T0142312

Lecture Notes in Electrical Engineering

Volume 490

** Indexing: The books of this series are submitted to ISI Proceedings, EI-Compendex, SCOPUS, MetaPress, Springerlink **

Lecture Notes in Electrical Engineering (LNEE) is a book series which reports the latest research and developments in Electrical Engineering, namely:

- Communication, Networks, and Information Theory
- Computer Engineering
- Signal, Image, Speech and Information Processing
- Circuits and Systems
- Bioengineering
- Engineering

The audience for the books in LNEE consists of advanced level students, researchers, and industry professionals working at the forefront of their fields. Much like Springer's other Lecture Notes series, LNEE will be distributed through Springer's print and electronic publishing channels.

For general information about this series, comments or suggestions, please use the contact address under "service for this series".

To submit a proposal or request further information, please contact the appropriate Springer Publishing Editors:

Asia:

China, *Jessie Guo, Assistant Editor* (jessie.guo@springer.com) (Engineering)

India, *Swati Meherishi, Senior Editor* (swati.meherishi@springer.com) (Engineering)

Japan, *Takeyuki Yonezawa, Editorial Director* (takeyuki.yonezawa@springer.com) (Physical Sciences & Engineering)

South Korea, *Smith (Ahram) Chae, Associate Editor* (smith.chae@springer.com) (Physical Sciences & Engineering)

Southeast Asia, *Ramesh Premnath, Editor* (ramesh.premnath@springer.com) (Electrical Engineering)

South Asia, *Aninda Bose, Editor* (aninda.bose@springer.com) (Electrical Engineering)

Europe:

Leontina Di Cecco, Editor (Leontina.dicecco@springer.com)
(Applied Sciences and Engineering; Bio-Inspired Robotics, Medical Robotics, Bioengineering; Computational Methods & Models in Science, Medicine and Technology; Soft Computing; Philosophy of Modern Science and Technologies; Mechanical Engineering; Ocean and Naval Engineering; Water Management & Technology)

(christoph.baumann@springer.com)
(Heat and Mass Transfer, Signal Processing and Telecommunications, and Solid and Fluid Mechanics, and Engineering Materials)

North America:

Michael Luby, Editor (michael.luby@springer.com) (Mechanics; Materials)

More information about this series at http://www.springer.com/series/7818

Asoke K. Nandi · N. Sujatha
R. Menaka · John Sahaya Rani Alex
Editors

Computational Signal Processing and Analysis

Select Proceedings of ICNETS2, Volume I

 Springer

Editors
Asoke K. Nandi
Department of Electronic and Computer
  Engineering
Brunel University London
Uxbridge
UK

N. Sujatha
Department of Applied Mechanics
Indian Institute of Technology Madras
Chennai, Tamil Nadu
India

R. Menaka
School of Electronics Engineering
VIT University
Chennai, Tamil Nadu
India

John Sahaya Rani Alex
School of Electronics Engineering
VIT University
Chennai, Tamil Nadu
India

ISSN 1876-1100   ISSN 1876-1119 (electronic)
Lecture Notes in Electrical Engineering
ISBN 978-981-13-4131-1   ISBN 978-981-10-8354-9 (eBook)
https://doi.org/10.1007/978-981-10-8354-9

Printed on acid-free paper

This Springer imprint is published by the registered company Springer Nature Singapore Pte Ltd. part of Springer Nature
The registered company address is: 152 Beach Road, #21-01/04 Gateway East, Singapore 189721, Singapore

Preface

This LNEE volume consists of papers presented at the Symposium-A entitled "Computational Signal Processing and Analysis" in the International Conference on "NextGen Electronic Technologies–Silicon to Software"—ICNETS^2-2017, which was held in VIT Chennai, India, during 23–25 March 2017.

The focus of this symposium was to bring together researchers and technologists working in different aspects of signal processing such as biomedical signal processing, image processing and video processing. One of the major objectives of this symposium is to highlight the current research developments in the areas of signal, image and video processing.

This symposium received over 64 paper submissions from various countries across the globe. After a rigorous peer review process, 37 full-length papers were accepted for presentation at the conference. This was intended to maintain the high standards of the conference proceedings. The presented papers were oriented towards addressing challenges involved in different application areas of signal processing. In addition to the contributed papers, renowned domain experts across the globe were invited to deliver keynote speeches at ICNETS^2-2017.

Acknowledgements

We would like to thank the VIT management for their support and encouragement. Editors are indebted to their respective university managements.

The success of the Symposium-A is due to Dr. S. R. S. Prabaharan, DEAN, SENSE, who has devoted his expertise and experience in promoting and coordinating the activities of the conference. We would like to express our sincere appreciation to the panel of reviewers who offered exemplary help in the review process. The quality of a refereed volume depends mainly on the expertise and dedication of the reviewers. We would like to express our gratitude to the keynote speakers who shared their expertise to the budding signal processing researchers.

The session chairs of different sessions played key roles in conducting the proceedings of each session in a well-organized manner.

We would like to recognize Springer LNEE for publishing the proceedings of Symposium-A as one volume. We would also like to thank the $ICNETS^2$-2017 Secretariat for dexterity. We would like to place on record the tireless work contributed by Symposium-A manager Dr. Jagannath. We acknowledge our publication committee Dr. Mohanaprasad, Dr. Annis Fathima and Dr. Velmathi for their efforts. Finally, we would like to show appreciation to our signal processing research group faculty members for their several months of hard work in making this symposium a prolific one.

Uxbridge, UK   Asoke K. Nandi
Chennai, India   N. Sujatha
Chennai, India   R. Menaka
Chennai, India   John Sahaya Rani Alex

Contents

About the Editors

Prof. Asoke K. Nandi received his Ph.D. degree from the University of Cambridge (Trinity College). He has held positions at the University of Oxford, Imperial College London, the University of Strathclyde and the University of Liverpool. In 2013, he moved to Brunel University London as the Head of Electronic and Computer Engineering. In 1983, his co-discovery of the three particles (W+, W− and Z0) was recognized by the Nobel Committee for Physics in 1984. He has made numerous fundamental contributions to signal processing and machine learning and has authored over 550 technical publications, with an h-index of 67. He is a fellow of the Royal Academy of Engineering, UK, and of seven other institutions including IEEE. Among the many awards he has received are the IEEE Heinrich Hertz Award (2012), the Glory of Bengal Award for his outstanding achievements in scientific research (2010), the Institution of Mechanical Engineers Water Arbitration Prize (1999) and the Mountbatten Premium from the Institution of Electrical Engineers (1998).

Dr. N. Sujatha graduated in Biomedical Optics from Nanyang Technological University (NTU), Singapore, and is an Associate Professor at the Department of Applied Mechanics, IIT Madras, India. She has also served as a Visiting Associate Professor at NTU, Singapore, in 2014–2015. Her major research interests are laser-based diagnostic imaging and diagnostic optical spectroscopy. She has co-authored a chapter and published over 50 international journal/conference papers, several of which have won best paper awards. She is a member of International Society for Optical Engineering, Optical Society of America, and is a fellow of the Optical Society of India.

Dr. R. Menaka received her doctoral degree in Medical Image Processing from Anna University, Chennai, and is currently an Associate Professor at the School of Electronics Engineering, VIT University, Chennai. She has served in both industry and academia for more than 25 years and has published over 35 research papers in various national and international journals and conferences. Her areas of interest include signal and image processing, neural networks and fuzzy logic.

John Sahaya Rani Alex is an Associate Professor at the School of Electronics Engineering, VIT Chennai. She has worked in the embedded systems and software engineering fields for 12 years, which includes 7 years' experience in the USA. Her research interests include spoken utterance detection and implementations of digital signal processing (DSP) algorithms in an embedded system. She is a member of IEEE and the IEEE Signal Processing Society.

Detecting Happiness in Human Face Using Minimal Feature Vectors

Manoj Prabhakaran Kumar and Manoj Kumar Rajagopal

Abstract Human emotions estimated from face become more effective compared to various modes of extracting emotion owing to its robustness, high accuracy and better efficiency. This paper proposes detecting happiness of human face using minimal facial features from geometric deformable model and supervised classifier. First, the face detection and tracking is observed by constrained local model (CLM). Using CLM grid node, the entire and minimal feature vectors displacement is obtained by facial feature extraction. Compared to entire features, minimal feature vectors is considered for detecting happiness to improve accuracy. Facial animation parameters (FAPs) helps in identifying the facial feature movements to forms the feature vectors displacement. The feature vectors displacement is computed in supervised bilinear support vector machines (SVMs) classifier to detect the happiness in human frontal face image sequences. This paper focuses on minimal feature vectors of happiness (frontal face) in both training and testing phases. MMI facial expression database is used in training, and real-time data are used for testing phases. As a result, the overall accuracy of happiness is achieved 91.66% using minimal feature vectors.

Keywords Constrained local model (CLM) · Facial animation parameters (FAPs) Minimal feature vectors displacement · Support vector machines (SVMs)

M. P. Kumar (✉) · M. K. Rajagopal
School of Electronics Engineering, Vellore Institute of Technology,
Chennai, Tamil Nadu, India
e-mail: manoj.prabhakaran2013@vit.ac.in

M. K. Rajagopal
e-mail: manojkumar.r@vit.ac.in

© Springer Nature Singapore Pte Ltd. 2018
A. K. Nandi et al. (eds.), *Computational Signal Processing
and Analysis*, Lecture Notes in Electrical Engineering 490,
https://doi.org/10.1007/978-981-10-8354-9_1

1

1 Introduction

Since 1990s, several researches are carried out on human emotion recognition for human–computer interaction (HCI), affective computing, etc. Emotion recognition in human has been established by the various modes of extraction [1]: physiological signal and non-physiological signal. From [1], the facial expression recognition is best out of the various modes of extracting emotion methods. From 1990 to till now, researchers are mostly concentrating on the robust automatic facial expression from image sequence compared to other modes of extracting emotions. In [2] has given the study of automatic facial expression system, through the photographic stimuli. In [3, 4] has established the automatic facial expression system from facial image sequence, which analyze the facial emotion through feature detection and tracking points.

From the literature survey [5–10], it is observed that the facial emotions are defined by the maximum number of facial feature points with action units (AUs) [11]. Therefore, usage of more feature points for facial emotion attains the complex data computation with less accuracy. To overcome this problem, the minimal feature points are selected for human facial expression. Facial action coding system (FACS) defines the combination of action units for facial emotion, using the entire feature points. Facial animation parameters (FAPs) [12] define facial emotion of action units within 10 groups, which use the entire feature points. Therefore, FAPs are considered for emotions' extraction using minimal feature vectors, which result in less data computational with high accuracy.

From [13] explain the importance of face modeling: the state of art with respect to different face models of face detection, tracking of automatic facial expression recognition.

In this paper, the detecting happiness is based on constrained local model (CLM) and bilinear support vector machines (SVMs). CLM [14] is developed for the face detection, tracking, and extracting the feature points. The extracted feature points form the minimal feature vectors displacements. The bilinear SVMs [15, 16] are formulated for classification of detecting happiness with help of FAPs [12]. The rest of the paper is as follows: The descriptions of detecting happiness are shown in Sect. 2. Section 3 describes the experimental results and discussion of proposed system. Section 4 summarizes the future work and conclusion.

2 System Description

The system description of detecting happiness is followed in three steps: face detection, tracking and feature extraction. From the facial feature vectors displacement, facial expressions are classified. Face detection and tracking, are carried out using deformable geometric grid node (CLM) [14]. Then feature vectors displacement is composed in supervised classifier (SVMs) [15] for defining the

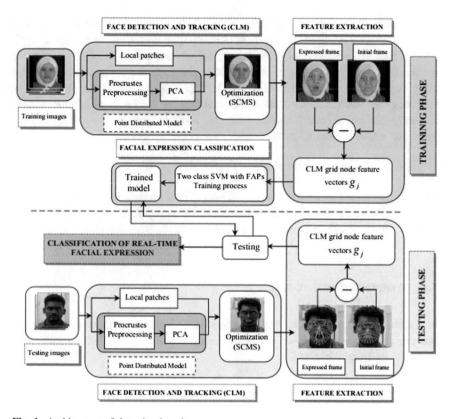

Fig. 1 Architecture of detecting happiness

happiness in human face. The proposed system architecture of detecting happiness is shown in Fig. 1.

2.1 Facial Detection and Tracking

In the proposed system, the face detection and tracking is carried out by constrained local model (CLM) (deformable geometric model fitting) [14], which represented two processes such as CLM model building and CLM search. The conceptual diagram of CLM model and search is as shown in Fig. 1.

2.1.1 Building a CLM Model

In CLM model building, there are two processes: shape model and patch model. In the shape model, first mark manually the landmark of feature points of face using point distribution model (PDM) [17]. PDM employed the non-rigid face shape of 2D+3D vector mesh. In PDM (Eq. (1)), building with principal component analysis (PCA) and Procrustes preprocessing. Principal component analysis (PCA) is applied for alignment of shape from the large database to get the mean value and eigen vectors shape of face. Before PCA, applying the Procrustes analysis for removing the scale, rotation, translations and gives the result of aligned shape. Similarly, the patch model applying logistic regression gives the result of mean value and eigen vectors of patch model.

$$x_i = sR(\tilde{x}_i) + T_{t_x,t_y} \Leftarrow T_{s,R,t_x,t_y}(\tilde{x}_i) \tag{1}$$

where x_i mentioned as ith landmark of 2D+3D PDM's location, $\tilde{x}_i$ identify as mean shape of 2D+3D PDM and pose parameters of PDM represent as $p = (s, R, t, q)$. s, R, t are denoted as shape, rotation, and translation.

2.2   Searching with CLM

In searching face with CLM, applying the linear logistic regressor algorithm for extracting the feature points of each face feature variation gives the response maps of ith image frames in Eq. (2).

$$p(l_i = \mathbf{aligned}|I, \mathbf{x}) = \frac{1}{1 + \exp\{\alpha\, C_i(I; \mathbf{x}) + \beta\}} \tag{2}$$

From the each feature point, crop a patch image of individual part (i.e., nose, left eye, right eye) and apply the linear logistic regressor [14], which is trained model to finding the local region of image and gives the result of response image. The quadratic function is fit on the response image of feature point position by optimization function is subspace constrained mean shift (SCMA) [14]. The mean shift algorithm [18] is applied for landmark location with aligned shape and patches in Eq. (3).

$$x_i^{(\tau+1)} \leftarrow \sum_{\mu_i \in \Psi_{x_i^c}} \frac{\alpha_{\mu_i}^i\, N\left(x_i^{(\tau)}; \mu_i, \sigma^2 I\right)}{\sum\limits_{y \in \Psi_{x_i^c}} \alpha_y^i\, N\left(x_i^{(\tau)}; y, \sigma^2 I\right)} \mu_i \tag{3}$$

Finally, combining a shape constraint model and local region of optimization function obtains the feature point of face, and fixed number of iteration gives the result of facial feature points tracking.

2.3 Classification

In classification, formulate the support vector machine (SVMs) with facial animation parameters (FAPs) of extracted feature points. Support vector machines (SVMs) [15, 16] are linear separating a maximum margin of hyperplane in a higher dimensionality space. Let $g_j = \{(\vec{x}_i, \vec{y}_i)\}$; $i = 1 \ldots k$; $\vec{x} \in \Re^n$; $y_i \in \{-1, +1\}$ is the training dataset of facial extraction of feature vectors displacement. Then maximum margin of separating hyperplane of linear data of the form is Eq. (4).

$$\begin{aligned} \vec{w}^T \cdot \vec{x} + b &\geq +1 \quad \text{for } (y_i = +1) \\ \vec{w}^T \cdot \vec{x} + b &\leq -1 \quad \text{for } (y_i = -1) \end{aligned} \tag{4}$$

$\vec{w}^T$ is weight vectors, where normal to the separating hyperplane and $\vec{w}^T$ is a bias. A decision function of separating hyperplane is as follows in Eq. (5).

$$f(\vec{x}) = \vec{w}^T \cdot \vec{x} + b \tag{5}$$

Subject to constraint inequalities is Eq. (6) the separating linear optimal hyperplane in form out Eq. (7) :

$$y_i\left(\vec{w}^T \cdot \vec{x}_i + b\right) - 1 \geq 0 \quad i = 1, \ldots N \tag{6}$$

$$\vec{w} = \sum_{i=1} \alpha_i \cdot \tilde{S}_i \tag{7}$$

The two class of linear SVMs of decision surface is as follows in Eq. (8):

$$f(x) = \sigma\left(\sum_{i=1} \alpha_i \, \Phi(\tilde{S}_i) \cdot \Phi(x)\right) \quad \text{or } y = \vec{w} \cdot x + b \tag{8}$$

From Eq. (8) gives the discriminating hyperplane of separating cluster in decision surface. For nonlinear case of SVMs, the training data are changed into linear separable data by using kernel function (polynomial, rbf), normalization and transformation of Φ mapping function [15, 16]. From Eq. (8), decision surface is classify the detecting happiness are seen detailed in Sect. 3.

3 Experimental Results and Discussion

3.1 Feature Vectors Displacement

The information of face detection, tracking and extraction are carried out for emotion in real-time human face using geometric deformable model (CLM) [14].

The extracted information of happiness is in frame-by-frame facial features movement to form the facial feature vectors displacement. The geometric information of feature vectors displacement is one node displacement $d_{i,j}$ defined as the consecutive frame-by-frame difference between the grid node displacements of first to ith node coordinates. The feature vectors displacement is in Eq. (9):

$$d_{ij} = \begin{bmatrix} \Delta x_{i,j} \\ \Delta y_{i,j} \end{bmatrix} = \begin{pmatrix} a_{11} - a_{12} & a_{13} - a_{14} & \cdots & a_{1,j+1} - a_{1,j+2} \\ a_{21} - a_{22} & a_{23} - a_{24} & \cdots & a_{2,j+1} - a_{2,j+2} \\ \vdots & & \ddots & \vdots \\ a_{i,j} - a_{i,j+1} & \cdots & & a_{n,m+1} - a_{n,m+2} \end{pmatrix} \tag{9}$$

$i = 1, \ldots F, j = 1, \ldots N$, where $\Delta x_{i,j}, \Delta y_{i,j}$ are x-axis, y-axis coordinates of grid node displacement of the ith node in jth frame image, respectively. F is the number of grid node ($F = 66$ nodes of CLM), and N is the number of the extracted facial images from the facial image sequence.

$$g_j = \begin{bmatrix} d_{1j} \, d_{2j} \ldots d_{Ej} \end{bmatrix}^T \quad j = 1, \ldots N \tag{10}$$

From Eq. (10), for every sequence of the happy face in dataset, an extracted feature vectors grid deformation vector g_j is created to form the displacements of the every geometric grid node $d_{i,j}$. In happy face, major muscle variation is happening in mouth region (Groups 8 and 2 of FAPs). From the extracted features from CLM, feature vectors displacement is computed. The entire and minimal feature vectors displacement of happy in CLM is shown in Fig. 2a, b; the blue color indicates as happy. The happy variations are more in outer lip and corner lip region with along x-axis direction defined from the FAPs [12].

In this system, the entire feature vectors displacement has high data computation and less accuracy of variation in happy. In order to achieve less data computation and high accuracy, minimal feature displacement is used and desired result is obtained. In Fig. 2a, the entire feature vectors displacement has feature variation in Group 8 (outer mouth lip region) and Group 2 (corner lip region) from the FAPs description. In our proposed, the minimal feature vectors displacements have the feature variation only in Group 2 (corner lip region) as shown in Fig. 2b. In this system, the geometric deformable grid node (CLM) has $L = 66 * 2 = 132$ dimensions. In the feature vectors displacement of image sequence, where computed the $d_{i,j}$ displacements of CLM grid node in order to form in start at neutral face to expressed face (i.e. Initial frame to peak response of frame) and the expressed face to neutral state. The CLM feature vectors displacement g_j is employing for the classification of happy face using two classes of SVMs in our proposed system. In our proposed system, the detecting happiness of CLM is developed in C++ with open framework tool and SVMs which was implemented in Intel i5 processor. In training and testing processes, MMI facial expression standard database [19] and real-time emotions of video rate is 30 frames/s are respectively and only frontal face image sequence are captured are shown in Fig. 3.

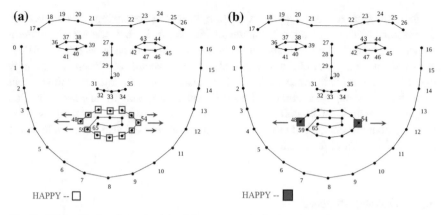

Fig. 2 CLM grid of entire and minimal feature vectors displacement of Happy

3.2 Training Process

In Happy, the major facial muscle movement in Group 8 and Group 2 of temporal segments in x-axis direction of the entire and minimal feature vectors displacement defined by FAPs [12]. From Fig. 4a, b are shown as expression value (i.e., offset-apex-onset region) of entire and minimal feature vectors displacements are respectively. In Happy, the major facial movement is horizontally expanded of both feature vectors. In that, the entire feature has taken all feature point for classification of happy. But it attained the high data computations with less accuracy. In order to achieve, the minimal feature vectors has only two feature points (49th and 55th of CLM grid node) for happy classification which attained the less data computations with high accuracy are shown in Fig. 4b. The reason for selecting minimal feature vectors, the two feature points have high variance compared to the outer lip mouth region (12 feature points) by FAPs.

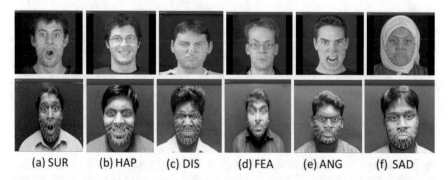

Fig. 3 Training and testing processes of MMI facial expression database (first row) and real-time (second row) facial expression datasets are respectively. **a** Surprise (SUR), **b** Happy (HAP), **c** Disgust (DIS), **d** Fear (FEA), **e** Anger (ANG), and **f** Sad (SAD).

In training process of happy classification, the entire and minimal feature vectors displacement of classification is shown in Fig. 4c, d. In that, trained 10 different subjects of happy (+ve class) and surprise (−ve class) were taken as bilinear SVMs are shown in Fig. 4c, d. In Fig. 4c, the entire feature vectors displacement of happy classification has attained the nonlinear data classification. In order to achieve linear classification of happy, applied the kernel function (polynomial, rbf), normalization and transformation of mapping function. In Fig. 4d, the minimal feature vectors displacement has conquered the linear separable datasets and also achieved less data computation with high accuracy.

3.3 Testing Process

In the testing process, the real-time facial data comprising of all basic six emotions were taken from 10 different subjects is shown in Fig. 3. Similarly, in the testing

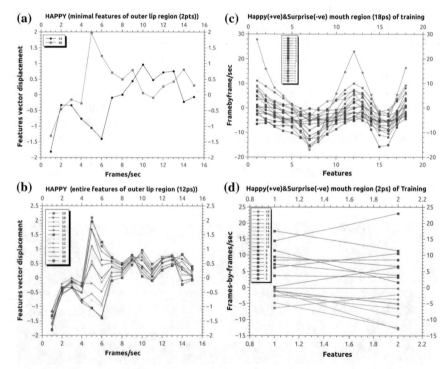

Fig. 4 **a** Entire feature vectors of Happy in outer lip corners (12 fps). **b** Minimal feature vectors of Happy in outer lip corner (48th and 54th fps). **c** Training process of happy (+ve) and surprise (−ve) of entire feature vectors of outer lip corners (12 fps) are in nonlinear case in bilinear SVMs. **d** Training process of happy (+ve) and surprise (−ve) of minimal feature vectors of outer lip corners (48th and 54th fps) are in linearly separable in bilinear SVMs

Table 1 Confusion matrix of Happy in bilinear SVMs classifier

EMO	HAP	SUR	SAD	FEA	ANG	DIS
HAP	10	0	0	0	0	0
SUR	0	10	0	0	0	0
SAD	0	0	10	0	0	0
FEA	1	0	0	9	0	0
ANG	2	0	0	0	8	0
DIS	2	0	0	0	0	8

process, where evaluated with the CLM face tracking and extracted features points to form the minimal features vectors displacement. The information of minimal feature vectors displacements was applied on the decision surface of trained model in the classification of happiness. The confusion matrix of Happy using bilinear SVMs is shown in Table 1. From the confusion matrix of happy, the overall accuracy is 91.66% achieved. The validation parameters are Precision is 3, Recall is 0.666, and F-measure values is 3 which is calculated from the confusion matrix of Happy.

4 Conclusion

In this paper, happiness is detected with minimal facial feature points using CLM and SVMs. In this system the minimal feature vectors are determined which contributes highly to detect happiness in human face. This leads to less computation and more accuracy. In this paper, the experiments are carried out with real time frontal facial expression and MMI Face expression database. Using minimal feature vector the accuracy of detecting happiness is 91.66% this work can be extends for the reminaing basic sets of emotion with multi-classification, different attributes (posed, spontaneous and wild), which helpful for developing HCI application.

Acknowledgements The authors would like to thanks my research colleague for real-time dataset from Vellore Institute of Technology, Chennai.

References

1. Kollias S, Karpouzis K (2005) Multimodal emotion recognition and expressivity analysis. In: 2005 IEEE international conference on multimedia and expo. IEEE, pp 779–783
2. Ekman P, Sorenson ER, Friesen WV et al (1969) Pan-cultural elements in facial displays of emotion. Science 164(3875):86–88

3. Mase K (1991) Recognition of facial expression from optical flow. IEICE Trans Inf Syst 74(10): 3474–3483
4. Samal A, Iyengar PA (1992) Automatic recognition and analysis of human faces and facial expressions: a survey. Pattern Recogn 25(1):65–77
5. Bartlett MS et al (2003) Real time face detection and facial expression recognition: development and applications to human computer interaction. In: Conference on computer vision and pattern recognition workshop (CVPRW'03), vol 5. IEEE, pp 53–53
6. Lucey P et al (2010) The extended Cohn-Kanade dataset (CK+): a complete dataset for action unit and emotion-specified expression. In: 2010 IEEE computer society conference on computer vision and pattern recognition-workshops. IEEE, pp 94–101
7. Kotsia I, Pitas I (2007) Facial expression recognition in image sequences using geometric deformation features and support vector machines. IEEE Trans Image Process 16(1):172–187
8. Zhang Y et al (2008) Dynamic facial expression analysis and synthesis with mpeg-4 facial animation parameters. IEEE Trans Circuits Syst Video Technol 18(10):1383–1396
9. Pantic M, Patras I (2006) Dynamics of facial expression: recognition of facial actions and their temporal segments from face profile image sequences. IEEE Trans Syst Man Cybern Part B (Cybern) 36(2):433–449
10. Okada T, Takiguchi T, Ariki Y (2010) Video searching system based on human face identification and facial expression recognition using MSM and AAM. Far East J Electron Commun 4(1):41–48
11. Tian Y-I et al (2001) Recognizing action units for facial expression analysis. IEEE Trans Pattern Anal Mach Intell 23(2):97–115
12. Tekalp AM, Ostermann J (2000) Face and 2-D mesh animation in MPEG-4. Signal Process Image Commun 15(4):387–421
13. Salam H (2013) Multi-object modelling of the face. Ph.D. thesis, Supelec
14. Saragih JM et al (2011) Deformable model fitting by regularized landmark mean-shift. Int J Comput Vis 91(2):200–215
15. Ventura D (2009) SVM example. Lectures notes, Mar 2009
16. Vapnik VJ, Vapnik V (1998) Statistical learning theory, vol 1. Wiley, New York
17. Cristinacce D, Cootes TF (2006) Feature detection and tracking with constrained local models. In: BMVC, vol 1, p 3
18. Cheng Y (1995) Mean shift, mode seeking, and clustering. IEEE Trans Pattern Anal Mach Intell 17(8):790–799
19. Valstar M, Pantic M (2010) Induced disgust, happiness and surprise: an addition to the MMI facial expression database. In Proceedings of the 3rd international workshop on EMOTION (satellite of LREC): Corpora for research on emotion and affect, p 65

Analysis of Myocardial Ischemia from Cardiac Magnetic Resonance Images Using Adaptive Fuzzy-Based Multiphase Level Set

M. Muthulakshmi and G. Kavitha

Abstract In this research work, cardiac magnetic resonance (CMR) images are analyzed to study the pathophysiology of myocardial ischemia (MI). It is a cardiac disorder that causes irreversible damage to heart muscles. The images considered for this study are obtained from medical image computing and computer-assisted intervention (MICCAI) database. Adaptive fuzzy-based multiphase level set method is utilized to extract endocardium and epicardium of left ventricle from short-axis view of CMR images. The segmentation results are validated with similarity measures such as Dice coefficient and Jaccard index. Further, five indices are derived from the segmentation results. The obtained results provide average Dice coefficient for endocardium and epicardium as 0.867 and 0.918, respectively. The mean Jaccard index for epicardium and endocardium is 0.855 and 0.766, respectively. It is observed that the proposed method segments the left ventricle more precisely from CMR images. The ischemic subjects show a reduced mean ejection fraction (32.52) compared to the normal subjects (59.04). The average stroke volume is found to be 70.16 and 64.05 ml for healthy subjects and ischemic subjects, respectively. Reduction in stroke volume and ejection fraction for ischemic subjects indicates lower quantity of blood drained by heart. It is also observed that there is an increase in myocardial mass for ischemic subjects (182.11 g) compared to healthy subjects (127.47 g). The thickened heart muscle contributes to the increased myocardial mass in abnormal subjects. Further, ischemic subjects show an increase in endocardium volume at end-diastolic and end-systolic phase when compared to normal subjects. Thus, the clinical indices evaluated from adaptive fuzzy-based multiphase level set method could differentiate the normal and ischemic subjects. Hence, this study can be a useful supplement in diagnosis of myocardial ischemic disorder.

M. Muthulakshmi (✉) · G. Kavitha
Department of Electronics Engineering, MIT Campus, Anna University, Chennai, India
e-mail: lakshmingm.2@gmail.com

G. Kavitha
e-mail: kavithag_mit@annauniv.edu

© Springer Nature Singapore Pte Ltd. 2018
A. K. Nandi et al. (eds.), *Computational Signal Processing and Analysis*, Lecture Notes in Electrical Engineering 490,
https://doi.org/10.1007/978-981-10-8354-9_2

Keywords Ischemia · Cardiac magnetic resonance images · Multiphase level set Adaptive fuzzy · Left ventricle

1 Introduction

Myocardial ischemia (MI) is an irreversible cardiovascular disorder. Cardiovascular disease (CVD) is the predominant cause of fatality globally. MI is characterized by weakened heart muscles [1]. The interruption of blood supply damages the heart muscles that inhibit its ability to pump blood. Eventually, this may be captured as abnormal heart rhythms, diastolic and systolic dysfunctions [2]. MI causes chest pain, discomfort in shoulder, arm, back, neck, and jaw. Mortality due to acute myocardial ischemia can be reduced by diagnosis and treatment at an earlier stage.

Various modalities used to diagnose CVD include echocardiography, magnetic resonance images (MRI), computed tomography (CT), single photon emission computed tomography (SPECT), positron-emitted tomography (PET) and integrated modalities. The effective noninvasive modality for CVD diagnosis is cardiac magnetic resonance (CMR) images [3]. CMR provides high soft-tissue contrast, multiplanar acquisition capability and lacks ionizing radiations. Left ventricle (LV) segmentation from CMR is essential for quantitative cardiac study. Segmentation of LV manually done by radiologists are complex, consumes more time, and prone to human errors [4]. The papillary muscles make automatic segmentation of LV difficult as their intensities are similar to myocardium. Intensity inhomogeneity and reduced contrast between other organs and myocardium pose additional challenges in segmentation. Clinical indices such as left ventricle volume, ejection fraction, and mass are evaluated with the outcomes obtained from segmentation of LV echocardiographic images [5]. These indices aid the diagnosis of myocardial ischemia, and they can be computed more precisely with the aid of efficient segmentation algorithm.

Previous works on left ventricle segmentation are based on local or global information [6], deformable models [7], atlas [8], and statistical models [9]. The local information-based methods better segregate region of interest based on intensity of pixels. However, they are less effective when tissues have overlapping intensities [10]. The region growing algorithms though work better for less gradient images; the drawback is that they leak into irrelevant adjacent regions. Atlas-based methods require prior information that depends on spatial probability pattern of different tissues. The training time of statistical models depends on the training population. Furthermore, model-based methods preserve anatomical spatial information. Past studies revealed that active contour models provide promising approach for left ventricle segmentation [11]. Here, a contour deforms its shape in accordance with internal and external forces. LV segmentation of CMR images is carried out with active contour model coupled with nonlinear shape priors [12]. Li et al. introduced multiphase level set method to segment X-ray, CT and MR images with intensity inhomogeneity [13]. Recently, two-step DRLSE is applied for LV and RV segmentation using CMR images [14]. In detection of cardiac ischemia, unsupervised support vector machine along with dictionary learning is carried out on CMR images dependent on blood oxygen level [15].

The limitation with majority of the segmentation methods based on active contour is that their precision depends on the appropriate placement of initial contour which requires manual intervention. In order to overcome this, Huang et al. initialized the contour for snake models utilizing fuzzy C-means clustering and graph-cut segmentation method [16]. Region-based level set method including fuzzy C-means clustering is applied to brain CT images for hemorrhage segmentation [17]. A fuzzy C-means clustering methodology that is adaptively regularized is implemented for brain tissue segmentation from MR brain images [18]. The initial contour obtained from adaptive fuzzy and the level set energy based on adaptive fuzzy membership function would provide more precise segmentation results.

In this work, a multiphase level set method based on adaptive fuzzy is employed for segmentation of endocardium and epicardium from CMR images. Fuzzy-based intensity descriptor is incorporated to define the energy of the multiphase level set function. The efficacy of the segmentation method is validated with similarity measures such as Jaccard index and Dice coefficient. From the segmented regions indices such as left ventricle end-diastole and end-systole volume, stroke volume, ejection fraction and myocardial mass are calculated. These indices could aid the diagnosis of cardiovascular disorders such as myocardial ischemia.

2 Material and Methods

2.1 Database

The short-axis cardiac magnetic resonance images used for the analysis are acquired from the medical image computing and computer-assisted intervention (MICCAI) left ventricle segmentation database [19]. The database contains cine-MR images of 45 patients from a range of pathology. The subjects are divided as normal, ischemic heart failure, non-ischemic heart failure and hypertrophy. Ground truths for evaluation purpose are provided by expert cardiologists. The description about age and gender of each subject is provided in the database.

2.2 Adaptive Fuzzy-Based Multiphase Level Set

Adaptive fuzzy-based multiphase level set (AFMLS) method is applied for segmentation of epicardium and endocardium of LV simultaneously. In multiphase level set method, k-level set contours $\Phi_1, \Phi_2,...,\Phi_k$ are used and their membership function is defined by $M_i(\Phi_1(y),\ldots, \Phi_k(y))$ [13]. The energy of an AFMLS function [13, 17] is given by

$$\varepsilon(\Phi, c, b) = \int \sum_{i=1}^{N} e_i(x)M_i(\Phi(x))\mathrm{d}x. \tag{1}$$

where e_i is the energy-based intensity descriptor, the k-level set functions $\Phi_k(x)$ is defined by adaptive fuzzy membership function output u_{ij}, cluster center [18], and N is the number of segmented regions.

$$u_{ij} = \frac{\left(\left(1 - K(x_i, v_j)\right) + \varphi_i\left(1 - K(\bar{x}_i, v_j)\right)\right)^{-1/(m-1)}}{\sum_{k=1}^{c}\left(\left(1 - K(x_i, v_k)\right) + \varphi_i(1 - K(\bar{x}_i, v_k))\right)^{-1/(m-1)}} \tag{2}$$

where φ is the adaptive regularization parameter, number of clusters denoted by c, K represents Gaussian radial basis kernel function, and m indicates the weighting exponent indicating the degree of fuzziness.

$$e_i = |I - bc_i|^2. \tag{3}$$

where $i = 1$ to N, original image is given by I, b represents bias field, and c denotes the cluster center. In this work, $N = 3$ is considered.

2.3 Similarity Measures

Segmentation outcomes are quantitatively evaluated using Dice coefficient and Jaccard index [10]. The similarity between the ground truth and computed segmentation results is evaluated by Dice coefficient and Jaccard index. The similarity measure has values in the range of 0–1. Higher value indicates better segmentation results. A_s is the segmented region using AFMLS method, and A_m is the ground truth.

3 Results and Discussion

The short-axis view CMR sequence of frames used in this work includes 9 normal and 12 ischemic subjects. Adaptive fuzzy-based multiphase level set (AFMLS) algorithm is applied for segmentation of epicardium and endocardium of left ventricle from CMR images. In this method, the value for level set parameters σ, timestep, μ, and υ are chosen as 7, 0.1, 1, and $0.01 * A^2$ where $A = 255$.

Figure 1a–h illustrates the endocardial and epicardial contours of left ventricle (LV) from end-diastole (ED) to end-systole (ES) phase for a normal subject.

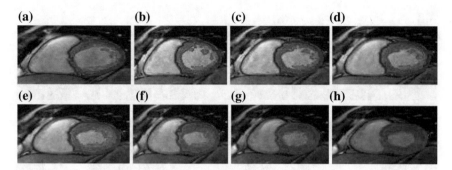

Fig. 1 AFMLS segmentation of left ventricle from ED phase to ES phase for normal subject. **a** ED image. **b–g** Progress from ED phase to ES phase. **h** ES image

Figure 1a corresponds to ED slice, and Fig. 1h corresponds to ES slice. The sequence of frames from ED to ES phase is shown in Fig. 1b–g. There is reduction in LV dimensions from ED phase to ES phase as the ventricle contracts. It is evident that the proposed AFMLS method could capture the variations in epicardial and endocardial geometry of LV from ED to ES phase.

The LV segmentation output at ED and ES phase using AFMLS method and ground truth for both normal and ischemic subjects is demonstrated in Figs. 2 and 3, respectively. Figure 2a–c shows the segmented endocardium during ED, epicardium during ED, and endocardium during ES for a healthy subject. The corresponding ground truth images for healthy subject are shown in Fig. 2d–f. The extracted endocardium at ED, epicardium at ED, and endocardium at ES for a ischemic subject are illustrated in Fig. 3a–c, respectively. Further, Fig. 3d–f illustrates the ground truth images for the same. Hence, it is evident that the proposed AFMLS algorithm is able to segment the endocardial and epicardial boundaries in both ischemic and normal subjects.

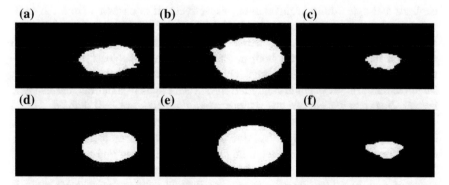

Fig. 2 **a** Segmented endocardium in ED phase. **b** Segmented epicardium in ED phase. **c** Segmented endocardium in ES phase. **d** Ground truth for endocardium in ED phase. **e** Ground truth for epicardium in ED phase. **f** Ground truth for endocardium in ES phase in normal subjects

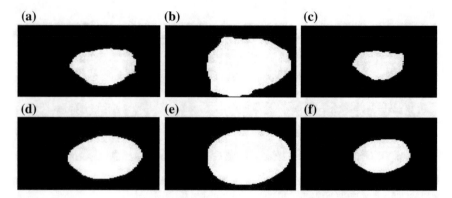

Fig. 3 a Segmented endocardium in ED phase. **b** Segmented epicardium in ED phase. **c** Segmented endocardium in ES phase. **d** Ground truth for endocardium in ED phase. **e** Ground truth for epicardium in ED phase. **f** Ground truth for endocardium in ES phase in ischemic subjects

The AFMLS algorithm is validated with the help of Dice coefficient and Jaccard index. The Dice metric calculates the overlapped area between the automatic segmentation result and the ground truth. The Dice coefficient obtained for different normal and ischemic subjects is shown in Fig. 4. The Dice coefficient for endocardium and epicardium segmentation of normal subjects is illustrated in Fig. 4a, b, respectively. Further, Fig. 4c, d shows the Dice coefficient for endocardium and epicardium segmentation of ischemic subjects. The mean Dice coefficient is obtained as 0.866 and 0.918 for endocardium and epicardium segmentation, respectively.

Figure 5 depicts the Jaccard index obtained for different normal and ischemic subjects. Figure 5a, b shows the Jaccard index for endocardium and epicardium segmentation of normal subjects, respectively. Further, the Jaccard index for endocardium and epicardium segmentation of ischemic subjects is illustrated in Fig. 5c, d, respectively. The average Jaccard index is 0.766 and 0.855 for endocardium and epicardium segmentation, respectively. It is observed from the similarity measures that the proposed AFMLS algorithm is able to segment the LV better from both normal and ischemic CMR images. Though the segmentation validation indices are high for both normal and ischemic subjects, the Dice coefficient and Jaccard index are relatively low for endocardium segmentation in ischemic subjects. This could be due to ill-defined edges in abnormal cardiac MR images. Fuzzy better clusters the regions when the edges are well defined.

The indices such as myocardial mass, ejection fraction, end-systole volume, end-diastole volume and stroke volume for normal and ischemic subjects are calculated for the segmented left ventricle [10]. Figure 6 shows end-diastole volume (EDV) for LV of normal and ischemic subjects, where the ventricle dilates. It is observed that there is an increase in the end-diastole volume for ischemic subjects compared to the normal subjects. This indicates an increase in quantity of blood intake by the heart. Figure 7 illustrates the end-systole volume (ESV) for LV of

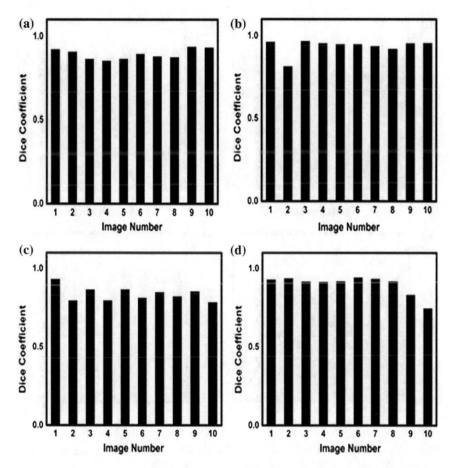

Fig. 4 Dice coefficient for **a** endocardium and **b** epicardium segmentation of normal subjects; **c** endocardium and **d** epicardium segmentation of ischemic subjects

normal and ischemic subjects, where the ventricle contracts. It is observed that the ESV shows a high range (90–180 ml) in ischemic subjects compared to normal subjects (40–80 ml). This analysis shows that the LV contraction is less in ischemic subjects and it is an indicator of systolic heart failure. In Fig. 8, the stroke volume (SV) for normal and ischemic subjects is shown. It is studied that the stroke volume reduces for ischemic subjects compared to normal subjects and it is more distributed in nature. The stroke volume is the difference in the EDV and ESV, which is an indicator of amount of blood drained by LV. Reduced SV indicates reduction in blood drained by the heart compared to normal subjects. It is associated with thickened myocardium and LV hypertrophy. Figure 9 shows the ejection fraction (EF) of left ventricle for normal and ischemic subjects. It is shown that the EF is low for ischemic subjects compared to normal subjects. The EF parameter well separates the ischemic and normal images. EF is the ratio of SV to EDV. Low EF

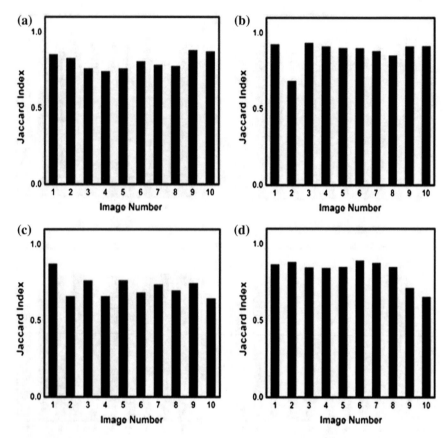

Fig. 5 Jaccard index for **a** endocardium and **b** epicardium segmentation of normal subjects; **c** endocardium and **d** epicardium segmentation of ischemic subjects

along with low SV signifies the abnormality in blood drained by heart. The myocardial mass (MM) for normal and ischemic subjects is shown in Fig. 10. It is observed that there is an increase in myocardial mass for ischemic subjects when compared to normal subjects. It is associated with thickened myocardium and LV hypertrophy.

The average and standard deviation values of the indices for the segmented LV region are given in Table 1. It is observed that the average values of ESV, EDV, and MM for ischemic subjects are higher than the normal subjects. Further, the mean values of SV and EF for ischemic subjects are lower than the normal subjects. The studies carried out reveal that there is significant reduction in EF for ischemic subjects and significant increase in ESV and MM for ischemic subjects. Thus, EF, ESV, and MM can be used as a measure in differentiating the ischemic from normal subjects.

Fig. 6 End-diastole left
ventricle volume for normal
and ischemic subjects

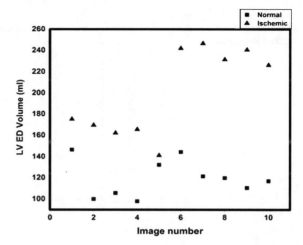

Fig. 7 End-systole left
ventricle volume for normal
and ischemic subjects

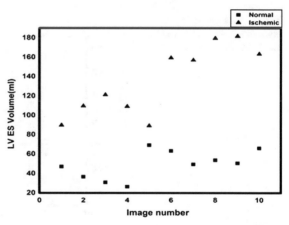

Fig. 8 Left ventricle stroke
volume for normal and
ischemic subjects

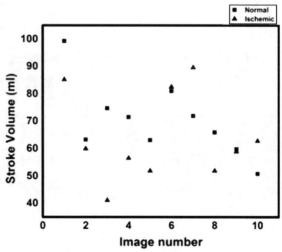

Fig. 9 Left ventricle ejection fraction for normal and ischemic subjects

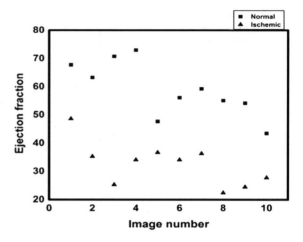

Fig. 10 Left ventricle myocardial mass for normal and ischemic subjects

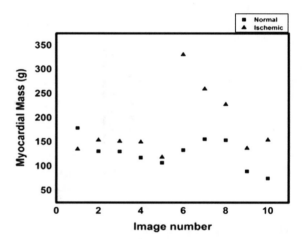

Table 1 Left ventricle indices for normal and ischemic subjects

Left ventricle indices	Mean ± standard deviation	
	Normal subjects (10)	Ischemic subjects (10)
End-diastole volume (EDV) (ml)	119.573 ± 17	200.128 ± 40
End-systole volume (ESV) (ml)	49.417 ± 14	136.082 ± 35
Stroke volume (SV) (ml)	70.156 ± 13	64.047 ± 16
Ejection fraction (EF)	59.044 ± 9	32.523 ± 7
Myocardial mass (MM) (g)	127.467 ± 31	182.111 ± 68

4 Conclusion

In this work, an attempt has been made to extract the left ventricle of normal and myocardial ischemic CMR images using multiphase level set method. Adaptive fuzzy-based energy descriptor which eliminates need for human intervention in initial contour placement has been considered for segmentation. The proposed AFMLS method is able to extract more precisely the endocardium and epicardium of LV from ED to ES phase for normal and ischemic subjects. The derived similarity measures such as Dice coefficient and Jaccard index show better adherence of segmentation results with ground truth. The reduction in stroke volume and ejection fraction for ischemic subjects compared to healthy subjects indicates abnormal cardiac behavior. The increase in myocardial mass and ESV predicts systolic heart failure. The ability of the proposed methodology to identify the abnormality at different cardiac phases makes it suitable to for study of cardiovascular disorders.

References

1. Radico F, Cicchitti V, Zimarino M, Caterina RD (2014) Angina pectoris and myocardial ischemia in the absence of obstructive coronary artery disease: practical considerations for diagnostic tests CME. J Am Coll Cardiol Interv 7:453–463
2. Detry J-MR (1996) The pathophysiology of myocardial ischaemia. Eur Heart J 17:48–52
3. Miller CA, Pearce K, Jordan P, Argyle R, Clark D, Stout M, Ray S, Schmitt M (2012) Comparison of real-time three-dimensional echocardiography with cardio-vascular magnetic resonance for left ventricular volumetric assessment in unselected patients. Eur Heart J Cardiovasc Imaging 13:187–195
4. Constantinides C, Chenoune Y, Kachenoura N, Roullot E, Mousseaux E, Herment A, Frouin F (2009) Semi-automated cardiac segmentation on cine magnetic resonance images using GVF-Snake deformable models. MIDAS J. http://hdl.handle.net/10380/3108
5. Marsousi M, Eftekhari A, Kocharian A, Alirezaie J (2010) Endocardial boundary extraction in left ventricular echocardiographic images using fast and adaptive B-spline snake algorithm. Int J Comput Assist Radiol Surg 5:501–513
6. Pednekar A et al (2006) Automated left ventricular segmentation in cardiac MRI. IEEE Trans Biomed Eng 53:1425–1428
7. Avendi MR, Kheradvar A, Jafarkhani H (2016) A combined deep-learning and deformable-model approach to fully automatic segmentation of the left ventricle in cardiac MRI. Med Image Anal 30:108–119
8. Zhuang X, Shen J (2016) Multi-scale patch and multi-modality atlases for whole heart segmentation of MRI. Med Image Anal 31:77–87
9. Ordas S, Oubel E, Leta R, Carreras F, Frangi AF (2007) A statistical shape model of the whole heart and its application to model-based segmentation. In: Proceedings for SPIE medical imaging, vol 6511
10. Petitjean C, Dacher J (2011) A review of segmentation methods in short axis cardiac MR images. Med Image Anal 15:169–184
11. Lynch M, Ghita O, Whelan PF (2006) Left-ventricle myocardium segmentation using a coupled level-set with a priori knowledge. Comput Med Imag Graph 30:255–262

12. Pham V-T, Tran T-T (2016) Active contour model and nonlinear shape priors with application to left ventricle segmentation in cardiac MR images. Opt-Int J Light Electron Opt 127: 991–1002
13. Li C et al (2008) A variational level set approach to segmentation and bias correction of images with intensity inhomogeneity. Med Image Comput Comput Assist Interv 11: 1083–1091
14. Zhang S, Limy C (2015) Distance regularized two level sets for segmentation of left and right ventricles from cine-MRI. Magn Reson Imag 34:699–706
15. Bevilacqua M, Dharmakumar R, Tsaftaris SA (2016) Dictionary-driven ischemia detection from cardiac phase-resolved myocardial BOLD MRI at rest. IEEE Trans Med Imag 35: 282–293
16. Huang QH et al (2012) A robust graph-based segmentation method for breast tumors in ultrasound images. Ultrasonics 52:266–275
17. Bhadauria HS, Singh A, Dewal ML (2013) An integrated method for hemorrhage segmentation from brain CT imaging. Comput Electr Eng 39:1527–1536
18. Elazab A et al (2015) Segmentation of brain tissues from magnetic resonance images using adaptively regularized kernel-based fuzzy C-means clustering. Comput Math Methods Med 2015(2015):1–12 (485495)
19. Radau P, Lu Y, Connelly K, Paul G, Dick AJ, Wright GA (2009) Evaluation framework for algorithms segmenting short axis cardiac MRI. MIDAS J Card MR Left Ventricle Segmentation Challenge. http://hdl.handle.net/10380/3070

Diagnosis of Schizophrenia Disorder Using Wasserstein Based Active Contour and Texture Features

M. Latha and G. Kavitha

Abstract Magnetic resonance (MR) brain images have a significant role in diagnosis of many neuropsychiatric disorders such as Schizophrenia (SZ). In this work, Wasserstein-based active contour and the texture features such as Hu moments and gray-level co-occurrence matrix (GLCM) are used to analyze Schizophrenic MR brain images. The images ($N = 40$) used for the analysis are obtained from National Alliance for Medical Image Computing (NAMIC) database. Initially, the normal and schizophrenic images are subjected to skull stripping using Wasserstein-based active contour method. The extracted brain from skull-stripping process is compared with Brain Extraction Tool (BET) and Brain Surface Extractor (BSE) methods. Seven features from Hu moment and twenty-two features from GLCM are extricated from the skull-stripped images. Further, these extracted features are analyzed to obtain discriminative information from normal and abnormal images. The result shows that the Wasserstein-based active contour method is able to separate the brain with an accuracy of 0.978, sensitivity of 0.934, and F-score of 0.958. The features extracted from Hu moments for abnormal images show higher magnitude value than normal images. Hu moments show significant percentage variation between normal and SZ subjects. Hu features such as $\phi3$, $\phi4$, and $\phi5$ yield higher variation of 26.3%, 21.4%, and 20.1%, respectively, between normal abnormal images. In GLCM-based features, the features such as sum of squares, autocorrelation, and maximal correlation coefficient show better variation of 19.2%, 18.4%, and 15.6% between normal and abnormal images. Hu moments show better percentage variance in normal and abnormal images compared to GLCM features. Hence, the combination of Wasserstein-based active contour and Hu moments could be used for better demarcation of normal and Schizophrenia subjects.

Keywords Schizophrenia · Wasserstein-based active contour · Texture feature GLCM · Hu moments · Magnetic resonance images

M. Latha (✉) · G. Kavitha
Department of Electronics Engineering, MIT Campus, Anna University, Chennai, India
e-mail: lathakaran@gmail.com

G. Kavitha
e-mail: kavithag_mit@annauniv.edu

© Springer Nature Singapore Pte Ltd. 2018
A. K. Nandi et al. (eds.), *Computational Signal Processing and Analysis*, Lecture Notes in Electrical Engineering 490,
https://doi.org/10.1007/978-981-10-8354-9_3

1   Introduction

Schizophrenia (SZ) is a chronic brain disorder. It is characterized by disturbances in cognition and emotional responsiveness resulting in disorganized speech, thinking, and behavior. It has its onset in late young or early adulthood and results in lifelong psychological, communal, and occupational disability. The symptoms include altered perceptual, cognitive, and emotional states in the auditory and sensory regions. The complications lead to social isolation, social dysfunction, decreased life expectancy, and increased risk of suicide [1]. The whole brain volume seems to be decreased, whereas ventricular volume is increased in SZ subjects [2].

Magnetic resonance (MR) imaging is used to study the brain anatomy. It is a noninvasive, high resolution, and flexible tool without the use of ionizing radiation. In MR images, analysis of different tissues and subcortical regions plays a pivotal role in the detailed investigation of various brain disorders such as Schizophrenia, Alzheimer, Autism, brain tumor, dementia, Huntington's disease, and multiple sclerosis [3].

Liu et al. [4] used Laws texture feature for automatic classification of Schizophrenia and Alzheimer's disease. Wavelet features are used to identify the first episode of SZ from MR brain images [5]. Most of the existing work focused on region of interest-based methods. The anatomical structural changes in the ventricle and corpus callosum are analyzed [2]. Recently, machine learning techniques are applied to classify the first episode and chronic SZ subjects [6]. Majority of the methods are stated to have hitches based on accuracy, modality, and demographics. Many studies could not find abnormality of the whole brain, and only region-based analysis is reported. The current methods rely on tools.

The MR brain image consists of brain and non-brain tissues. Skull-stripping process is adopted to improve the accuracy of diagnosis, by detaching the non-brain tissues. Skull stripping is a challenging task. The presence of noise and non-homogeneous intensity degrades the quality of image and increases the complexity of skull stripping. An isolated brain obtained from skull stripping is useful for processing, such as registration, inhomogeneity correction, segmentation, and tissue classification [7].

Skull stripping is done using morphology, intensity, and deformable model-based methods [8]. In morphology-based method, initial ROI is selected by thresholding and edge detection method. The thresholding-based method is subtle to noise and intensity inhomogeneities existing in MR images. It fails to segment the region as the structural element depends on size and object shape. The intensity-based methods are sensitive to intensity bias due to poor resolution, noise, and artifacts. A combination of anisotropic diffusion filter along with the boundary detection based on Marr-Hildreth method and morphological operations are used in Brain Surface Extraction (BSE) to disparate the brain region from other tissues [9]. Deformable-based methods are contracted or stretched with respect to the image boundary. They discern both the inner and outer boundaries concurrently. They use a closed curve to separate the cerebral brain into non-brain and brain regions and gives promising results. Skull stripping using Brain Extraction Tool (BET) employs deformable model to position the plane of the brain by forces that are adaptive [10].

Jiang et al. [11] integrated region and boundary-based methods for the extraction of brain. Geometric active contour method is used for skull stripping, and it eliminates the boundary leakage problem [12]. Multispectral adaptive region growing method was adopted for skull stripping and the results are compared with tool-based methods [13]. Nonparametric region-based active contour model, with local histogram distance minimization by Wasserstein distance, is used to segment the images [14].

Zhang et al. [15] used Hu moment invariants and machine learning for pathological brain detection. The same features are applied to recognize the tooth in dental radiograph images [16]. Hu moments are used to determine the degree of energy intense of spectrogram in speech emotion and also to recognize a specified shape in different objects [17, 18]. GLCM features are applied for breast cancer detection [19]. The same features could be used to classify cervical cancer, thyroiditis, and liver tumor [20–22].

In this analysis, Wasserstein-based active contour method is applied to remove the non-brain tissues from normal and Schizophrenic MR images. These results are compared with BET and BSE tools. The results are validated with the similarity and performance measures. Then, the texture features such as Hu moments and GLCM are extracted from skull-stripped brain image. These features are further analyzed to study the differentiation between normal and SZ subjects.

2 Methodology

2.1 Wasserstein-Based Active Contour

The level set method is used to compute and analyze the motion of an object with the velocity. It is nonconvex and needs a reasonable initialization in order to avoid local minima. This could be eliminated by nonparametric region-based active contour method. It is based on Wasserstein distance measurement. The local histogram information is used in order to divide the image into two regions, in each of which the difference of the cumulative distribution function from its median is minimal. The regularization term and region term in energy are used to correct the partition boundaries length and distinguish different regions with the help of histograms of pixel intensity, respectively [14].

The convex total variational model is given as follows:

$$\min_{0 \leq u \leq 1, P_1, P_2} \left\{ E_2(.,.,.|I) = \int |\nabla u(x)| \, \mathrm{d}x + \lambda \int_\Omega W_1(P_1, P_x)u(x)\mathrm{d}x + \lambda \int_\Omega W_1(P_2, P_x)(1 - u(x))\mathrm{d}x \right\}.$$

$$(1)$$

where λ is a multiplier, W_1 is Wasserstein distance, measure the distance of two histograms P_1 and P_2. $P_x(y)$ is the local histogram of image pixel. $G_x(y)$ and $G_y(y)$ are the cumulative distribution function. The energy E_2 is varied with respect to G_1 and G_2 for fixed u. G_1 and G_2 are calculated based on the weighted factor of image region $u(x)$ median of G_x and G_y, respectively.

In order to achieve a fast solution, an auxiliary variable v is added to it. The energy model can be given by:

$$\min_{u, 0 \leq v \leq 1} \int_\Omega |\nabla u(x)| dx + \frac{1}{2\phi} \int_\Omega (u(x) - t(x))^2 dx + \lambda \int_\Omega r(x, G_1, G_2) t(x) dx. \quad (2)$$

$$r(x, G_1, G_2) = \int_0^L |G_1(y) - G_x(y)| - |G_2(y) - G_x(y)| dy. \quad (3)$$

ϕ is a parameter (greater than 0) used to penalize the error between $u(x)$ and $t(x)$. On convergence, $u = t$ is obtained. Here u and r are initialized. Then u and t are updated based on the equations to estimate r.

$$u(x) = t(x) - \phi \operatorname{div} q(x). \quad (4)$$

$$q^{n+1} = \frac{q^n + dt \nabla (\operatorname{div} q^n - t/\phi)}{1 + dt |(\operatorname{div} q^n - t/\phi)|}. \quad (5)$$

where $dt \leq 1/8$. The solution can be derived as:

$$t(x) = \text{maximum} \{ \text{minimum} \{ u(x) - \phi \lambda r(x, G_1, G_2), 1 \}, 0 \} \quad (6)$$

The equations are iterated (60) until converge is obtained. Here, $\phi = 100$, $\lambda = 1/5$ are considered.

2.2 Validation Measures

The images that are skull stripped are compared against ground truth to prove the validity of the algorithm. The segmentation results are analyzed using statistical and similarity measures. The similarity measures such as Dice coefficient (DC), Jaccard Coefficient (JC), and Hausdroff Distance (HD) are used to validate the degree of correspondence between ground truth images and skull-stripped images. The statistical measures such as accuracy and sensitivity and F-score are used to analyze the performance of the segmentation algorithm [13].

2.3 Texture Features

Texture features are used to categorize the structural pattern change in the images. Hu moments and GLCM are used to extract the texture information from the images.

Hu moments are represented as a statistical expectation of variable. Moments and their invariants are used to characterize the patterns in an image [15]. Six absolute (does not depend on location, dimension, and orientation) and one skew (distinguish mirror images) orthogonal invariant are derived by Hu [16].

Gray-level co-occurrence matrix measures the relative occurrence of pixel intensities in a particular spacing and direction between two pixels [19]. Second-order statistical features are computed from GLCM. The texture variations are calculated using spatial dependence matrix based on gray tone $P(i, j)$ separated with a distance (d) for pixels. In this present work, angle (θ) is considered as $0°$ with pixel distance of 1. Twenty-two statistical texture features are extracted from GLCM [21].

3 Results and Discussion

The image dataset used in this work includes normal and abnormal Schizophrenia subjects obtained from National Alliance for Medical Image Computing (NAMIC) database (http://insight-journal.org/midas/collection/view/190). The images that are considered for the analysis are skull stripped by means of Wasserstein-based active contour method and compared with skull-stripping tools such as BET and BSE. Then, the extracted brain is validated with different measures. The texture features such as Hu moment and GLCM-based features are extracted from the skull-stripped image and are analyzed.

The MR brain images ($N = 40$) in axial views are subjected to skull stripping process. It is observed that in some cases BET and BSE-based method was not able to properly segment the brain regions. BET over estimates the brain by comprising non-brain tissues and BSE included some non-brain portions and it also failed to deracinate the brain portions. These methods require parameter changes. The Wasserstein-based active contour method could segment the brain region in majority of the cases. Figure 1a represents the typical abnormal SZ image. The contour formed over the abnormal image is depicted in Fig. 1b. The extracted brain image is depicted in Fig. 1c. The ground truth is represented in Fig. 1d. From Fig. 2, the similarity measures such as DC (0.96), JC (0.92), and HD (0.62) seems to be high for Wasserstein-based active contour compared to BET and BSE tools. This shows that the segmented image obtained from this active contour method is more similar to the ground truth.

(a)   **(b)**   **(c)**   **(d)**

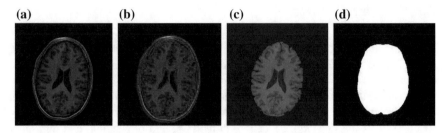

Fig. 1 Skull stripping using Wasserstein-based active contour, **a** typical MR image, **b** contour on segmented image, **c** segmented whole brain, and **d** ground truth

Fig. 2 Similarity measures obtained using Wasserstein-based active contour, BET, and BSE

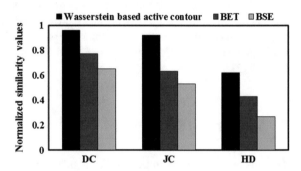

Table 1 Comparison of performance measures for Wasserstein based active contour, BET, and BSE

	Normalized values		
Method ($N = 40$)	Accuracy	Sensitivity	F-Score
Wasserstein-based active contour	0.978	0.934	0.958
BET	0.899	0.634	0.768
BSE	0.872	0.527	0.662

Table 1 indicates the comparison of performance measures for Wasserstein-based active contour with existing brain extraction approaches. It is observed that the performance measures are high in the considered method. The accuracy, sensitivity, and F-score are 0.978, 0.934, and 0.958, respectively. The high accuracy indicates the adherence of resulted segmented brain region with reference to ground truth. The BET and BSE methods showed low accuracy, sensitivity, and F-score. Hence, the Wasserstein-based active contour technique is used to segregate the brain better in the considered subjects compared to other two methods.

Then, Hu moments and GLCM-based features are extracted from the skull-stripped images. The comparison of Hu moments extracted from normal and abnormal images is represented in Table 2. It is observed that Hu moments for abnormal images show higher magnitude value than normal images. This shows

Table 2 Comparison of Hu moments for normal and abnormal images

Hu moment ($N = 40$)	Normalized values	
	Normal	Abnormal
$\phi1$	0.33	0.42
$\phi2$	0.29	0.32
$\phi3$	0.12	0.38
$\phi4$	0.12	0.33
$\phi5$	0.03	0.23
$\phi6$	0.09	0.23

Fig. 3 Percentage variation between normal and abnormal images using Hu moments

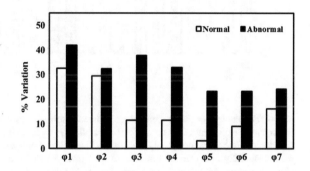

that the Hu moments are able to detect the high-intensity variation in SZ images. The percentage variation among normal and SZ images using Hu moments is depicted in Fig. 3. It is observed that the features such as $\phi3$ (26.3%), $\phi4$ (21.4%), and $\phi5$ (20.1%) show significant difference between normal and abnormal images.

GLCM features do not show better discrimination between normal and abnormal images. GLCM features such as autocorrelation, maximum probability, cluster prominence, sum of square, information measures of correlation 2, inverse difference normalized and moment normalized are high for abnormal images. Autocorrelation shows that pixels in abnormal images are more correlated to each other and has more orderliness. Cluster prominence measures lack of symmetry in the abnormal image. Sum of square indicates the intensity present in abnormal image. The average percentage variation between normal and abnormal image using GLCM is represented in Fig. 4. In GLCM-based features, the features such as sum of squares, autocorrelation, and maximal correlation coefficient show better variation of 19.2%, 18.4%, and 15.6% among normal and SZ images. In GLCM, the difference observed between the normal and SZ subjects seems to be less compared to Hu moments. Hence, Hu moments could be used to study the neuropsychiatric disorder such as schizophrenia. In further, this work can be extended by considering more number of subjects with different texture features and classifiers to study the differentiation between the normal images and SZ subjects.

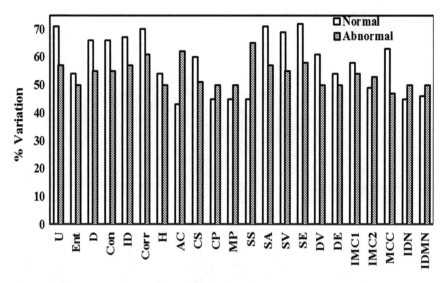

Fig. 4 Percentage variation percentage between normal and abnormal images using GLCM features. *U* Uniformity; *Ent* entropy; *D* dissimilarity; *Con* contrast; *ID* inverse difference; *Corr* correlation; *H* homogeneity; *AC* autocorrelation; *CS* cluster shade; *CP* cluster prominence; *MP* maxi probability; *SS* sum of squares; *SA* sum average; *SV* sum variance; *SE* sum entropy; *DV* difference variance; *DE* difference entropy; *IMC1* information measures of correlation 1; *IMC2* information measures of correlation 2; *MCC* maximal correlation coefficient; *IDN* inverse difference normalized; *IDMN* inverse difference moment normalized

4   Conclusion

In this work, the normal and SZ subjects are analyzed from MR brain images using Wasserstein-based active contour with Hu moments and GLCM features. First, the MR brain images are subjected to skull-stripping process by means of Wasserstein based active contour method. Then the performance of this active contour method is compared with BET and BSE. The analysis showed that the Wasserstein based active contour method gives high-performance measures than BET and BSE methods. It gives the average values of 0.96, 0.92, and 0.62 for DC, JC, and HD similarity measures, respectively. The Wasserstein based active contour could skull strip the MR image with an accuracy of 0.978, sensitivity of 0.934, and F-score of 0.958. Hence, Wasserstein based active contour method could skull strip the brain MR image superior than other tool-based methods. The Hu moments show higher variation between normal and abnormal images compared to GLCM-based features. The Hu moments such as $\phi3$ (26.3%), $\phi4$ (21.4%), and $\phi5$ (20.1%) show significant difference between normal and abnormal images. Hence, the combined framework of Wasserstein based active contour method and Hu moments could be used to aid the diagnosis of Schizophrenic subjects.

References

1. Pawan KS, Sarkar R (2015) A simple and effective expert system for schizophrenia detection. Int J Intell Syst Technol Appl 14(1):27–49
2. Del Re EC, Konishi J, Bouix S, Blokland GA (2015) Enlarged lateral ventricles inversely correlate with reduced corpus callosum central volume in first episode Schizophrenia: association with functional measures. Brain Imaging Behav. https://doi.org/10.1007/s11682-015-9493-2.j
3. Andre GRB, Traina AJM, Ribeiro MX, Paulo MAM, Balan CT (2012) Smart histogram analysis applied to the skull-stripping problem in T1-weighted MRI. Comput Biol Med 42 (5):509–522
4. Liu Y, Teverovskiy L, Carmichael O, Kikinis R, Shenton M, Carter CS, Davis Stenger VA, Davis S, Aizenstein H, Becker JT, Lopez OL, Meltzer CC (2004) Discriminative MR image feature analysis for automatic Schizophrenia and Alzheimer's disease classification. Medical Image Computing and Computer-Assisted Intervention—MICCAI 2004 (Lecture Notes in Computer Science), vol 3216, pp 393–401
5. Dluhos P, Schwarz D, Kasparek T (2014) Wavelet features for recognition of first episode of Schizophrenia from MRI brain images. Radioengineering 23(1):274–281
6. Goulda IC, Shepherda AM, Laurensa KR, Cairns MJ, Carra VJ, Greena MJ (2014) Multivariate neuroanatomical classification of cognitive subtypes in Schizophrenia: a support vector machine learning approach. NeuroImage Clin 6:229–236
7. Somasundaram K, Kalavathi P (2011) Skull stripping of MRI head scans based on chan-vese active contour model. Int J Knowl Manag e-Learn 3(1):7–14
8. Kalavathi P, Prasath VB (2016) Methods on skull stripping of MRI head scan images—a review. J Digit Imaging 29(3):65–79
9. Smith SM (2002) Fast robust automated brain extraction. Hum Brain Mapp 17:143–155
10. Shattuck DW, Sandor-leahy SR, Schaper KA, Rottenberg DA, Leahy RM (2001) Magnetic resonance image tissue classification using a partial volume model. NeuroImage 13:856–876
11. Jiang S, Zhang W, Wang Y, Chen Z (2013) Brain extraction from cerebral MRI volume using a hybrid level set based active contour neighborhood model. Biomed Eng Online 12(31):1–18
12. Zhang H, Liu J, Zhu Z, Haiyun L (2011) An automated and simple method for brain MR image extraction. BioMed Eng Online. https://doi.org/10.1186/1475-925X-10-81
13. Roura E, Oliver A, Cabezas M, Vilanova JC, Rovira À, Ramio-Torrenta L, Llado X (2014) MARGA: multispectral adaptive region growing algorithm for brain extraction on axial MRI. Comput Methods Programs Biomed 113:655–673
14. Kangyu N, Bresson X, Chan T, Esedoglu S (2009) Local histogram based segmentation using the Wasserstein Distance. Int J Comput Vis 84:97–111
15. Zhang Y, Jianfei Y, Shuihua W, Zhengchao D, Preetha P (2015) Pathological brain detection in MRI scanning via Hu moment invariants and machine learning. J Exp Theor Artif Intell. https://doi.org/10.1080/0952813X.2015.1132274
16. Pattanachai N, Covavisaruch N, Sinthanayothin C (2014) Tooth recognition in dental radiographs via Hu's moment invariants. In: IEEE international conference on mechatronics and automation, pp 1581–1586
17. Sun Y, Wen G, Wang J (2015) Weighted spectral features based on local Hu moments for speech emotion recognition. Biomed Signal Process Control Biomed Signal Process Control 18:80–90
18. Zhang HF, Zhang X (2011) Shape recognition using a moment algorithm. In: International conference on multimedia technology, pp 3226–3229
19. Beura S, Majh B, Ratnakar D (2015) Mammogram classification using two dimensional discrete wavelet transform and gray-level co-occurrence matrix for detection of breast cancer. Neurocomputing 154:1–14
20. Mariarputham EJ, Stephen A (2015) Nominated texture based cervical cancer classification. Comput Math Methods Med 586928:1–10

21. Shin YG, Yoo J, Kwon HJ, Hong JH, Lee HS, Yoon JH, Kim EK, Moon HJ, Han K, Kwak JY (2016) Histogram and gray level co-occurrence matrix on gray-scale ultrasound images for diagnosing lymphocytic thyroiditis. Comput Biol Med 75:257–266
22. Xian G (2010) An identification method of malignant and benign liver tumors from ultrasonography based on GLCM texture features and fuzzy SVM. Expert Syst Appl 37:6737–6741

Anticipatory Postural Adjustments for Balance Control of Ball and Beam System

Vinayak Garg, Astik Gupta, Amit Singh, Yash Jain,
Aishwarya Singh, Shashanka Devrapalli and Jagannath Mohan

Abstract Designing a robust controller for nonlinear unstable system involves major constraints such as nonlinearity, disturbance and uncertainty. Therefore, a control strategy realization for such unstable systems generally explores interests among scientific community. The objective of the present study was to design a control strategy of anticipatory postural adjustments for balancing of ball and beam system. Due to the characteristics like unstable, nonlinear and double-integrating, the ball and beam system is commonly considered as a benchmark control operation for estimating several control strategies. Although a PID, i.e. proportional–integral–derivative controller, can be developed for stable system, for double-integrating unstable system, it is very less common. In order to control the ball's position, a PID controller as non-model-based control strategy was presented in this study. To obtain the dynamics of ball and beam system, Euler–Lagrangian approach is used. The system's prototype was built using UV sensor as feedback network; servo-motor as an actuator; and Arduino microcontroller as PID controller. Simulation models such as set-point tracking and disturbance rejection analysis of ball and beam system are performed using MATLAB/Simulink in order to evaluate the performance of proposed system. The outcomes revealed that the proposed system provides increased stability and reduced percentage of error. Along with these, transient response is also improved and the system is more robust against disturbance.

Keywords Ball and beam system · Unstable double-integrating system
Euler–Lagrangian approach · PID controller · Stability

V. Garg · A. Gupta · A. Singh · Y. Jain · A. Singh · S. Devrapalli · J. Mohan (✉)
School of Electronics Engineering, VIT University, Chennai, Tamil Nadu, India
e-mail: jagan.faith@gmail.com

© Springer Nature Singapore Pte Ltd. 2018
A. K. Nandi et al. (eds.), *Computational Signal Processing and Analysis*, Lecture Notes in Electrical Engineering 490,
https://doi.org/10.1007/978-981-10-8354-9_4

33

1 Introduction

Accurate and reliable control of instrumentation system is a challenging assignment among researchers. Many control strategies have been established from the ancient years. The characteristics, unstable, nonlinear and double-integrating system [1, 2] of ball and beam balancing system, are the contributing factor for using it as a benchmark control operation and estimation of several control strategies. The primary model of ball and beam system can be implemented to a balancing issue of different control systems like landing stabilization of an aircraft and the balancing problem of robots for transport of goods [3, 4]. The important motivating factor of ball and beam balancing system is that it is a double-integrating unstable system, whose control strategy is not well defined [5].

Euler–Lagrangian approach is used to obtain the dynamics of ball and beam system. The prototype of ball and beam system was built using UV sensor as feedback network; servomotor as an actuator; and Arduino microcontroller as proportional–integral–derivative (PID) controller. The evaluation of K_p, K_i and K_d, i.e. the controller parameters, is performed by various methods, which includes the time-domain, frequency-domain and Ziegler–Nichols method.

2 System Dynamics

The ball and beam system is made up with a support frame to keep up the beam from one side where the UV sensor is placed. The other side of the beam is attached to a lever arm whose other end is connected to a servomotor. The ball is free to rotate along the length of the beam which in other words means that its degree of freedom is 2, i.e. rotational and translational. The position of the ball calculated by the UV sensor is given as the feedback to the system, and based on the difference between the desired and actual position of the ball, i.e. error, the corrective measure is taken by rotating the motor in the suitable direction by the suitable angle which is decided by the controller. The fact is that open-loop system is unstable, and feedback becomes necessary for the proposed system. The mathematical modelling of the system is done using the simplified Lagrangian equation for the motion of the ball. The required parameters for calculation of the transfer function are provided in Table 1.

The mathematical model of the ball and beam system is derived using Lagrangian approach [6] (Eqs. 1–6):

The simplified Lagrangian equation is given as

$$\left(m + \frac{J}{R^2}\right)\ddot{r} + \frac{J}{R\ddot{\alpha}} - mr\dot{\alpha}^2 + mg\sin\alpha = 0 \tag{1}$$

Linearizing above about beam angle, $\sin\alpha = \alpha$, for small α, gives

Table 1 Parameters of ball and beam system

Symbol	Description	Values
M	Mass of the ball	0.11 kg
R	Radius of the ball	0.015 m
d	Lever arm offset	0.03 m
G	Gravitational acceleration	-9.8 m/s^2
L	Length of the beam	1 m
J	Ball's moment of inertia	$9.99\ e^{-6}$ kg m^2
r	Ball position coordinate	–
α	Beam angle coordinate	–
θ	Servomotor angle coordinate	–

$$\left(m + \frac{J}{R^2}\right)\ddot{r} = mg\alpha \tag{2}$$

Relating beam angle to servomotor angle by

$$\alpha = \left(\frac{d}{L}\right)\theta \tag{3}$$

Substituting above equation in (2) and taking Laplace transform give

$$\left(m + \frac{J}{R^2}\right)s^2 r(s) = mg\left(\frac{d}{L}\right)\theta(s) \tag{4}$$

Writing the above equation in form of a transfer function

$$\frac{r(s)}{\theta(s)} = \frac{mgd}{L\left(m + \frac{J}{R^2}\right)}\frac{1}{s^2} \tag{5}$$

Substituting the value of all the parameter gives

$$\frac{r(s)}{\theta(s)} = \frac{0.21}{s^2} \tag{6}$$

3 Controller Design

The transfer function given by Eq. (6) shows the system to be an undamped second-order system. Based on the fact, theoretically this system can be categorized as "marginally" stable, but in practical conditions it turns out to be unstable. Hence, the appropriate controller to be designed makes the system stable and meets the performance requirements.

3.1   Time-Domain Specifications

In this section, we set the required time-domain specifications in order to tune the controller. As far as the time-domain specifications are concerned, an upper limit on the settling time and the maximum overshoot were set to less than 3 s and 5%, respectively.

Firstly, calculating the steady-state error (e_{ss}),

$$e_{ss} = \lim_{s \to 0} \frac{sR(s)}{1 + G(s)H(s)} \tag{7}$$

Considering a unity feedback, i.e. $H(s) = 1$, and unit step input, i.e. $R(s) = \frac{1}{s}$, the steady-state error becomes zero, as shown in Eq. (8)

$$e_{ss} = 0 \tag{8}$$

Since the error is zero, we can set the integral gain, K_i, to zero as well. The closed-loop transfer function of the system is given by Eq. (9). The poles of this system lie on the imaginary axis which implies that the system is marginally stable.

$$\frac{C(s)}{R(s)} = \frac{0.21}{0.21 + s^2} \tag{9}$$

Therefore, we can choose the proportional gain, K_p, as any number greater than zero. With K_i being zero, the derivative gain, K_d, also needs to be evaluated which can be found by using time-domain specification. Let K_p be assumed as 15.

For the above system, calculating the steady-state error given by Eq. (10)

$$e_{ss} = \lim_{s \to 0} \frac{1}{s^2 + 0.2\,sK_d + 3} \tag{10}$$

Equating with the standard system equation of second-order system, we get $\omega_n = \sqrt{3}$ and $2\varsigma\omega_n = 0.2\,K_d$. We also know the maximum overshoot for the system is 5%; from this, ς can be calculated from Eq. (11) and found to be $\varsigma = 0.69$.

$$MO\% = 100\,e^{\frac{-\varsigma\pi}{\sqrt{1 - \varsigma^2}}} \tag{11}$$

Hence, the derivative gain is found as $K_d \geq 11.95$. This is the boundary value of K_d, and it can be increased to decrease the maximum overshoot of the system. The increase in the value of K_p will result in reduction of settling time of the system but a slight increase in the rise time.

The open-loop transfer function of the system with controller designed using time-domain specifications is depicted in Eq. (12). Equation (13) shows the closed-loop transfer function with unity feedback.

$$G'(s) = \frac{3.15 + 2.51s}{s^2} \tag{12}$$

$$\frac{C'(s)}{R'(s)} = \frac{3.15 + 2.51s}{s^2 + 2.51s + 3.15} \tag{13}$$

To check the stability of the system after cascading the controller with the plant, Routh stability test is performed. Considering the denominator of Eq. (13), i.e. the characteristic equation, Routh stability test is performed and the result shows there is no sign change found in the first column of the Routh table. The result concluded that the system is stable after cascading the controller.

3.2 Frequency-Domain Specifications

From the frequency response of the open-loop transfer function of the system, we can find out the phase margin offered by the system, and from there, a check can be made whether the system is stable or not. From Eq. (6) which is the open-loop transfer function, the magnitude (Eq. 14) and phase response of the system can be calculated.

$$|G(j\omega)|_{dB} = -40 \log \omega - 0.7 \tag{14}$$

Looking at the open-loop transfer function, it is fairly easy to conclude that the phase response of the system is constant ($\angle G(j\omega) = -180°$) for all values of ω. It can be seen that the phase margin of the system is zero since at the gain crossover frequency; the phase of the system is $-180°$. For the system to be stable, the phase margin has to be positive which means that this system is either unstable or marginally stable. Therefore, to rectify this issue, a lead controller is to be added.

Equation (15) shows the open-loop transfer function of phase lead compensator.

$$G_c(s) = K\left(\frac{1 + Ts}{1 + \alpha Ts}\right) \tag{15}$$

In this study, the following assumptions are made: damping ratio (ζ) is set to 0.7 (approximately) with the phase margin (Φ) as 100 times the damping factor. The bandwidth is assumed to 1.75 rad/s. The centre frequency is half of the bandwidth and found as $\omega \simeq 1$ rad/s. α and T can be written as a function of phase margin Φ and centre frequency, which is given in Eqs. (16) and (17).

$$\alpha = \frac{1 - \sin \Phi}{1 + \sin \Phi} \tag{16}$$

$$T = \frac{1}{\omega\sqrt{\alpha}} \tag{17}$$

α and T are found to be 0.0311 and 5.67, respectively. The value of K is set to 1, and Eqs. (18) and (19) give the controller's transfer function and the system's open-loop transfer function after cascading the lead controller, respectively.

$$Gc(s) = \frac{1+5.67s}{1+0.176s} \tag{18}$$

$$G''(s) = \frac{1.254s + 0.21}{0.0176s^3 + s^2} \tag{19}$$

The compensated system stability is verified with Routh stability test. As shown in Fig. 1, an additional phase of 80 degree has made the system stable.

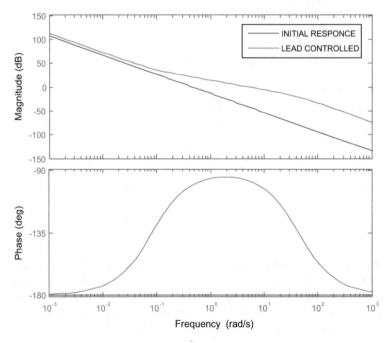

Fig. 1 Bode plot of the uncompensated system and compensated system designed using frequency-domain specifications

Table 2 Formulae for the controller parameters in the Ziegler–Nichols closed-loop system

Control	K_p	T_i	T_d
P	0.5 K_{cr}		0
PI	0.45 K_{cr}	0.33 T_{cr}	0
PID (tight)	0.6 K_{cr}	0.5 T_{cr}	0.125 T_{cr}
PID (little overshoot)	0.33 K_{cr}	0.5 T_{cr}	0.33 T_{cr}
PID (no overshoot)	0.2 K_{cr}	0.3 T_{cr}	0.5 T_{cr}

3.3 Ziegler–Nichols Tuning Method

The Ziegler–Nichols tuning method as the name suggests was proposed by John G. Ziegler and Nathaniel B. Nichols and is another way to tune the PID controller. The rule says that integral and derivative gains (K_i and K_d) are initialized to zero, and then, the proportional gain (K_p) is incremented from zero to its "ultimate gain (K_{cr})" at which the output has stable oscillations. The period of oscillation is defined as "critical time period (T_{cr})". From these K_{cr} and T_{cr} values, K_p, K_i and K_d can be calculated using Table 2.

It is important to keep in mind that the system is already undamped and critically stable. Choosing PID controller with little overshoot, the following values are obtained: $K_p = 8.8235$, $K_i = 4.81282$ and $K_d = 19.2513$. Therefore, the transfer function of the open-loop compensated system after cascading the PID controller is expressed in Eq. (20).

$$G'''(s) = \frac{4.043s^2 + 1.853s + 1.011}{s^3} \tag{20}$$

The compensated system has characteristic equation expressed as $s^3 + 4.043s^2 + 1.853s + 1.011 = 0$. From the characteristic equation, the stability test is performed and the result shows that the system becomes stable after cascading the PID controller with the original system.

4 Simulation Results and Discussion

For any dynamic system like ball beam system, stability is not the only criterion which determines the performance of the system. Many other specifications such as the settling time and maximum overshoot also need to be checked before deciding on which controller design method should be implemented for the final project. A comparative study is done for all the design procedures described/used in this study to find out which method provides the best performance with respect to stability as well as time-domain specifications.

The system responses of transfer functions obtained from the three methods are simulated using MATLAB/Simulink. In the PID controller, different gain values

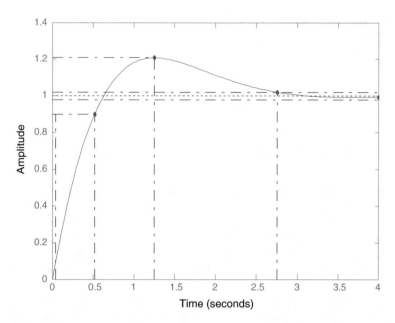

Fig. 2 Step response of the compensated system from time-domain specifications

obtained by different methods can be input and the system response can be observed. Figure 2 depicts the step response obtained from the compensated system designed from the time-domain specifications.

From Fig. 2, the rise time calculated is 0.477 s, overshoot is 20.8%, and the settling time is 2.76 s. Comparing with the initial conditions, the settling time and overshoot do not match which are because of the assumption and the trial-and-error method used to derive the value of K_p.

Figure 3 depicts the compensated system's step response designed from the frequency-domain specifications. The rise time of the above system is 1.14 s, overshoot is 10.8%, and the settling time is 12 s. Compared to the previous case, there is an improvement in performance in terms of overshoot, but the system's settling time has increased to 12 s. Figure 4 depicts the step response of the system consisting of the controller designed using the Ziegler–Nichols method.

The rise time is 0.42 s, overshoot is 11.6%, and the settling time is 9.28 s. The performance of this system can be considered almost equivalent to the previous system. The overshoot and the settling time are close to each other, but the rise time is decreased almost 3 times. Figure 5 shows the comparison plot of step responses obtained from T-D, F-D and Z-N methods. Table 3 depicts the performance parameters obtained from the three methods.

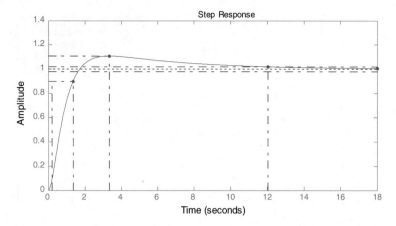

Fig. 3 Step response of the compensated system from frequency-domain specifications

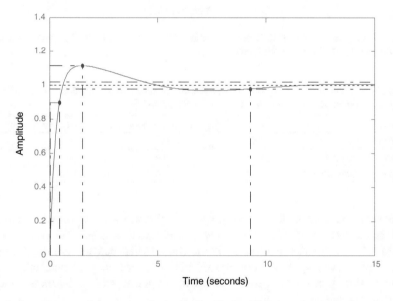

Fig. 4 Step response of the system obtained using Ziegler–Nichols tuning method

5 Conclusion

In this paper, the control strategies of anticipatory postural adjustments for balance control of ball and beam system are designed using time-domain, frequency-domain and Ziegler–Nichols methods. The controller parameters are obtained from the methods and being incorporated in the original ball and beam system. The comparison study of the three methods is also been made in the view of performance

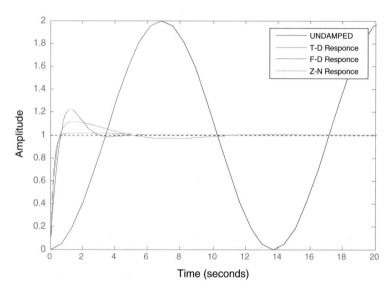

Fig. 5 Comparison plot of step responses obtained from time-domain (T-D), frequency-domain (F-D) and Ziegler–Nichols (Z-N) methods

Table 3 Summaries of the performance parameters obtained from time-domain specification, frequency-domain specification and Ziegler–Nichols method

Methods	Rise time (s)	Overshoot (%)	Settling time (s)
Time-domain	0.477	20.8	2.76
Frequency-domain	1.14	10.8	12
Ziegler–Nichols Method	0.42	11.6	9.28

parameters. To analyse the proposed system's performance, the set-point tracking and the disturbance rejection analysis were performed out in simulations supported by MATLAB/Simulink platform. The results revealed that the proposed system retains good stability of the position of ball with reduced error percentage. The system also provides improved transient response with robustness against disturbance.

References

1. Ker CC, Lin CE, Wang RT (2007) Tracking and balance control of ball and plate system. J Chin Inst Eng 30(3):459–470
2. Lin CE, Ker CC (2008) Control implementation of a magnetic actuating ball and plate system. Int J Appl Electromagn Mech 27(1–2):133–151
3. Yu W (2009) Nonlinear PD regulation for ball and beam system. Int J Electr Eng Educ 46:37–59

4. Chang YH, Chang CW, Tao CW, Lin HW, Taur JS (2012) Fuzzy sliding-mode control for ball and beam system with fuzzy ant colony optimization. Expert Syst Appl 39(3):3624–3633
5. Herman W, Mohd F (2009) A study of different controller strategies for a ball and beam system. J Teknol 50(D):93–108
6. Carlos G, Gerson B (2014) Modelling the ball and beam system from Newtonian mechanics and from Lagrange methods. In: Proceedings of Latin American and Caribbean Conference for Engineering and Technology (LACCEI), Ecuador, vol 1, pp 1–9

A New Extended Differential Box-Counting Method by Adopting Unequal Partitioning of Grid for Estimation of Fractal Dimension of Grayscale Images

Soumya Ranjan Nayak, Jibitesh Mishra and Rajalaxmi Padhy

Abstract Fractal dimension (FD) is most useful research topic in the field of fractal geometry to evaluate surface roughness of digital images by using the concept of self-similarity, and the FD value should lie between 2 and 3 for surfaces of digital images. In this regard, many researchers have contributed their efforts to estimate FD in the digital domain as reported in many kinds of the literature. The differential box-counting (DBC) method is a well-recognized and commonly used technique in this domain. However, based on the DBC approach, several modified versions of DBC have been presented like relative DBC (RDBC), improved box counting (IBC), improved DBC (IDBC). However, the accuracy of an algorithm for FD estimation is still a great challenge. This article presents an improved version of DBC algorithm by partitioning the box of grid into two asymmetric patterns for more precision box count and provides accurate estimation of FD with less fit error as well as less computational time as compared to existing method like DBC, relative DBC (RDBC), improved box counting (IBC), and improved DBC (IDBC).

Keywords Fractal dimension · DBC · RDBC · SDBC · IBC · IDBC

S. R. Nayak (✉) · R. Padhy
Department of Information Technology, College of Engineering and Technology,
Bhubaneswar 751003, Odisha, India
e-mail: nayak.soumya17@gmail.com

R. Padhy
e-mail: padhyrajalaxmi123@gmail.com

J. Mishra
Department of Computer Science and Application, College of Engineering and Technology,
Bhubaneswar 751003, Odisha, India
e-mail: jmishra@cet.edu.in

© Springer Nature Singapore Pte Ltd. 2018
A. K. Nandi et al. (eds.), *Computational Signal Processing and Analysis*, Lecture Notes in Electrical Engineering 490,
https://doi.org/10.1007/978-981-10-8354-9_5

1 Introduction

Fractal geometry was popularized by Mandelbrot [1] to interpret and characterize complex shapes and surfaces with self-similarity in nature, like mountain, coastlines, and clouds. The concept self-similarity is most important to describing the textures images and useful for estimation of fractal dimension (FD) of complicated objects found in nature, which is failed to analyze by the traditional Euclidean geometry. Fractal dimension can be used as a most important feature in the field of image processing to characterize roughness, smoothness, and self-similarity of natural images. Pentland [2] popularized about smoothness versus roughness of image surface, where the smooth image indicates the fractal dimension of two and rough image as three. In this regard, various researchers have developed many box-counting algorithms. Fractal geometry has provided a numerical model for various complex and irritated objects that initiate in nature [1–3], in terms of coastline, mountains, and clouds. Fractal dimension (FD) is also useful in several fields of application that including analysis of texture and texture segmentation [4–6], shape measurement and texture classification [7], the study of the image as well as graphic analysis in other fields [8–10]. Various popular fractal dimension theories have been proposed and suggested by many researchers for grayscale and color images. Gangepain has presented the well-known reticular box-counting algorithm [11], and Keller et al. presented probabilistic theories [12].

Sarkar and Chaudhuri [4, 13] studied five different algorithms in box counting on synthetic images and suggested the proficient differential box-counting (DBC) algorithm to computing fractal dimension of images. Their results have shown that Keller et al. [12] give the satisfactory result, after reaching a particular level of noise (s), and image intensity surface with the slope of the curve tends to zero. As the natural scenes rather reveal some statistical self-similarity, not deterministic self-similarity, therefore both the methods do not estimate the dynamic range of FD; the reduction factor r was introduced so that if an object is broken down with a factor r in all n dimensions, then we can call as statistically alike to the original one to satisfy Eq. (1) that described in Sect. 2.

DBC algorithm was considered to be an efficient method presented in [14, 15]. However, in the case of DBC method there are three major drawbacks reported by Li et al. [16], i.e., selection of proper height of the box, calculation of exact box number and partition of image plane. In order to avoid this situation, the same author presented another method called an improved box-counting method [16] in order to avoid the three above-mentioned drawbacks of DBC method. Again some demerits found in DBC method are reported by Liu et al. [17] that over-counting and under-counting problem found while executing the method. To avoid such situation, the same author [17] presented another method called improved differential box-counting method. Woraratpanya [18] proposed triangle box counting (TBC) to improve the exactness, though this algorithm is only meant for binary images. Recently, an improved triangle box-counting method (ITBC) was proposed by Kaewaramsri and Woraratpanya [19] based on DBC approach to solve over-counting problem by implementing triangle box partition. Nayak et al. [20] presented simple and proficient improved version of DBC algorithm for more accurate FD estimation of grayscale images. Another modified DBC method also presented by same author Nayak et al. [21] to avoid shortcomings like over-counting and under-counting and simultaneously provides less fitting error for grayscale images.

2 Basic Principle of Fractal Dimension and Related Work

There are several dimensions to express fractal dimensions in the field of fractal geometry, such as Hausdorff dimension, information dimension, packing dimension, box-counting dimension. Among these dimensions, box-counting dimension is most popular and widely used techniques in fractal geometry to express the concept of self-similarity. Fractal dimension can be evaluated by using equation below:

$$D = \log(N)/\log(1/r) \tag{1}$$

2.1 DBC Approach

Sarkar and Chaudhuri [13] projected the differential box-counting (DBC) method for estimating FD of grayscale images. In order to implement this method, they represent grayscale image in $3D$ space, where $2D$ space like (x, y) represents an image plane and third coordinates like z represent the gray level. For this experimental analysis, they took a square image of size $M \times M$ and partitioned into $L \times L$ grids. Each and every grid comprises a stake of boxes of size $L \times L \times H$, where H indicates the height of every box, and this height can be calculated in terms of $L \times G/M$, where G represents the total number of gray levels. Let the maximum and minimum gray values of (i,j)th grid fall in Lth and Kth box, respectively, and then the box count $n_r(i,j)$ can be calculated as follows:

$$n_r(i,j) = L - K + 1 \tag{2}$$

By taking involvement from all blocks, N_r is counting for different values of L as follows:

$$N_r = \sum_{i,j} n_r(i,j) \tag{3}$$

2.2 RDBC Approach

Based on original DBC, Jin and Jayasooriah [22] presented an improved version of DBC called relative DBC (RDBC) by adopting same maximum and minimum intensity point on the grid and taking the scale limit such as higher and smaller limits of scale ranges for accurate evaluation of FD of texture images. Finally, N_r can be evaluated as follows:

$$N_r = \sum_{i,j} \text{ceil}[k \times ((K - L)/L')] \qquad (4)$$

where k represents the coefficient in z-direction and ceil(.) is used to set the nearest integer.

2.3 IBC Approach

In similar like DBC and RDBC, Li et al. [16] presented another improved DBC mechanism by adopting three major parameters like the selection of box height, box number estimation, and partition of intensity surface. They are selecting box height by using the formula as follows:

$$r' = \frac{L}{1 + 2a\sigma} \qquad (5)$$

where a is a positive integer and set the appropriate value as 3, σ represents standard deviation, and $2a\sigma$ represents image roughness. Finally, $n_r(i,j)$ are evaluated as follows

$$n_r(i,j) = \begin{cases} \text{ceil}\left(\frac{K-L}{r'}\right) & \text{if } K \neq L \\ 1 & \text{otherwise} \end{cases} \qquad (6)$$

N_r can be calculated by taking the contribution of all grids.

2.4 IDBC Approach

Liu et al. [17] proposed improved version of DBC algorithm called improved differential box-counting method (IDBC) for estimating FD of grayscale image. In their proposed method, three modifications have been done such as modifying box counting, recalibrating box of block in spatial plane of (x, y) direction, and choosing proper grid box size and final $n_r(i,j)$ was calculated by taking the maximum contribution from original grid and shifted grid by using Eq. (2)

$$n_r(i,j) = \begin{cases} \text{ceil}\left(\frac{I_{max} - I_{min} + 1}{s'}\right) I_{max} I_{min} \\ 1 \quad \text{otherwise} \end{cases} \tag{7}$$

N_r can be calculated by taking the contribution of all grids, and finally, the FD can be estimated for all these methods from the least square regression line of $\log(N_r)$ verses $\log(1/r)$.

After analyzing above-mentioned methods, we conclude that the improvement technique is implemented to evaluate FD based on two major issues, namely over-counting and under-counting problems. Most of the techniques [22, 16] attempt to solve only under-counting by implementing box height. On the other hand, the technique [19] attempts to solve over-counting problems and few techniques like [17, 21] tried to solve both the problems. All these methods provide more fitting error as well as more computational time; hence, no proper box-counting technique is presented. Therefore, this article presented more accurate estimation by partitioning the box of the grid into two asymmetric patterns for more precision box count and yields more accurate estimation in terms of less fit error as well as less computational time.

3 Proposed Methodology

In this section, our improved proposed methods are discussed. The improvement can be achieved by partitioning the grid box into two asymmetric patterns for more precision box count, which is represented in Fig. 1.

Fig. 1 Sktech of partitioning grid into two asymmetric pattern

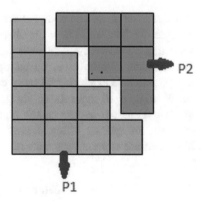

3.1 Algorithm Details

In this section, the detailed algorithm of our proposed method is summarized below.

1. Read the image of size $M \times M$.
2. Divide the image into the box of size $L \times L$, where L varies from 2 to $M/2$.
3. Estimate height of grid $H = L \times G/M$, and each box size should be $L \times L \times H$, where G represents total gray level.
4. Partitioning the box of grid into two asymmetric patterns such as $p1$ and $p2$ based on step 3.
5. Determine $n_r(i,j)$ for each pattern in Fig. 1 based on Eq. (8), where max and min represents maximum and minimum intensity point at each pattern and NP represents total no of pixel points present at each pattern.

$$
\begin{aligned}
P_1 &= \text{ceil}(((\max - \min) + 1)/H) \times (\text{NP}/(L \times L)) \\
P_2 &= \text{ceil}(((\max - \min) + 1)/H) \times (\text{NP}/(L \times L))
\end{aligned}
\tag{8}
$$

6. Final $n_r(i,j)$ can be calculated by taking maximum contribution from each pattern using Eq. (9).

$$
n_r(i,j) = \max\{P_1, P_2\}
\tag{9}
$$

7. For the distinct value of L, the entire amount of boxes required to wrap entire image surface is evaluated using Eq. (3).
8. Plot the least square linear fit of $\log(N_r)$ versus $\log(1/r)$ to estimate FD from negative slope of the fitted straight line.

4 Experimental Setup

In order to evaluate and compare the performance of our proposed method, we are considering four well-known methods such as DBC, RDBC, IBC, and IDBC. All these algorithms are implemented in MATLAB (R2014a, 64 bit) with the standard environment (Windows 8, 64-bit operating system, Intel (R) i7–4770 CPU @ 3.40 GHz). The experimental setup carried out on three experiments, which have a set of original 16 real brodatz images [23] presented in Fig. 2, one set of twelve generated smooth texture image represented in Fig. 6, one set of fifty generated synthetic images by different gray level shifts, the precision of FD estimation algorithm are express by means of fitting error as follows in Eq. (10)

$$
\text{Error fit} = \frac{1}{n} \sqrt{\sum_{i=1}^{n} \frac{dx_i + c - y_i}{1 + d^2}}
\tag{10}
$$

Fig. 2 Sixteen Brodatz database texture images

4.1 Test on Real Texture Images

In this section, we are using a set of 16 real texture images [23] from the Brodatz database of size 256×256 for our experimental analysis which is presented in Fig. 2. In this experimental setup, the FD generated from DBC algorithm are in the range of 2.19 to 2.67 similarly, the other measure like RDBC, IBC, IDBC and proposed method are range from 2.28 to 2.73 and 2.29 to 2.73 and 2.30 to 2.72 and 2.34 to 2.77 which are represented in Table 1 and error fit are listed in Table 2 and graphical representation are presented in Figs. 3 and 4 respectively. The average error fit is estimated from each method like DBC, RDBC, IBC, IDBC, and proposed method are 0.0418, 0.0669, 0.0681, 0.0453, 0.0419 respectively are listed in Table 5 and presented in Fig. 5. From this experimental result, it is crystal clear that the fitting error from proposed method yields less average error fit as compared to other existing four chosen methods as well as less fitting error for individual image presented in Fig. 4, and provides less computational time presented in Table 4 and Fig. 6.

4.2 Test on Synthetic Images1

In this section, we have generated fifty transformed images based on Eq. (11). Resultant generated images are achieved from original a100 Brodatz images.

$$\text{RTI}(i,j) = \text{round}\left(\frac{I(i,j) - \min(i,j)}{\max(i,j) - \min(i,j)} \times W\right) + (G/2 - 2) \tag{11}$$

where RTI is the resultant transformed image generated from original image $I(i,j)$, $round(.)$ is used for rounding to nearest integer, $\max(i,j)$ and $\min(i,j)$ are

Table 1 Computational FD of Brodatz images

Image name	Fractal dimension				
	DBC	RDBC	IBC	IDBC	Proposed
a8	2.27	2.36	2.39	2.40	2.43
a11	2.59	2.67	2.68	2.66	2.71
a23	2.58	2.61	2.62	2.63	2.67
a38	2.51	2.58	2.60	2.57	2.61
a55	2.67	2.73	2.73	2.72	2.77
a56	2.53	2.60	2.61	2.63	2.65
a62	2.50	2.55	2.57	2.55	2.58
a69	2.51	2.53	2.55	2.56	2.60
a71	2.54	2.54	2.56	2.58	2.62
a89	2.42	2.49	2.52	2.50	2.54
a90	2.30	2.43	2.45	2.47	2.51
a91	2.19	2.28	2.29	2.30	2.34
a93	2.60	2.65	2.67	2.66	2.70
a98	2.42	2.50	2.51	2.46	2.50
a99	2.41	2.48	2.49	2.48	2.52
a100	2.57	2.66	2.67	2.61	2.66

Table 2 Computational EF of Brodatz images

Image name	Error fit				
	DBC	RDBC	IBC	IDBC	Proposed
a8	0.0686	0.0757	0.0779	0.0665	0.0623
a11	0.0525	0.0550	0.0561	0.0449	0.0412
a23	0.0655	0.0677	0.0681	0.0585	0.0559
a38	0.0453	0.0478	0.0496	0.0356	0.0332
a55	0.0497	0.0517	0.0521	0.0400	0.0376
a56	0.0658	0.0683	0.0688	0.0581	0.0573
a62	0.0664	0.0678	0.0695	0.0559	0.0528
a69	0.0558	0.0545	0.0556	0.0459	0.0422
a71	0.0633	0.0617	0.0630	0.0554	0.0519
a89	0.0648	0.0671	0.0687	0.0563	0.0524
a90	0.0570	0.0675	0.0695	0.0630	0.0584
a91	0.0665	0.0756	0.0751	0.0644	0.0590
a93	0.0492	0.0474	0.0498	0.0383	0.0335
a98	0.0647	0.0691	0.0698	0.0546	0.0499
a99	0.0665	0.0690	0.0699	0.0589	0.0552
a100	0.0534	0.0569	0.0576	0.0422	0.0387

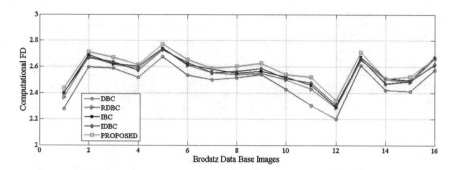

Fig. 3 Computational FD of sixteen Brodatz database images

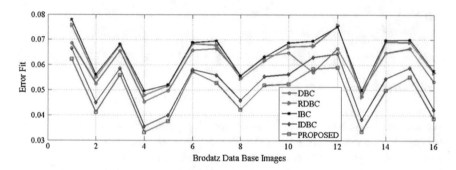

Fig. 4 Computational error fit of sixteen Brodatz database images

Table 3 Computational FD of twelve generated synthetic images

Image name	Fractal dimension				
	DBC	RDBC	IBC	IDBC	Proposed
s1	2.42	2.49	2.52	2.50	2.54
s2	2.42	2.49	2.52	2.50	2.54
s3	2.42	2.49	2.52	2.50	2.54
s4	2.42	2.49	2.52	2.50	2.54
s5	2.42	2.49	2.52	2.50	2.54
s6	2.42	2.49	2.52	2.50	2.54
s7	2.43	2.49	2.52	2.50	2.54
s8	2.43	2.49	2.52	2.50	2.54
s9	2.43	2.49	2.52	2.50	2.54
s10	2.43	2.49	2.52	2.50	2.54
s11	2.44	2.49	2.52	2.50	2.54
s12	2.44	2.49	2.52	2.50	2.54

Table 4 Computational time estimation

Images	Computational time				
	DBC	RDBC	IBC	IDBC	Proposed
Brodatz	69.803	42.184	398.499	112.897	53.051
Synthetic1	24.246	18.834	129.339	38.002	17.682
Synthetic2	16.817	10.204	102.438	26.156	12.655

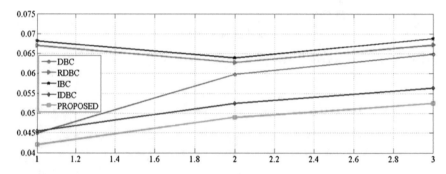

Fig. 5 Computational average error fit

represented maximum and minimum intensity point of original image, W is weighted coefficient varies from $1 \leq W \leq 50$, and G represents gray level; we estimate FD based on different W values, as the value of W increases simultaneously FD also increases and vice versa. Figure 7 represents the estimated FD of fifty generated synthetic images, from this experimental analysis.

We have seen that our proposed methodology yields better FD estimation than other competent methods. As we mention earlier that when the value of W increased means the generated FD also increased accordingly, while method like DBC, RDBC and IBC are failed to estimate accurate FD. When W value ranges from 1 to 21, the FD value from DBC yields FD less than 2 and, and similarly in case of RDBC the FD value also suddenly decreased W27 to W28 images. But in case of IBC the FD value was evaluated as greater than three in the first image that is W1 image and gradually decrease when the W value is increased, while our proposed method provides consistent result in terms of increasing order as increased of W presented on Fig. 5, from above point of view it is crystal clear that our proposed method provides the wider range of FD.

4.3 Test on Synthetic Images2

In this experimental setup, a set of 12 synthetic images are generated, which are presented in Fig. 8. In this process, each image is generated from original image by incrementing each intensity point (by formula $2 \times (k - 2)$, where k lies from 1 to

12 in such a way that at the kth of 12th image, the maximum gray level should not exceed 255, and therefore alternatively we can say that the entire pixel surface is recalibrated to particular position of gray level against the original one in z-direction. Theoretically, it was clear that if we either step up or step down the intensity value with a constant number, then the FD should stay same as original one because theoretically both have the equal degree of roughness. From this experimental result presented in Table 3, we have seen that except DBC all other methods provide same estimated value. In DBC, the gray level is established on the particular area in the z-direction; the first box is in zero gray level position, and the utmost box is in 255 gray level position; this may cause over-counting but in case of proposed method. In this experimental analysis, our proposed method provides accurate estimation with less fit error and less computational time listed in Tables 4 and 5 respectively.

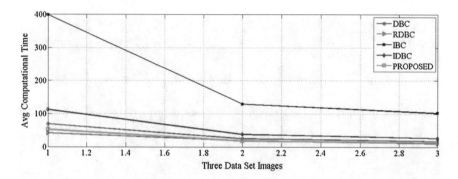

Fig. 6 Estimation of average computational time

Table 5 Average error fit estimation

Images	Average error fit				
	DBC	RDBC	IBC	IDBC	Proposed
Brodatz	0.0448	0.0669	0.0681	0.0453	0.0419
Synthetic1	0.0597	0.0627	0.0638	0.0524	0.0488
Synthetic2	0.0648	0.0671	0.0687	0.0563	0.0524

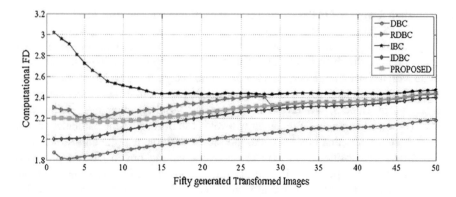

Fig. 7 Fractal dimension estimation of fifty generated synthetic images

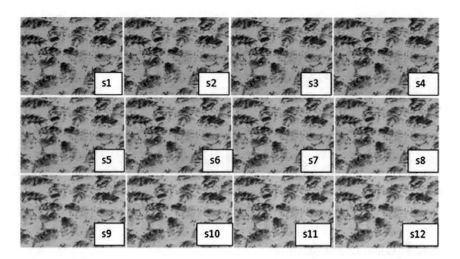

Fig. 8 Twelve generated synthetic images

5 Conclusion

In this article, we presented improved differential box-counting by partitioning the box of the grid into two asymmetric patterns for accurate estimation of FD of grayscale images. Three experiments were carried out to test our proposed method, and the experiment results are compared with our well-known methods such as DBC, RDBC, IBC, and IDBC. The result is generated from Brodatz database, and synthetic images show that the proposed method eradicated with maximum efficiency in terms of less fitting error for each image and also produced average less fitting error as compared to other chosen methods. It shows that our modified method covers all objects with the wider range of fractal dimension as compared to existing methods. It is a robust and more precise method.

Acknowledgements The authors are sincerely thankful to the Department of Information Technology, College of Engineering and Technology, Bhubaneswar. And we are also thankful to all the authors of references.

References

1. Mandelbrot BB (1982) The fractal geometry of nature. Freeman, San Francisco
2. Pentland AP (1984) IEEE Trans Pattern Anal Mach Intell 6
3. Peitgen HO, Jurgens H, Saupe D (1992) Chaos and fractals: new frontiers of science, 1st edn. Springer, Berlin
4. Chaudhuri BB, Sarkar N (1995) IEEE Trans Pattern Anal Mach Intell 17
5. Liu S, Chang S (1997) IEEE Trans Image Process 6 (1997)
6. Ida T, Sambonsugi Y (1998) IEEE Trans Circ Syst Video Technol 8
7. Neil G, Curtis KM (1997) Pattern Recogn 30
8. Lin KH, Lam KM, Siu WC (2001) IEEE Proceedings on vision, image and signal processing 148
9. Asvestas P, Matsopoulos GK, Nikita KS (1998) J Vis Commun Image Represent 9
10. Bisoi AK, Mishra J (2001) Pattern Recogn Lett 22
11. Gangepain J, Carmes CR (1986) Wear 109
12. Keller J, Crownover R, Chen S (1989) Comput Vis Graph Image Process 45
13. Sarkar N, Chaudhuri BB (1994) IEEE Trans Syst Man Cybern 24
14. Wenlu X, Weixin X (1997) Chin Signal Process J 13
15. Yu L, Zhang D, Wang K, Yang W (2005) Pattern Recogn 38
16. Li J, Du Q, Sun C (2009) Pattern Recogn 42
17. Liu Y, Chen L, Wang H, Jiang L, Zhang Y, Zhao J, Wang D, Zhao Y, Song Y (2014) J Vis Commun Image Represent 25
18. Woraratpanya K, Kakanopas D, Varakulsiripunth R (2012) ASEAN Eng J Part D 1
19. Kaewaramsri Y, Woraratpanya K (2015) Recent advances in information and communication technology. In: Proceedings of the 11th international conference on computing and information technology, July 2–3, Bangkok, Thailand
20. Nayak SR, Mishra J, Padhy R (2016) Signal processing, communication, power and embedded system. In: Proceeding of the IEEE international conference, Oct 3–5, Paralakhemundi, India
21. Nayak SR, Mishra J, Padhy R (2016) Int J Comput Sci Inf Secur 14
22. Jin XC, Jayasooriah SH (1995) Pattern Recogn Lett 16
23. Brodatz P (1966) Texture: a photographic album for artists and designers, New York

Computing Shortest Path for Transportation Logistics from High-Resolution Satellite Imagery

Pratik Mishra, Rohit Kumar Pandey and Jagannath Mohan

Abstract One of the most key facets of transportation logistics systems is the traffic management where city planning, road monitoring and speed of transportation play significant role. Regardless of the firm technologies available, road extraction from high-resolution satellite imagery has been an interesting research field focused in recent years. The study deals with extraction of topographical features like roads from high-resolution satellite imagery and computation of shortest path for transportation logistics. In this study, the topographical features and model from high-resolution satellite imagery were analysed. Then the comparison study of various road extraction algorithms was performed. After performing the preliminary processing like histogram visualization and grey-level thresholding, path opening and closing morphological filter was used. The response of the filter was then used to extract the curvilinear structure which represents the road. Post-processing morphological operation like thinning was also applied for removing distorted artefacts. Finally, shortest path between the source and the destination was approximately commuted using quasi-Euclidean distance method. The algorithm was extensively tested using several satellite imageries, and some of the selected results were presented in the paper. It is evident that one type of topographical features is not enough to obtain good results. The road extraction would be combined with other algorithms based on the requirement of applications to yield optimal solution for finding the shortest path. The proposed algorithms would also have the application of map generation for speedy transportation.

Keywords High-resolution satellite imagery · Mathematical morphology
Road extraction · Transportation logistics

P. Mishra · R. K. Pandey · J. Mohan (✉)
School of Electronics Engineering, VIT University, Chennai, Tamil Nadu, India
e-mail: jagannath.m@vit.ac.in

© Springer Nature Singapore Pte Ltd. 2018
A. K. Nandi et al. (eds.), *Computational Signal Processing and Analysis*, Lecture Notes in Electrical Engineering 490,
https://doi.org/10.1007/978-981-10-8354-9_6

1 Introduction

Roads have assumed a focal part of our life since the start of the human advancement. Prior to the development of ocean, air and rail travel and transportation, roads were the single method for transporting merchandise and individuals starting with one area then onto the next [1]. Indeed, even in present-day society, roads stay a standout amongst the most imperative methods of travel and transportation. Since roads assume a vital part in our day-by-day lives, data relating to the area of road is basic. This data not just permits people to settle on educated choices with respect to their surroundings as a rule, additionally expands productivity in the selection of courses for transporting merchandise and individuals. At present, data in regard to road areas and their qualities is put away carefully inside geographic databases [2]. This advanced representation of the road system is sufficiently adaptable to empower various Geographic Information Systems (GISs), viz. administrations which incorporate satellite route, course arranging, transportation framework demonstrating, social insurance availability arranging, arrive cover grouping and even foundation administration. Road information sets can be gotten by either ground reviewing and outlining roads and depicting or by extricating the road systems from satellite imagery [2, 3].

Road network extraction from the satellite imagery can be delegated manual, semi-mechanized or completely computerized prepare [4]. Manual extraction includes a prepared human administrator outlining roads from remotely detected imagery, while semi-computerized extraction requires some human contribution to manage an arrangement of robotized procedures lastly; the mechanized extraction handle requires no human intervention. Completely mechanized road extraction frameworks contain an assortment of algorithms, which can be generally ordered into three levels of processing [1]:

- Preprocessing of raw image,
- Refinement of features extracted from preprocessing, and
- Production of the road networks.

The greater part of the conditions of India and in this way the entire of India should be carefully mapped digitally. A large portion of the existing administrations of GPS is given in view of the information separated and embedded physically on to the databases of online guide assets like Google Earth and Wikimapia. The extraction and overhauling of road system databases are significant to numerous Geographic Information System (GIS) applications like route, urban arranging and so on. The improvement of develop procedures for road extraction and change identification in the light of imagery may give an optimal solution for this issue. An effective road database separating and redesigning framework ought to incorporate the accompanying three primary capacities [2]:

- Frame of a road network from high-resolution satellite imagery,
- Trace of roads using image processing techniques, and
- Upgrading of the road database.

The first and foremost step in satellite image analysis is the collection of data. For this purpose, various satellites have been deployed in space which function day and night and send real-time data back to earth. Once the image is captured, it can be used for various purposes like road tracking, disaster management, pollution control, forest cover estimation. Road tracing in a given satellite image can be done in numerous ways. In this paper, we have used histogram processing to segment the road [5]. Then we have performed morphological operations to remove discontinuities in the segmented image. Then we have taken two random points on the skeletal framework of the traced road, which is free from any discontinuity, and have found out the shortest path between them using quasi-Euclidean technique.

2 Methodology

The work for tracing for road networks and calculating the shortest path between two points from a satellite image is based on grey-level thresholding, morphological operation and calculating the quasi-Euclidean distance. The workflow diagram for the present work is given in Fig. 1.

2.1 Grey-Level Thresholding

The acquired satellite image is converted into a grey image and the histogram plot is found out. From the histogram plot, the road region is segmented on the basis of pixel intensity. The histogram plot is basically divided into four parts based on pixel intensity [6]. The road region lies in the fourth part. From this methodology, estimated road component is identified [7, 8]. The pixels which lie in part four are allotted value one, and remaining pixels are allotted value zero.

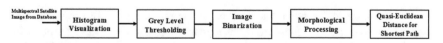

Fig. 1 Workflow diagram of proposed work

2.2 Morphological Operation

Morphological operations involve erosion and dilation. In our work, we have used dilation technique. In dilation, the segmented region gets thickened. In our case, the segmented road region gets thickened. An appropriate structuring element was applied on the binary image to get the desired result.

2.3 Skeletal Framework of the Road

The dilated road region was converted into a skeletal framework by using an appropriate structuring element [9].

2.4 Computing the Shortest Path Between Two Points

After segmenting the road region and applying morphological operation, two random points on the skeletal network of the road are chosen. The respective coordinates of the two points were taken as input. Then the quasi-Euclidean distance was found out by the value of the coordinates which are named to be (x_1, y_1) and (x_2, y_2).

Quasi-Euclidean distance "d" is given as

$$d = \begin{cases} |x_1 - x_2| + (\sqrt{2} - 1)|y_1 - y_2|, & |x_1 - x_2| > |y_1 - y_2| \\ (\sqrt{2} - 1)|x_1 - x_2| + |y_1 - y_2|, & \text{otherwise.} \end{cases}$$

After finding the quasi-Euclidean distance, the shortest path was overlaid between the two random points on the skeletal road network [1].

3 Result and Discussion

In our study, we took several satellite images as input. Figure 2a shows the road map of satellite image taken from "Wikimapia" where the road-path does not contain over bridge. Figure 3a shows the road map of satellite image taken from "Wikimapia" where the road-path contains over bridge. Figures 2b and 3b are the equivalent grayscale images of cases having with over bridge and without over bridge, respectively. The histogram of the grayscale images was plotted. From the histogram plot, the range of values in which road component falls was taken and that was selected as threshold for image binarization. We have successfully traced the road region from high-resolution satellite image (Fig. 4a). Our algorithm has

(a)   (b)

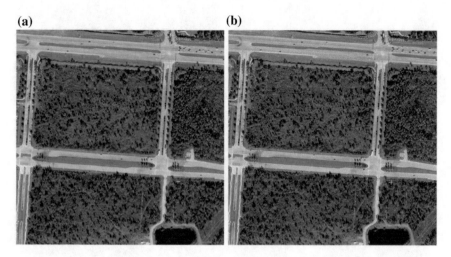

Fig. 2 Road map of satellite image. **a** Road network without over bridge. **b** Grayscale image

(a)   (b)

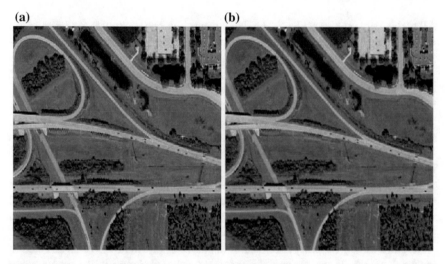

Fig. 3 Road map of satellite image. **a** Road network having over bridge. **b** Its grayscale image

been tested on various satellite images and functioned as per our expectation in each case. After extracting the road region, we took any two points on the detected road region and found the shortest distance between the chosen points (Fig. 4b). Tracking of roads from satellite image has been extensively done in recent times. Our study not only deals with tracking of roads but also involves further work on the segmented road component.

In today's world, time plays a crucial role in every aspect of life. In corporate world, transportation and various other fields' punctuality is a sought-after entity.

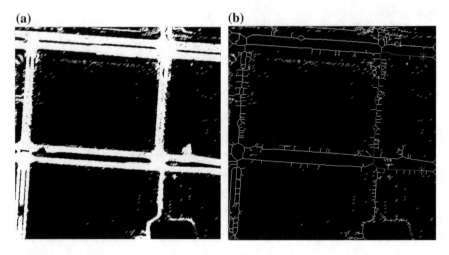

Fig. 4 a Segmented image of Fig. 2a. **b** Extracted road with shortest distance between two random points

All entities are interdependent. The delay caused by one entity will be reflected on other entities as well. In our case, we have targeted transportation sector. In businesses across the world, people want faster outcomes. For example, UBER cab service's mission is to transport customers from one place to another in minimum time. That's where the concept of shortest path arises. As the path is short, the time taken to cover the distance will ultimately be less. So, it becomes very essential for the software to identify the shortest route between the source and the destination. Our study focuses on this aspect also. The present study tracks the road, computes the shortest distance between two coordinates and also gives the shortest route between them.

This study has certain limitations. If a satellite image contains a flyover/over bridge, then the part of the road on which the shadow of the flyover falls could not be tracked since that region has different colour representation. Rai and Kumra [10] revealed the same inference which is corroborative to the present study.

Regions having intensity level nearer to our threshold value also gets detected in the final output. If we change the threshold value to improve the result, other undesired components also get added up with the outcome. So, we adjusted our threshold value to the best of the limits. Figure 5 shows the discontinuity between roads in the places where the over bridges are constructed (indicated in Fig. 5).

Further study can be carried out in this field. The above-mentioned limitations can be worked upon. One possible solution is to assign a different colour to the shadowed region but that will hamper the uniformity of the output. Work can be done on finding all possible routes between the source and destination. After that the user can take the route as per their convenience.

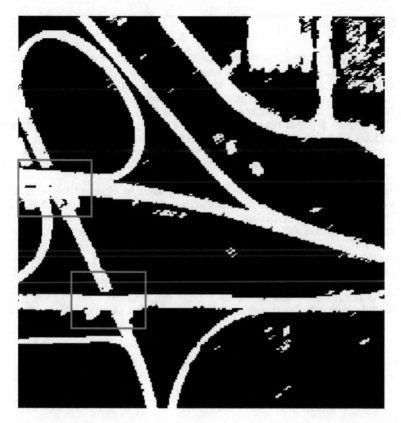

Fig. 5 Segmented image of Fig. 3a with discontinuities between roads indicating the over bridges

4 Conclusion

One of the key aspects of scientific research is to solve day-to-day problems. In today's world, numerous vehicles ply on roads and it is a direct consequence of increase in population. To be on time, people have to take help of technology. So to make things simpler, the proposed algorithm traces the road and finds the shortest distance between the source and destination, and not only this, it can also show the shortest path in a distinguished manner for efficient navigation. Our study has also opened up opportunities for further research in the field of high-resolution satellite imagery.

References

1. Wang W, Yang N, Zhang Y, Wang F, Cao T, Eklund P (2016) A review of road extraction from remote sensing images. J Traffic Transp Eng (English Edition) 3(3):271–282
2. Das S, Mirnalinee TT, Varghese K (2011) Use of salient features for the design of a multistage framework to extract roads from high-resolution multispectral satellite images. IEEE Trans Geosci Remote Sens 49(10):3906–3931
3. Cem U, Beril S (2012) Road network detection using probabilistic and graph theoretical methods. IEEE Trans Geosci Remote Sens 50(11):4441–4453
4. Movaghati S, Moghaddamjoo A, Tavakoli A (2010) Road extraction from satellite images using particle filtering and extended Kalman filtering. IEEE Trans Geosci Remote Sens 48 (7):2807–2817
5. Mnih V, Hinton GE (2010) Learning to detect roads in high-resolution areal images. Computer Vision—ECCV (Lecture Notes in computer science), vol 6316, pp 210–223
6. Sujatha C, Selvathi D (2015) Connected component-based technique for automatic extraction of road centerline in high resolution satellite images. EURASIP J Image Video Process 2015 (1):1–16
7. Gonzalez RC, Woods RE (2008) Digital image processing, 3rd edn. Prentice Hall Inc.
8. Jain AK (1989) Fundamentals of digital image processing, 3rd edn. Prentice Hall Inc.
9. Alex DM (2013) Robust and efficient method to extract roads from satellite images. Int J Latest Trends Eng Technol 2(3):26–29
10. Rai PK, Kumra VK (2011) Role of geoinformatics in urban planning. J Sci Res 55:11–24

Combined Analysis of Image Processing Transforms with Location Averaging Technique for Facial and Ear Recognition System

A. Parivazhagan and A. Brintha Therese

Abstract In the current biometric human recognition scenario, novel ideas are evolving to solve the errors in facial and ear recognition system. In this proposed work, a blooming new technique called location averaging technique is combined with few image processing transforms, i.e., location averaging technique is combined with FFT, DCT, and DWT for human face and ear recognition. Location averaging technique is a feature extraction/reduction process; it transforms the whole size of an image into a single column vector. It helps to accumulate more number of images for recognition system. Location averaged FFT, location averaged DCT, and location averaged DWT are the three methods proposed for face and ear recognition system. The standard face and ear database images are used for analyzing the accuracy, runtime, and mismatching. The maximum accuracy value of about 99% is achieved in shortest run time with less mismatching.

1 Introduction

Biometrics face and ear recognition is a growing field in human recognition system. Using ear for human recognition is a blooming technology. In human biometric identification, researches are growing all around the world to get good accuracy. The location averaging technique is focusing on averaging the intensity values with intensity location values; it is concentrating on each and every element of an image, so accuracy of the result is satisfactory. The location averaging is combined with FFT, DCT, and DWT to complete the recognition system. It is a process of obtaining values from spatial domain and analyzing it in frequency domain; it act as a bridge between both spatial and frequency domain analysis. In Fourier transform,

A. Parivazhagan (✉) · A. Brintha Therese
School of Electronics Engineering, VIT University, Chennai Campus, Chennai, India
e-mail: parivazhagan.a2013@vit.ac.in

A. Brintha Therese
e-mail: abrinthatherese@vit.ac.in

© Springer Nature Singapore Pte Ltd. 2018
A. K. Nandi et al. (eds.), *Computational Signal Processing and Analysis*, Lecture Notes in Electrical Engineering 490,
https://doi.org/10.1007/978-981-10-8354-9_7

the location averaging technique values are represented in frequency domain, and in discrete cosine transform, it is represented by cosine-based basic functions in frequency domain, and in discrete wavelet transform, it is represented by both frequency and location information. The remaining topics of this paper are organized as, 2. Location averaging technique, 3. Location averaged FFT, 4. Location averaged DCT, 5. Location averaged DWT, 6. Results and Discussions, and 7. Conclusions, and References.

2   Location Averaging Technique

The location averaging technique [1, 2] shown in Fig. 1 is a column-wise averaging process used for image reduction and recognition. It will reduce the whole size of an image into its column vector size. Hence, the reduced image vector can be used for recognition purpose. All the cell elements (C_{kt}) (intensity values) of the image are multiplied with its location value (L_{kt}), and it is divided by its column size (C_s). This is called as location averaged value of the cell (X_{kt}). The averaging process is performed for this location averaged value of the cells, i.e., addition of all the location averaged value of the cells present in a row, and it is divided by the column size (C_s). This is called as location averaged value of the row (Y_k). A single row is transformed into a single element; first row values are transformed to location averaged value of the first row (Y_1), similarly for all rows. So finally, an image matrix is transformed into a column vector.

$$X_{kt} = (C_{kt} * L_{kt})/C_s$$

$$Y_k = (X_{11} + X_{12} + X_{13} + \ldots + X_{1N})/C_s; \quad \text{Here } k = 1;$$

$$\text{Location averaged value for } \acute{N} \text{ Rows} = \begin{bmatrix} Y_1 \\ \cdots \\ Y_N \end{bmatrix}$$

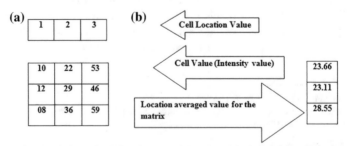

(a) Sample image's matrix   (b) Location averaged value (vector) for the Sample image's matrix

Fig. 1 Location averaging technique block diagram

C_{kt} Cell value (intensity value);
L_{kt} Cell location value;
C_s Column size;
k Row;
t Column;
N Last element;
X_{kt} Location averaged value of the cell;
Y_k Location averaged value of the row;
Y_1 Location averaged value of the first row;
Y_N Location averaged value of the 'Nth' row.

Here, the test image and database images are given as input to the location averaging technique, and they are transformed to location averaged test image (Lt) and location averaged database images (Ld).

3 Location Averaged FFT

The location averaging technique [1, 2] is performed on test image and database images, and its outputs are given as inputs to the FFT [3–5]. This combined process is called as location averaged FFT, shown in Fig. 2. In frequency domain, signals are analyzed with respect to frequency and any image can be transformed to frequency domain from spatial domain. The spatial domain data, i.e., the location averaged test image and location averaged database images are given as input to the FFT.

$$F(x, y) = \sum_{m=0}^{M-1} \sum_{n=0}^{N-1} f(m, n) e^{-j2\Pi\left(\frac{xm}{M} + \frac{yn}{N}\right)} \qquad (1)$$

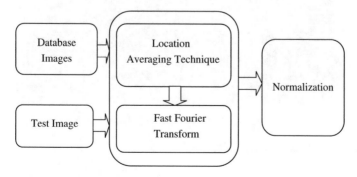

Fig. 2 Location averaged FFT block diagram

The frequency transformation techniques (1) are applied on the image data from the location averaged test image and database images, and after this, a normalization process is performed. In normalization process, the Euclidean distance technique is used to find the minimum differences between the test image and the database images, to detect the correct matched image. And the matched images results are shown in Fig. 3 (Euclidean distance $= \sqrt{[(\text{Database images} - \text{Test image})^2]}$).

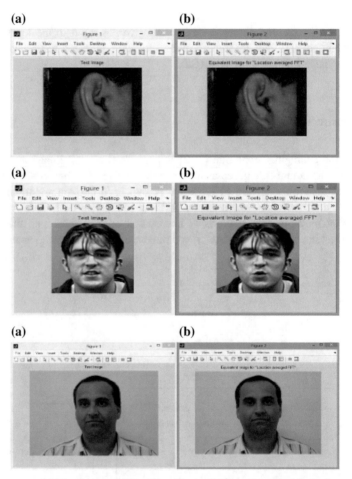

(a) Test image (b) Database image (Output Matched image)

Fig. 3 Results of location averaged FFT

4 Location Averaged DCT

The location averaging technique's results that are given as input to the DCT [6–9] are shown in Fig. 4. The cosine-based basic functions (2) are applied on the location averaged test image and location averaged database images. DCT transforms the given input data to linear combination of weighted-based function. These functions are called as frequency components of the given data. The results from this step are given to the normalization process. In normalization, the Euclidean distance method is used, to find the correct database image that is matched with the test image, and are shown in Fig. 5.

$$Y_{m,n} = \sum_{i=0}^{N-1} \sum_{j=0}^{N-1} f_{i,j} \cos \frac{(2i+1)m\Pi}{2N} \cos \frac{(2j+1)n\Pi}{2N};$$

$$0 \le m \le N-1; \quad 0 \le n \le N-1 \tag{2}$$

5 Location Averaged DWT

The DWT operations are applied on the results from the location averaging technique, shown in Fig. 6.

$$w_\phi(J_o, K) = \frac{1}{\sqrt{M}} \sum_n s(n) \phi_{Jo,K}(n) \tag{3}$$

$$w_\Psi(J, K) = \frac{1}{\sqrt{M}} \sum_n s(n) \Psi_{J,K}(n) \tag{4}$$

$$\frac{1}{\sqrt{M}} = \text{normalizing term}$$

ϕ—scaling function; Ψ—wavelet function; $n = 0, 1 \ldots M - 1$.

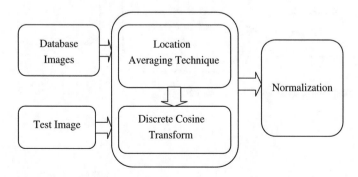

Fig. 4 Location averaged DCT block diagram

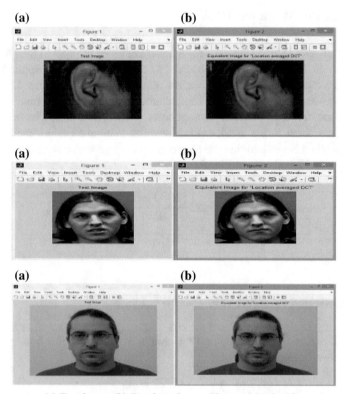

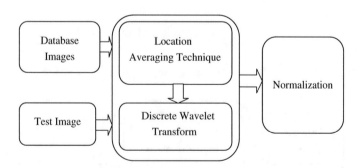

(a) Test image (b) Database image (Output Matched image)

Fig. 5 Results of location averaged DCT

Fig. 6 Location averaged DWT block diagram

Wavelets [10–12] are functions that are well localized on time and frequency around a certain point; it is suitable for non-stationary signals, and it is designed for low-frequency components to get good frequency resolution and high temporal resolution for high-frequency components. Depending on time scale representation, it performs multi-resolution sub-band decomposition of signals. The wavelet transformation techniques (3) and (4) are applied on the location averaged test image and location averaged database images. The results are then given to the normalization process. In that, the Euclidean distance phenomenon is performed on it to get the correct database image that is matched with the test image. The matched images outputs are shown in Fig. 7.

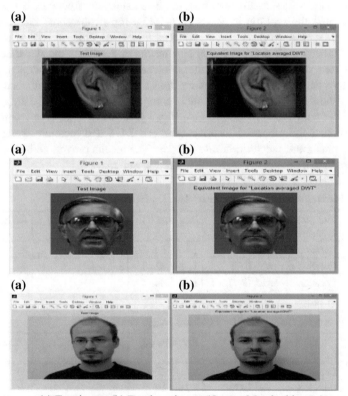

(a) Test image (b) Database image (Output Matched image)

Fig. 7 Results of location averaged DWT

6 Results and Discussions

Location averaged FFT, location averaged DCT, and location averaged DWT techniques are performed on 100 subjects of Libor Spacek's facial images database, 50 subjects of FEI face database, and 50 subjects of IIT Delhi ear database, and it produces excellent results that are presented in Tables 1, 2 and 3 and are also shown in Figs. 8 and 9. Accuracy, runtime, and recognition percentage are the parameters used to analyze the face and ear images. The highest accuracy of about 99% is achieved. For Libor Spacek's facial images database and FEI face database, the location averaged DCT yields minimum runtime of about 1.53 and 3.21 s, and it is lesser than other two methods. The accuracy for all the three methods is between 90 and 99% of both the facial databases. The runtime variations are between 1.53 and 3.41 s for all the three methods of both the face databases. Accuracy is between 72 and 74%, and the runtime is about 1.22–1.45 s for all the three methods of ear database. Location averaged DCT produces lesser run time of 1.22 s, with good accuracy. The overall process yields best results of accuracy and run time.

From the results, it is explained that the location averaging technique can act as an effective pair with image processing transforms for face and ear recognitions. Table 4 shows the proposed model comparison with the existing model. Comparison with PCA, modular PCA, Eigen face, and Fisher face techniques, it represents that the proposed location averaged DCT and location averaged FFT yield the maximum recognition rate of 99%. The results proved and validated the new technique's efficiency, originality, and perfection.

Table 1 Face recognition output for Libor Spacek's facial images database

S. No.	Proposed recognition techniques	Run time (s)	Mismatching (%)	Accuracy (%)
1	Location averaged FFT	1.54–1.64	1	99
2	Location averaged DCT	1.53–1.63	1	99
3	Location averaged DWT	1.77–1.86	2	98

Table 2 Face recognition output for FEI facial images database

S. No.	Proposed recognition techniques	Run time (s)	Mismatching (%)	Accuracy (%)
1	Location averaged FFT	3.24–3.30	10	90
2	Location averaged DCT	3.21–3.28	10	90
3	Location averaged DWT	3.30–3.41	10	90

Table 3 Ear recognition output for IIT Delhi ear images database

S. No.	Proposed recognition techniques	Run time (s)	Mismatching (%)	Accuracy (%)
1	Location averaged FFT	1.25–1.32	26	74
2	Location averaged DCT	1.22–1.28	26	74
3	Location averaged DWT	1.35–1.45	28	72

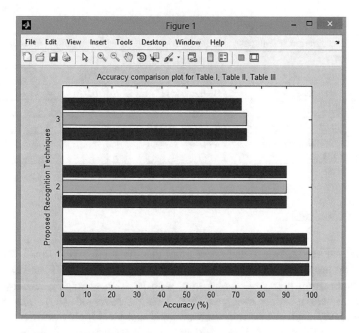

Fig. 8 Accuracy comparison plot for Tables 1, 2, and 3

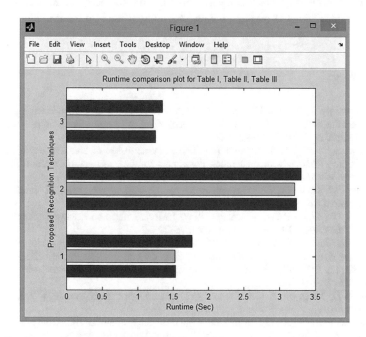

Fig. 9 Runtime comparison plot for Tables 1, 2, and 3

Table 4 Proposed system comparison with existing system

S. No.	Recognition techniques	References	Recognition performance (%)
1	PCA	Martinez and Kak [13]	70
2	Modular PCA	Zhang [14]	76.07
3	Eigen face	Belhumeur [15]	76.07
4	Fisher face	Belhumeur [15]	55.4
5	Location averaged FFT	Proposed system	99
6	Location averaged DCT	Proposed system	99

7 Conclusions

Face and ear recognition system of human identification is performed with three recognition processes, and it produces effective results. FFT, DCT, and DWT give good combination pair while combining with location averaging technique. The proposed novel location averaged FFT, location averaged DCT, and location averaged DWT methods performance are good and yield highly positive results. It represents that the three methods yield output with good accuracy, among these the location averaged DCT and location averaged FFT produce results in shortest runtime with less mismatching and yield accuracy of about 99%. Location averaged DWT also produces good accuracy in minimum runtime. Results show that all these three methods can act as an effective tool for recognition in biometric identification.

References

1. Parivazhagan A, Brintha Therese A (2015) A novel, location averaging, linear equation and exponential function techniques for face recognition in human identification system. Int J Appl Eng Res (IJAER) 10(87):21–26
2. Parivazhagan A, Brintha Therese A (2016) A combination of location averaging feature reduction technique with recognition algorithms for face recognition system. Int J Intell Electron Syst 10(2):1–10
3. Cooley JW, Tukey JW (1965) An algorithm for the machine calculation of complex Fourier series. Math Comput 19(90):297–301
4. Cooley JW, Lewis PAW, Welch PD (1967) Historical notes on the fast Fourier transform. IEEE Trans Audio Electroacoust Au-15(2):76–79
5. Uzun IS, Amira A, Bouridane A (2005) FPGA implementations of fast Fourier transforms for real-time signal and image processing. IEE Proc-Vis Image Signal Process 152(3):283–296
6. Cho NI, Lee SU (1991) Fast algorithm and implementation of 2-D discrete Cosine transform. IEEE Trans Circuits Syst 38(3):297–305
7. Ahmed N, Natarajan T, Rao KR (1974) Discrete Cosine transform. IEEE Trans Comput 100(1):90–93
8. Ezoji M, Faez K (2011) Use of matrix polar decomposition for illumination-tolerant face recognition in discrete cosine transform domain. IET Image Process 5(1):25–35

9. Huang S-M, Chou Y-T, Yang J-F (2014) Low-resolution face recognition in uses of multiple-size discrete cosine transforms and selective Gaussian mixture models. IET Comput Vis 8(5):382–390

10. Mallat S (2009) A wavelet tour of signal processing the sparse way, 3rd edn. Academic Press

11. Akansu AN, Serdijn WA, Selesnick IW (2010) Emerging applications of wavelets: a review. Phys Commun 3:1–18

12. Lesson 12 Multi-resolution analysis: discrete wavelet transforms, Version 2 ECE IIT, Kharagpur

13. Martinez AM, Kak AC (2001) PCA versus LDA. IEEE Trans Pattern Anal Mach Intell 23(2):228–233

14. Zhang T, Fang B, Tang YY, He G, Wen J (2008) Topology preserving non-negative matrix factorization for face recognition. IEEE Trans Image Proc 17(4):574–584

15. Belhumeur PN, Hespanha JP, Kriegman DJ (1997) Eigenfaces vs. Fisherfaces: recognition using class specific linear projection. IEEE Trans Pattern Anal Mach Intell 19:711–720

Robust Detection and Tracking of Objects Using BTC and Cam-Shift Algorithm

S. Kayalvizhi and B. Mounica

Abstract Face detection is used in several applications in the field of object recognition and pattern recognition tools. It is a demand nowadays that face detection to be performed using the compressed data. In this paper, we discussed on the face detection method applied on compressed images and video streams where only little decompression is required to retrieve the important data. This approach is faster and consumes less computational time and processing power when compared to the pixel domain-based algorithms. We used the Block Truncation Coding (BTC) algorithm for compression process. Viola and Jones proposed a fast and accurate method to detect the object. Haar-like features are used to detect the variation between the black and light portion of the image. Cam-shift algorithm is used to develop the face and head tracking. The object search is done using the back-projection procedure through probability distribution maximum obtained.

Keywords Viola–Jones algorithm · Cam-shift · Haar feature selection BTC algorithm

1 Introduction

Face is the most important feature that is considered when it comes to photos, news mows, and videos. In this research, we implemented a novel sensor fusion-based face tracking system which can track the faces in compressed videos and help in automatic video indexing. Face tracking is used in several applications like the surveillance, HCI, video compression, biometry, human–robot communication [1].

S. Kayalvizhi (✉) · B. Mounica
Department of Electronics and Communication Engineering,
SRM University, Chennai, India
e-mail: kayalvizhi.s@ktr.srmuniv.ac.in

B. Mounica
e-mail: battinamounica_bat@srmuniv.edu

© Springer Nature Singapore Pte Ltd. 2018   79
A. K. Nandi et al. (eds.), *Computational Signal Processing
and Analysis*, Lecture Notes in Electrical Engineering 490,
https://doi.org/10.1007/978-981-10-8354-9_8

This method can perform well in compressed domains where the face can be detected in compressed images and video streams by decompressing only little data. This approach is fast and robust when compared to the pixel domain-based algorithms. Image and video data are the trend in Internet now and with the growing size the demand also increases. In order to meet these expectations, images and videos can be stored with less storage space. BTC was initially introduced for compression of unaltered image data. It uses one-bit adaptive moment by retaining quantizers with statistical moments like standard deviation and standard mean. Face tracking is being employed in control robots. Its usage is still being experimented in some of the applications like health-related measurements like body temperature, posture tracking, supervised exercise and its integration with other technology. There are some existing algorithms to track the facial image in videos but they have its own pros and cons [2]. In this research, we studied in detail about the Viola–Jones method of object detection. It was trained to detect various objects. The main purpose of proposing this algorithm was to detect facial image [3]. It has the capacity to detect the image extremely fast. The methods of tracking face images on a course of period can drift its interest from the main object. To minimize the drift and to work efficiently, we are using Cam-shift algorithm [4].

2 Proposed System

In our research work, we used the Block Truncation Coding (BTC) for image compression. BTC uses the one-bit adaptive moment-preserving quantizers by retaining the main goal for visual tracking is the robustness and the real time of tracking. The matrix is formed during the process of tracking once the rectangle region is tracked in the first frame [5]. According to the Viola–Jones method, the detector is modified than the actual input image. The detector is run on the image several times with different size. Initially, both the methods look time-consuming but the scale invariant detector requires the same amount of calculation no matter what big is the size. This detector used in Viola and Jones is made up of integral image and uses some simple rectangular features reminiscent Haar-like features (Fig. 1).

The cam-shift method is used to track the head and face part in perceptual user interfaces. It is mainly used to find the peak of probability density. The one-dimensional histogram is used by the algorithm. The HSV color space consists of the hue (H) channel in the histogram. The object search is done using the probability distribution maximum obtained from the histogram back-projection procedure.

Fig. 1 Block diagram of proposed system

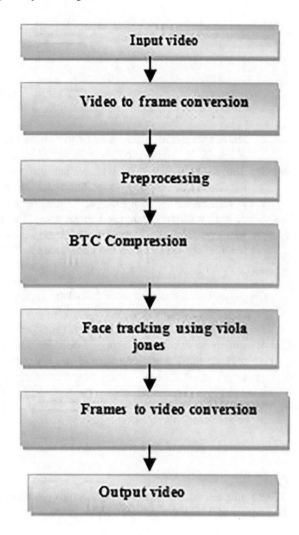

2.1　BTC Algorithm

The steps involved in this method are [6, 7]:

Step 1:　The input image is divided into rectangular regions without any overlapping. In order to simplify the computation task, let us consider the regions to of square size as $m \times m$.

Step 2:　Now each pixel of two-level quantizers will be represented with the mean x and standard deviation σ values.

$$X = \frac{1}{n} \sum_{i=1}^{n} x_i \tag{1}$$

$$\sigma = \sqrt{\frac{1}{n} \sum_{i=1}^{n} (x_i - \bar{x}_i)^2} \tag{2}$$

Here, x_i is the ith pixel value of the image block and the total number of pixels in the image block is represented by n.

Step 3: The two quantizers of BTC values are represented by x and σ. The threshold value is chosen as x. The threshold value for a two-level bit plane is constructed by comparing each pixel value x_i with the threshold value. A binary block (B) is used to represent the pixels. If the gray level is greater than or equal to x, we considered the pixel value as "1," and if gray level is less than x, then the pixel value is "0."

$$B = \begin{cases} 1, & x_i \geq \bar{x} \\ 0, & x_i < \bar{x} \end{cases} \tag{3}$$

In this way, all the blocks are converted into a bit plane.

Step 4: Finally to decode the image, a decoder will reconstruct the image block with all the 1's replaced with H and 0's with L as below:

$$H = \bar{x} + \sigma \sqrt{\frac{p}{q}} \tag{4}$$

$$L = \bar{x} - \sigma \sqrt{\frac{q}{p}} \tag{5}$$

Here p represents the number of 0's in the compressed bit plane and q represents the number of 1's in the compressed bit plane.

2.2 Viola–Jones Face Detection Algorithm

For a method to properly track the face, there are few challenges that are to be faced:

1. The pose of the object in the given video frame: In a moving posture in the video, the appearance of the face varies since the face is moving.
2. Occlusions: When the video is capturing the face image, there are possibilities that another object can come hiding the face and making it difficult for the tracker to capture the face image. This is called as occlusion.

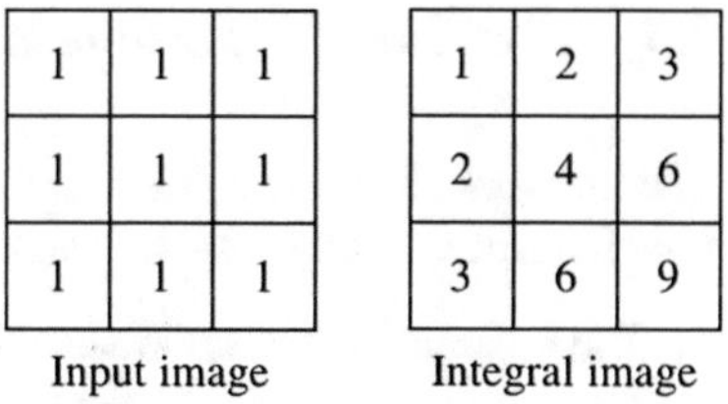

Fig. 2 Integral image

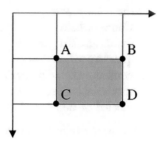

Input image   Integral image

Fig. 3 Sum calculation

3. Noise: During the process of capturing the face image, there can be noise which affects the clarity and quality of the image captured. The amount of noise in an image depends on the quality of the sensor being used for acquiring the video image.
4. Ambient illumination: There are chances that the image can change due to the direction, intensity, ambient light in appearance and color of the face image in a video. We convert each pixel equal to the sum of all pixels above and to the left of the concerned pixels as shown in Fig. 2.

The sum of gray rectangle is shown in Fig. 3.

$$\text{Sum of grey rectangle} = D - (B + C) + A$$

As both the rectangles B and C contain rectangle A in it, the sum of A will also be added to the calculation. In Viola–Jones method, the face detector will use two or more rectangles to analyze the given sub-window. The various types of features are shown in Fig. 4.

Fig. 4 Various types of features

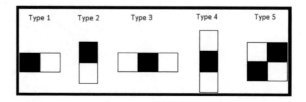

2.3 The Cam-Shift Algorithm

The purpose of inventing this method was to track the face and head in perceptual user interfaces. The Cam-shift algorithm consists of the below steps:

1. The calculation region is considered as the whole frame.
2. The initial location is selected from the two-dimensional mean shift search window.
3. The color probability distribution is calculated in the 2D region centered in the window where the area is slightly larger than that of the mean shift window size.
4. The mean shift parameter is applied to find the maximum density probability in order to perform the convergence or to decide the number of iterations. The zero moment and middle position are stored.
5. In the next video frame, the search window is placed in the middle position selected in step 4, and then goes to step 3.

The back-projection procedure is used to calculate the probability distribution.

$$M_{00} = \sum_x \sum_y I(x, y), \tag{6}$$

$$M_{01} = \sum_x \sum_y y I(x, y), \tag{7}$$

$$M_{10} = \sum_x \sum_y x I(x, y) \tag{8}$$

$$X_c = \frac{M_{10}}{M_{00}}, \tag{9}$$

$$Y_c = \frac{M_{01}}{M_{00}}, \tag{10}$$

$I(x, y)$ is the value of the discrete probability distribution at the point (x, y) on the back projected image. M_{00}—zero moment; M_{01} and M_{10}—first moments; (x_c, y_c)—search window mean location (centroid).

The objects' scale and shape are calculated as below:

$$M_{20} = \sum_x \sum_y x^2 I(x, y), \tag{11}$$

$$M_{02} = \sum_x \sum_y y^2 I(x, y), \tag{12}$$

$$M_{11} = \sum_x \sum_y xyI(x, y) \tag{13}$$

$$\mu_{20} = \frac{M_{20}}{M_{00}} - x^2 \tag{14}$$

$$\mu_{02} = \frac{M_{02}}{M_{00}} - y^2 \tag{15}$$

$$\mu_{11} = 2\left(\frac{M_{11}}{M_{00}} - x_c y_c\right) \tag{16}$$

The first two eigenvalues from cam-shift detector can be computed as:

$$w = \frac{\sqrt{\mu_{21} + \mu_{02} - \sqrt{\mu_{11}^2 + (\mu_{20} - \mu_{02})^2}}}{2} \tag{17}$$

$$l = \frac{\sqrt{\mu_{21} + \mu_{02} - \sqrt{\mu_{11}^2 + (\mu_{20} - \mu_{02})^2}}}{2} \tag{18}$$

The minor and the major semi-axes w and l of the distribution centroid are found through the following equations:

$$\theta = 1/2 \arctan\left(\frac{\mu_{21}}{\mu_{20} - \mu_{02}}\right) \tag{19}$$

If in case the target histogram contains a lot of information about the background image or about an adjacent object, then the target position and scale cannot be estimated accurately. In order to outperform this problem, a simple technique is used; the weights of target's histogram are reduced. In other terms, the ratio between the background histograms bin and target histograms bin is used. This is called as histogram weighing. This can be written as:

$$q_r = \begin{cases} \frac{q_0}{q_b}, & q_b \neq 0 \\ q_0, & \text{otherwise} \end{cases} \tag{20}$$

q_r is the weighed histogram (ratio histogram); q_0 represents the target histogram; and q_b the background histogram (Figs. 5 and 6; Table 1).

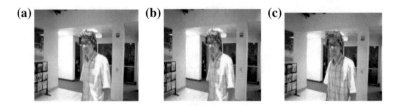

Fig. 5 Screen shots of **a** input image, **b** compressed image, **c** tracking image

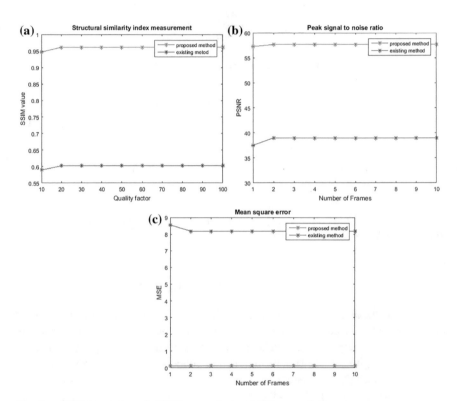

Fig. 6 **a** SSIM comparison, **b** PSNR comparison, **c** MSE comparison

Table 1 Results comparison

Method	SSIM	PSNR	MSE
FCT	0.602	38.99	8.18
BCT with Viola–Jones	0.961	57.65	0.1225

3   Conclusion

Though there are several papers to handle the real-time face detector problem, this paper is unique by its features as it can directly detect the frontal face image in the compressed domain. The computational performance of the proposed system is considerably low where it is suitable to be implemented in low computational resource devices. Image compression using BTC is investigated in this research. In order to provide a robust method along with less complexity and high accuracy, we have combined BTC along with Viola–Jones and Cam-shift algorithm. By using this novel method, we can detect the features like skin tone, or a contrast background or any other feature of the face.

References

1. Fonseca P, Nesvadha J (2004) Face detection in the compressed domain. In: 2004 international conference on image processing, 2004, ICIP'04, vol 3, pp 2015–2018
2. Sakure NS, Bankar RT, Salankar SS (2016) Camparative analysis of face tracking. Int J Adv Res Electron Commun Eng (IJARECE) 5(3)
3. Viola P, Jones M (2001) Rapid object detection using a boosted cascade of simple features. In: Proceedings of the 2001 IEEE computer society conference on computer vision and pattern recognition, CVPR 2001, vol 1, pp I-511–I-518
4. Varfolomieiev A, Antonyuk O, Lysenko O (2009) Camshift object tracking algorithm implementation on DM6437 EVM. In: Proceedings of 4th European DSP in education & research in 2009
5. Mohammed D, Abou-Chadi F (2011) Image compression using block truncation coding. Multidiscip J Sci Technol J Sel Areas Telecommun (JSAT), Febr Ed
6. Gupta A, Kumar S, Raja A (2014) Enhancement image compression using BTC algorithm. Int J Adv Res Comput Sci Softw Eng 4(2)
7. Baraniuk RG (2007) Compressive sensing [lecture notes]. IEEE Signal Process Mag 24(4): 118–121

Effect of Dynamic Mode Decomposition-Based Dimension Reduction Technique on Hyperspectral Image Classification

P. Megha, V. Sowmya and K. P. Soman

Abstract Hyperspectral imaging has become an interesting area of research in remote sensing over the past thirty years. But the main hurdles in understanding and analyzing hyperspectral datasets are the high dimension and presence of noisy bands. This work proposes a dynamic mode decomposition (DMD)-based dimension reduction technique for hyperspectral images. The preliminary step is to denoise every band in a hyperspectral image using least square denoising, and the second stage is to apply DMD on hyperspectral images. In the third stage, the denoised and dimension reduced data is given to alternating direction method of multipliers (ADMMs) classifier for validation. The effectiveness of proposed method in selecting most informative bands is compared with standard dimension reduction algorithms like principal component analysis (PCA) and singular value decomposition (SVD) based on classification accuracies and signal-to-noise ratio (SNR). The results illuminate that the proposed DMD-based dimension reduction technique is comparable with the other dimension reduction algorithms in reducing redundancy in band information.

Keywords Least square denoising · Dimension reduction · Principal component analysis · Singular value decomposition · Dynamic mode decomposition Alternating direction method of multiplier

P. Megha (✉) · V. Sowmya · K. P. Soman
Centre for Computational Engineering and Networking (CEN), Amrita School
of Engineering, Amrita Vishwa Vidyapeetham, Coimbatore 641112, India
e-mail: meghaparakatt@gmail.com

V. Sowmya
e-mail: v_sowmya@cb.amrita.edu

K. P. Soman
e-mail: kp_soman@amrita.edu

© Springer Nature Singapore Pte Ltd. 2018
A. K. Nandi et al. (eds.), *Computational Signal Processing
and Analysis*, Lecture Notes in Electrical Engineering 490,
https://doi.org/10.1007/978-981-10-8354-9_9

1 Introduction

Hyperspectral imaging is one of the recently developed techniques in remote sensing which provides very high spectral resolution. It collects information of a target across the electromagnetic spectrum [1]. Set of images captured by the hyperspectral sensor is layered one over the other and represented as image cube. Each image corresponds to one particular wavelength in the electromagnetic spectrum. Figure 1 shows single band of Indian pines and SalinasA hyperspectral data. Hyperspectral images (HSIs) are largely used in surveillance, mineralogy, agriculture, and gas industry as it offers the detailed picture of surveying area [8]. In all these applications, the main task is to classify the hyperspectral image. Although the richness of information provided by the hyperspectral image is very useful in classifying and identifying materials, its high dimension often leads to highly complex computation [6]. The presence of noise adds burden to the processing of a hyperspectral image.

Denoising is a preprocessing step which helps to increase the classification accuracy by removing noise from HSI bands [13]. Here, least square-based denoising is used since it works much faster than other denoising algorithms, thereby decreasing the time required for denoising [13]. Classification process may involve large number of training samples if we use hyperspectral data as such. Moreover, adjacent bands of hyperspectral image show high correlation between them. Many of the existing dimension reduction techniques are unsuitable for hyperspectral images since they destroy the pattern of HSI [5]. Various algorithms have been proposed to reduce the dimension of hyperspectral data by selecting only the most relevant bands [7]. Principal component analysis (PCA) is one such standard dimension reduction technique used to project hyperspectral data into a lower dimensional space [5]. Singular value decomposition (SVD) is another technique used for the dimension reduction of HSI. This paper presents a new hyperspectral dimension reduction (DR) technique based on dynamic mode decomposition (DMD), a technique originally evolved in fluid mechanics community. DMD was introduced as a tool to understand and analyze the dynamicity of flow fields [11]. Attempts have been made

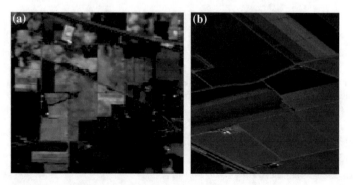

Fig. 1 Original hyperspectral image, **a** Indian pines, **b** SalinasA

to incorporate this technique in video processing for foreground–background separation, target detection, etc. [4]. Here, the method is used in static hyperspectral images to find the spectral variation. The effect of DMD-based algorithm is then compared with the standard dimension reduction techniques like PCA and SVD, through classification accuracies obtained and signal-to-noise ratio (SNR) calculation of selected bands. In HSI, pixel vectors with similar characteristics are classified as same group. A sparsity-based basis pursuit classifier is employed in this work. The L-1 norm optimization problem in basis pursuit is solved using alternate direction method of multipliers (ADMMs) [1].

The organization of the paper is as follows: A brief description of methods used for preprocessing is given after introductory part. The next section explains the proposed DMD-based dimension reduction technique, followed by result and analysis.

2 Mathematical Background

2.1 Least Square Denoising

Least square (LS) denoising is a one-dimensional signal denoising method proposed by Ivan Selesnick [12]. This technique is mapped into two-dimensional image denoising to remove noise present in the image [13]. LS problem for denoising approach is formulated as

$$\min_x \|y - x\|_2^2 + \lambda \|Dx\|_2^2 \tag{1}$$

where y is the noisy signal, x is the denoised signal, and D is the second-order differential matrix. The first part of the equation takes care of denoising, and the second part is to regularize denoising process which depends on the regularization parameter [13]. Solving this optimization problem leads to the following solution

$$x = (I + \lambda D^T D)^{-1} y \tag{2}$$

In the case of an image, this solution is applied first column wise and then row wise. Each band of hyperspectral image undergoes least square denoising.

2.2 Dimension Reduction

Hyperspectral images have attracted more researchers in recent years. But the abundance of spectral information present in the hyperspectral image dramatically increases the computational burden [7]. Dimension reduction of a hyperspectral

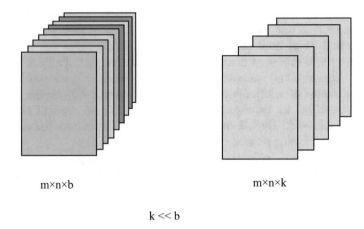

m×n×b   m×n×k

k << b

Fig. 2 Dimension reduction

image cube is the process of selecting relevant bands from the entire dataset without losing information. Figure 2 depicts the concepts of dimension reduction of hyperspectral images.

(1) Principal Component Analysis: PCA is one of the standard dimension reduction techniques used for HSI. It assumes that neighboring bands of HSI show high degree of correlation and convey redundant information. PCA algorithm finds the optimum linear combination of bands which shows maximum variation in pixel values [10]. It rearranges the original bands in such a way that first band has maximum variance and last band has minimum variance. Therefore, first few PCA bands contain majority of the information present in the original hyperspectral data [10].

(2) Singular Value Decomposition: SVD is another form of matrix analysis which gives low-dimensional representation of a high-dimensional matrix. SVD will help us to eliminate less important part of the matrix. It decomposes the matrix $[r \times c]$ into U, Σ and V^T with the following properties.

1. U is a $r \times m$ column-orthonormal matrix.
2. Σ is a diagonal matrix consists of singular values as diagonal elements.
3. V is a $c \times m$ column-orthonormal matrix. So, V^T is row-orthonormal.

where m is the rank of the matrix.

(3) QR Decomposition: QR decomposition or QR factorization is matrix orthogonalization method in which a matrix X $[m \times n]$ is decomposed into an orthogonal matrix Q and an upper triangular matrix R [2, 9].

$$X = QR \tag{3}$$

To reorder the diagonal values in R in descending fashion a permutation matrix P is used such that

$$XP = QRP \tag{4}$$

where Q is a $m \times m$ matrix and R is a $m \times n$ matrix.

(4) Dynamic Mode Decomposition: Dynamic mode decomposition is a powerful tool for extracting dynamicity of a nonlinear system [11]. The advantage of using DMD algorithm is that it is not necessary to have an explicit knowledge of dynamic operator to determine eigenvalues and eigenvectors of it. Here, DMD is applied on a 2D matrix formed by vectorizing each band and appending it as columns.

$$X_1^{m-1} = \begin{bmatrix} & \cdot & \cdot & & & \cdot \\ & \cdot & \cdot & & & \cdot \\ x_1 & x_2 & \cdot & \cdot & \cdot & x_{m-1} \\ & \cdot & \cdot & & & \\ & \cdot & \cdot & & & \cdot \end{bmatrix}$$

where m is the number of bands. We then assume that a linear operator A connects x_i to x_{i+1}. Our aim is to find the dynamic characteristics like eigenvalue and eigenvector to extract the dynamical process described by matrix A. We also assume that beyond a critical number of snapshots, columns of matrix X become dependent [11]. Once that point is reached, we can express the vector x_i as linear combination of previous vectors, which are linearly independent

$$x_m = a_1 x_1 + a_2 x_2 + \cdots + a_{m-1} x_{m-1} + r \tag{5}$$

where r is the residual vector. In matrix form,

$$x_m = X_1^{m-1} a + r \tag{6}$$

If we follow the concept from [6], we can write

$$A X_1^{m-1} = X_2^m \tag{7}$$

X_2^m can also be written as

$$X_2^m = X_1^{m-1} S + r e_{m-1}^T \tag{8}$$

where S is the low-rank approximation of A. Eigenvalues of S are same as that of nonzero eigenvalues of A [11]. Sometimes data from an experiment may be contaminated by the presence of noise or other uncertainties. So, we choose another matrix $\widetilde{S}$ which is related to S via matrix similarity to implement more robust results [11]. The computation of $\widetilde{S}$ is as follows: take the SVD of data sequence.

$$X_1^{m-1} = U\Sigma V^H \tag{9}$$

Substituting (9) in (8) and rearranging, we will get

$$\widetilde{S} = U^H X_2^m V \Sigma^{-1} \tag{10}$$

With this low-rank approximation, we can find the dynamic properties described by A. For using this concept to dimension reduction, we find eigenvectors of $\widetilde{S}$. By taking QR decomposition of eigenvector matrix of $\widetilde{S}$, we are able to get the permutation matrix P which stores the order in which eigenvectors are arranged [2]. When the permutation matrix is multiplied with X_1^{m-1}, the columns of the matrix, which represents the vectorized form of bands are arranged from most informative bands to least informative bands.

2.3 Basis Pursuit

In HSI, each pixel vector is classified into different classes based on ground truth. Normally, 10% or 20% of the data is used for training and rest of the data is used for testing. Basis pursuit is a sparsity-based algorithm in which each class is represented in low-dimensional subspace [3]. L1-norm optimization problem in basis pursuit is solved using ADMM. ADMM solve the convex optimization problems by breaking them into smaller problems. These smaller problems converge faster making ADMM better than other solvers [13].

3 Methodology

In this work, two hyperspectral image datasets, Indian pines and SalinasA are used for the experiment. Both the images are collected by NASA's Airborne Visible Infrared Imaging Spectrometer (AVIRIS) over agricultural sites at Indiana and SalinasA valley, California. Indian pines has ground pixel size of 145 × 145 and 220 spectral bands. SalinasA has ground pixel size of 86 × 83 and 224 spectral bands. In both datasets, spectral bands covering the range from 0.4 to 2.5 μm.

The main objective of this project is to incorporate DMD algorithm for the dimension reduction of hyperspectral images. The effectiveness of proposed method is compared with standard dimension reduction methods like PCA and SVD. In the first stage, each band of HSI image undergoes LS denoising giving denoised image cube. Before applying dimension reduction techniques, 3D HSI data is converted into a 2D matrix by vectorizing each band and appending as columns of a matrix. To analyze the effect DR techniques, we first classify hyperspectral data without dimension reduction. In the second stage, three different dimension reduction techniques PCA, SVD–QR, and DMD are applied separately on 2D matrix. After removing the redundant information data is dimensionally reduced by 50%. The effect of all dimension reduction techniques is validated through classification. The assessment of classification is done through visual interpretation and calculation of classification accuracies.

The accuracies obtained from PCA, SVD, and DMD are compared. The bands selected by the different algorithms are analyzed with the help of a numerical approach called SNR values analysis. SNR value is a ratio of signal power to noise power. But the problem in calculating SNR value of hyperspectral image is that the lack of reference image. Linlin Xu has proposed a method for calculating SNR value even if there is no reference image present [13, 14].

$$\text{SNR} = 10 \log_{10} \frac{\sum_{ij} \hat{x}_{ijb}^2}{\sum_{ij} \left(\hat{x}_{ijb} - m_b\right)^2} \tag{11}$$

where $\hat{x}_{ijb}$ is the denoised signal, m_b is the mean value of $\hat{x}_{ijb}$ where pixels are homogeneous [13].

4 Result and Analysis

4.1 Classification Accuracies

Hyperspectral image data after being denoised are validated through classification accuracies. The same classifier is applied on dimensionally reduced dataset. Five parameters are taken for the validation process. They are kappa coefficient, overall accuracy, average accuracy, elapsed time, and classwise accuracy [13]. Table 1 shows the classification accuracies obtained on Indian pines dataset before and after applying DR algorithms. From the table, it is clear that accuracies are almost same even after reducing 50% of bands. Kappa coefficient, a metric that compares an

Table 1 Classification result on Indian pines dataset

	Without dimension reduction	With dimension reduction		
		PCA	SVD	DMD
Kappa coefficient	0.98	0.97	0.98	0.98
Overall accuracy (%)	98.56	97.58	98.79	98.43
Average accuracy (%)	97.88	93.5	97.42	96.17
Elapsed time (min)	27.5	23	24.6	23

Table 2 Classification results on SalinasA dataset

	Without dimension reduction	With dimension reduction		
		PCA	SVD	DMD
Kappa coefficient	0.99	0.97	0.99	0.99
Overall accuracy (%)	99.9	98.35	99.9	99.9
Average accuracy (%)	99.9	98.46	99.9	99.9
Elapsed time (min)	5.06	4.04	3	3.5

observed accuracy with expected accuracy is 0.98 without dimension reduction. The results tabulated in Table 1 shows that SVD and DMD methods are able to get the same kappa coefficient even after removing redundant band information. PCA has 0.01 decrease in kappa coefficient. Average accuracy and overall accuracy also follow the similar trend, i.e., SVD and DMD results are comparable with the original result, whereas PCA shows a significant decrease. Figure 3 shows the

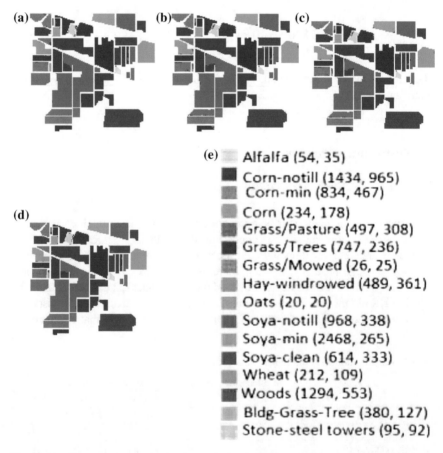

Fig. 3 Classification maps obtained before and after dimension reduction for Indian pines dataset, **a** without dimension reduction, **b** PCA, **c** SVD–QR, **d** DMD

comparison of classification maps obtained for Indian pines dataset under different dimension reductions. Table 2 contains the analysis of classification accuracies obtained on SalinasA dataset. It can be observed from the table that kappa coefficient obtained after applying SVD and DMD dimension reduction is same as the kappa coefficient obtained before dimension reduction. Overall and average accuracies also follow the same trend, i.e., same as accuracies obtained before dimension reduction. But PCA shows a decrease in accuracies. From these tables, it can be inferred that SVD–QR and proposed DMD-based algorithms work much better than PCA-based dimension reduction. So, in the next step of analysis PCA is discarded. In the next stage, we compare bands selected by SVD–QR and DMD based on SNR value analysis.

4.2 Comparative Analysis of SNR Values

Table 3 shows the comparison of best six SNR values of bands selected by SVD–QR and DMD. The most informative bands selected by DMD are different from the most informative bands selected by SVD–QR. From Table 3, it is clear that SNR values of bands selected by DMD are higher than that of bands selected by SVD–QR. The highest SNR values obtained for DMD are 55.77 and 23.71 dB for Indian pines and SalinasA dataset, respectively. In the case of SVD–QR, the best SNR values are 55.70 and 23.13 dB, which means that the algorithm that is used in fluid mechanics for finding the dynamicity of flow fields can be treated as a new effective method for hyperspectral dimension reduction (Fig. 4).

Table 3 SNR values in dB

SNR Values in dB				
	Indian pines		SalinasA	
	SVD–QR	DMD	SVD–QR	DMD
1	55.76	55.77	23.13	23.71
2	55.05	55.76	23.05	23.22
3	52.92	55.59	22.27	23.18
4	52.49	55.05	21.84	23.05
5	52.41	54.97	21.76	22.10
6	51.98	53.49	21.62	22.00

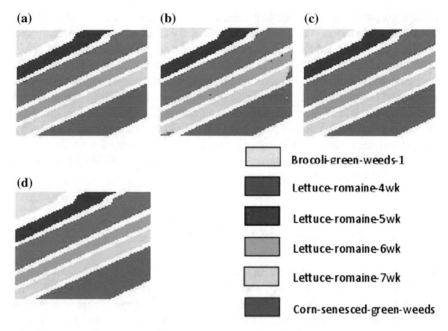

Fig. 4 Classification maps obtained before and after dimension reduction for SalinasA dataset, **a** without dimension reduction, **b** PCA, **c** SVD–QR, **d** DMD

5 Conclusion

This paper has proposed a new algorithm based on dynamic mode decomposition for the dimension reduction of hyperspectral images. Quantitative presentation demonstrates that proposed technique is very effective in removing redundancy between bands. Calculation of SNR values shows that proposed technique is comparable with existing dimension reduction algorithms. Dimensionally, reduced data is classified using basis pursuit classifier which uses ADMM to solve the optimization problem. The classification results show that even after eliminating redundant bands, selected by proposed algorithm, we are able to get the same accuracies obtained before dimension reduction. The SNR values of bands selected by proposed technique are slightly higher than that of bands selected by the other dimension reduction technique.

References

1. Aswathy C, Sowmya V, Soman K (2015) ADMM based hyperspectral image classification improved by denoising using Legendre Fenchel transformation. Indian J Sci Technol 8(24):1
2. Bhushan DB, Sowmya V, Manikandan MS, Soman K (2011) An effective pre-processing algorithm for detecting noisy spectral bands in hyperspectral imagery. In: 2011 International symposium on ocean electronics, pp 34–39
3. Chen S, Donoho D. Basis pursuit, vol 1, pp 41–44
4. Erichson NB, Brunton SL, Kutz JN (2015) Compressed dynamic mode decomposition for real-time object detection. arXiv preprint arXiv:1512.04205
5. Fong M (2007) Dimension reduction on hyperspectral images. Univ. California, Los Angeles, CA
6. Koonsanit K, Jaruskulchai C, Eiumnoh A (2012) Band selection for dimension reduction in hyper spectral image using integrated information gain and principal components analysis technique. Int J Mach Learn Comput 2(3):248
7. Li Y (2014) Dimension reduction for hyperspectral imaging using laplacian eigenmaps and randomized principal component analysis: midyear re-port
8. Lodha SP, Kamlapur S (2014) Dimensionality reduction techniques for hyperspectral images. Int J Appl Innov Eng Manag (IJAIEM) 3(10)
9. Reshma R, Sowmya V, Soman K (2016) Dimensionality reduction using band selection technique for kernel based hyperspectral image classification. Procedia Comput Sci 93:396–402
10. Rodarmel C, Shan J (2002) Principal component analysis for hyper-spectral image classification. Surv Land Inf Sci 62(2):115
11. Schmid PJ (2010) Dynamic mode decomposition of numerical and experimental data. J Fluid Mech 656:5–28
12. Selesnick I (2013) Least squares with examples in signal processing. Connexions 4
13. Srivatsa S, Ajay A, Chandni C, Sowmya V, Soman K (2016) Application of least square denoising to improve ADMM based hyperspectral image classification. Procedia Comput Sci 93:416–423
14. Xu L, Li F, Wong A, Clausi DA (2015) Hyperspectral image denoising using a spatial–spectral monte carlo sampling approach. IEEE J Sel Top Appl Earth Obs Remote Sens 8(6):3025–3038

Performance Evaluation of Lossy Image Compression Techniques Based on the Image Profile

P. Poornima and V. Nithya

Abstract Digital images have become part of our everyday life. It is important to store these images in an efficient manner in the given storage space and therefore digital image compression has become a large focus in the recent years. Image compression deals with the rebate of bits required to epitomize the image. It is also important to ensure that reduction in size of image is without degradation of image quality and loss of information. As a result of this, image compression techniques have been developed with lot of new algorithms and also with variations of the already existing ones. For a given application, the image compression algorithm is selected with an objective either to provide better compression or to get a good quality of reconstructed image. The profile of the image such as aspect ratio, color, pixel intensities, smooth regions, edges, shading, pattern, and texture is not considered while selecting an algorithm. In this paper, an attempt is made to provide the users a ready reckoner to select a compression technique based on the profile of the image. In this regard, three popular techniques such as set partitioning in hierarchical tree (SPIHT), compressive sensing (CS), and fractal coding (FC) in the family of lossy image compression are selected for analysis. In addition to this, the best choice of wavelet filter banks, measurement matrix, and reconstruction algorithm are also identified to provide improved performance. The performances of these algorithms are evaluated with the metrics such as compression ratio (CR), mean square error (MSE), and peak signal-to-noise ratio (PSNR).

P. Poornima (✉) · V. Nithya
Department of Electronics and Communication Engineering, SRM University, Kattankulathur, Chennai, India
e-mail: poornima_p@srmuniv.edu.in

V. Nithya
e-mail: nithya.v@ktr.srmuniv.ac.in

© Springer Nature Singapore Pte Ltd. 2018
A. K. Nandi et al. (eds.), *Computational Signal Processing and Analysis*, Lecture Notes in Electrical Engineering 490, https://doi.org/10.1007/978-981-10-8354-9_10

101

1   Introduction

Digital image processing is the process of examining and manipulating an image in order to improve its quality. In wireless transmission, it is important to reduce the file size as it would shorten the time required to broadcast an image or a video to the destination. The two main classifications of image compression techniques are the lossless and lossy techniques. Lossy compression algorithms are broadly classified into transformation schemes, non-transformation schemes, and the distributed coding. Transformation schemes include discrete cosine transform (DCT) and discrete wavelet transform (DWT). Non-transformation schemes are further classified into vector quantization (VQ) and fractal coding (FC). Distributed coding includes distributed source coding (DSC) and compressive sensing (CS) [1]. Different algorithms are developed in the literature by using a single image to analyze the quality of image and compression ratio. Arora et al. [2] have discussed the major lossy and lossless image compression techniques. Pokle et al. [3] have discussed that depending on the application appropriate compression algorithm has to choose. Singh et al. [4] results pointed that fractals provide better result with better PSNR and low MSE. Nahar [5] has surveyed the theory of compressive sensing, and its applications in various fields of wireless video and image transmission. All the above work in the literature mainly focused on the selection of an algorithm for high compression rate or image quality based on single image. This paper addresses the image compression algorithm should be elected based on image characteristics such as color, shape, smooth and edge regions, pixel intensities and texture. Our work concentrated on the comparative study on three well-known algorithms that can be chosen for image compression based on image profile. The highlights of the chosen algorithms are (1) SPHIT which is based on DWT and has multi-resolution properties that benefit in good compression rate. (2) Compressive sensing technique hinges on sparsity and results in good recovered image quality, and (3) fractal coding computes image similarity at regular intervals and provides the best compression rate. Therefore, this paper will be a quick guide to the readers to select an algorithm based on the image profile to achieve good image quality or compression ratio.

The remainder of the paper is structured as follows. Section 2 briefs about three compression algorithms considered for analysis. The simulation results obtained with the wide variety of images are discussed in Sect. 3. Section 4 presents the conclusion.

2   Image Compression Algorithms

Image compression is the process of downsizing the size in bytes of an image or video without deteriorating the quality [6]. From Fig. 1 [1], it is clear that lossy image compression techniques are broadly classified into three categories,

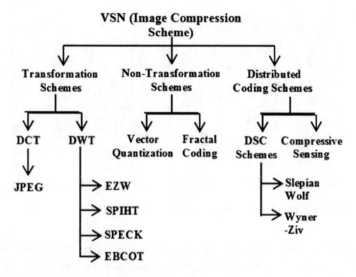

Fig. 1 Classification of lossy image compression [1]

namely transformation schemes, non-transformation schemes, and the distributed coding schemes. After detailed study from the literature, in our work, we have considered the outperforming compression technique in each category for further analysis. Also, to further improve the performance wavelet filter banks, measurement matrix, and reconstruction algorithm are also widely studied and the one that produces better results are utilized during simulation. The following section briefly discusses the three compression algorithm chosen for study.

2.1 Transform-Based Algorithm—Discrete Wavelet Transform—SPHIT

SPHIT exploits spatial orientation trees and exhibits self-similarity characteristics with the coefficients belonging to the same spatial location in different sub-bands in the pyramid structure. From these pyramids, the higher levels represent parents to the children on lower levels [7].

Procedure for SPIHT Algorithm:

Step 1: Define the decomposition levels (n) and set the threshold value and initially set the list of significant pixel (LSP) as an empty list.

Step 2: List of insignificant pixels (LIPs) are those which have magnitudes lesser than the determined threshold value.

Step 3: All the entries in the list of insignificant pixels (LIPs) are examined for their significance with the help of the determined threshold value.

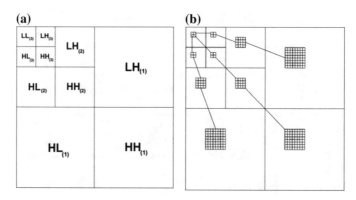

Fig. 2 Hierarchy of coefficient and tree of wavelet coefficients [8]

If an entry satisfies the determined threshold, a bit "1" is given as output and one more bit for the sign of coefficient that is represented by "1" for positive nor "0" for negative. The significant entry is assigned to the LSP. If an entry does not satisfy the threshold then the bit "0" is the output.

Step 4: The wavelet coefficients that have magnitudes lower than the thresholds are designated as list of insignificant sets (LIS). LIS is refined and to test their significance. If it found significant then they are partitioned into subsets and those subsets which have one coefficient and declared to be significant and will be moved to the LSP else they will be moved to the LIP.

Step 5: During the refinement pass, final output is obtained by the last (nth) MSB of the coefficients in the LSP which contains list of pixels which have magnitude greater than the threshold.

Step 6: Decrease the value of n, the sorting and refinement passes are applied once again and will be continued until either the expected rate is reached.

Figure 2 [8] illustrates (a) pyramid decomposition with coefficients hierarchy; (b) tree structure of wavelet coefficients.

2.2 Non-Transform-Based Algorithm—Fractal Coding

The fractal coding is performed using quadtree decomposition and Huffman encoding and decoding. In this method of compression, the image is first branched into numerous blocks named as range. The original image is further prorated as massive blocks labeled domain blocks. The domain blocks with their rotation property search for the finest match for whole range block. The correct domain and related information for every range block that are required for image retrieval are

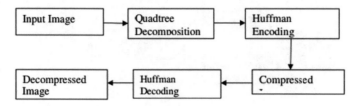

Fig. 3 Block diagram of fractal coding [9]

stored. Hence, the compression is achieved by storing only the parameters. Numbers of iterative operations are performed by decoder in order to reproduce the original image. Figure 3 [9] depicts the block diagram of fractal coding.

Procedure for Fractal Algorithm

Step 1: Original image is divided into small blocks with a threshold value as 0.2, minimum dimension as 2 and maximum dimension as 64 by using quadtree decomposition.

Step 2: From quadtree decomposition store the values of x and y coordinates, mean value, and the block size of the image.

Step 3: Save the fractal coding data and then apply the Huffman coding to complete the encoding procedure.

Step 4: For the encoded image, apply Huffman decoding to recover the image.

Step 5: Calculate compression ratio, MSE, and PSNR for the image.

2.3   Distributed Coding Schemes—Compressive Sensing

It is a signal processing technique that is capable of acquiring and reconstructing a signal in much more efficient way than the established Nyquist sampling theorem. It exploits the sparse nature of signal and computes sparsity of the signal [10]. The two main principles are sparsity and incoherence. The main idea of CS is the potential to recover the image with fewer measurements. Figure 4 [11] represents the block diagram of CS. The signal is first transformed into a sparser domain (DWT) and then multiplied with a measurement matrix (Gaussian Matrix, Toeplitz Matrix, or Bernoulli Matrix). The measured coefficients are obtained and are then used to reconstruct the image by using OMP reconstruction algorithm.

Mathematical expression is given by:

$$Y = \Phi X \tag{1}$$

Y = Measured Vectors, Φ = Measurement Matrix and X = Signal of Interest.

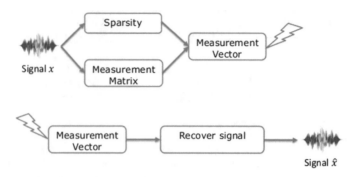

Fig. 4 Block diagram of compressive sensing [11]

Procedure for Compressive Algorithm

Step 1: An original image $(N \times N)$ is decomposed using Discrete Wavelet Transform (DWT) into four sub-bands coefficients namely LL1, HL1, LH1, HH1.

Step 2: The low-frequency components (LL1) contains maximum information of the image and, hence, it is kept unchanged.

Step 3: The high-frequency components (HL1, LH1, and HH1) are around zero and therefore are considered as sparse matrices.

Step 4: Sparse matrices are multiplied with measurement matrix to obtain the matrices of the measured coefficients.

Step 5: Orthogonal matching pursuit (OMP) reconstruction algorithm to reconstruct the three high-frequency components.

Step 6: Apply inverse discrete wavelet transform (IDWT) to the three high-frequency components and to the unchanged low-frequency components to reconstruct the image.

3 Results and Discussion

The analysis of the image compression techniques was carried out using MATLAB to evaluate the performance metrics. Different images of size 128 * 128 which vary in their intensities, regions, shapes, and texture were taken for study, and the MSE, PSNR, CR were evaluated for different compression algorithms.

3.1 Simulation Parameters

The various simulation parameters considered for the image compression algorithms are listed below:

SPHIT: We have used Haar Transform as it is orthogonal, fast, real, and easy for manual calculations.

Compressive Sensing: We have analyzed with different wavelet families, namely Haar, coif3 and sym5 and with different measurement matrices such as Gaussian, Toeplitz, and Bernoulli and found that sym5 wavelet family with Gaussian matrix produces good PSNR. Hence, we have used "sym5" wavelet family, Gaussian matrix as the measurement matrix and OMP as the reconstruction algorithm in our analysis.

Fractal Coding: We have set the threshold as 0.2, maximum dimension as 64, and minimum dimension as 2.

3.2 Performance Metrics

Mean square error (MSE) is calculated between the reconstructed and the decompressed image by the squared accumulated error.

$$\text{MSE} = \frac{1}{MN} \sum_{i=0}^{M-1} \sum_{j=0}^{N-1} \|X(i,j) - Y(i,j)\|^2 \tag{2}$$

where M, N = image size >0, X = original image, and Y = compressed image.

Peak signal-to-noise ratio (PSNR) can be measured as the quantum of peak error.

$$\text{PSNR} = 10 \log_{10}\left(\text{MAX}^2/\text{MSE}\right) \tag{3}$$

Compression ratio (CR) is the ratio of comparing the image size represented in bytes before and after compression.

$$\text{Compression Ratio} = \text{Original Image Size}/\text{Compressed Image Size} \tag{4}$$

3.3 Analysis of Compression Ratio and PSNR

The general way of classifying the features of the image is through the elements in visual interpretation to identify the homogeneous characteristics. The digital photographic images are decomposed into smooth regions and edge regions. The smooth regions are areas which have smooth color variations and lack sharp details. The edge region areas contain sharp details in the image.

- From the analysis of three algorithms, it is clear that for the images with fewer variations for large area can attain high compression ratio by adopting the fractal coding technique which has the property of self-similarity and similar image information can be stored in less space.

- Very good quality of reconstructed image is obtained by adopting compressive sensing technique and SPHIT.

Image with more number of sharp details in an image can achieve

- Less compression ratio from all the three techniques fractal coding technique
- Good image quality can be realized from compressive sensing when compared with SPIHT substantiating the fact that good quality of image is obtained when the compression ratio is low.

From Table 1, it is evident that nature image has more smooth regions such as blue sky that is spread out for a vast area, and has the highest compression ratio by the fractal coding technique, and high quality of reconstructed image by compressive sensing technique. Similarly, the lighthouse image having the vast sky and the rocks display high compression ratio and PSNR value.

From Table 2, it is obvious that the Lena image which has a blend of details, color, texture, and shading has less compression ratio by fractal coding when compared to smooth images. The image quality of the reconstructed image is fairly high by compressive sensing. Similarly, the other images (koala bear, Tulips, Penguins) also have more details in the image and show a low compression ratio and a reasonable PSNR value.

Table 1 Compression algorithm simulation results for smooth images

Smooth images	Compression algorithms and performance metrics			
	Compression algorithm	Compression ratio	MSE	PSNR (dB)
Water.jpg	SPHIT	1.173	137.446	26.749
	CS	5.968	0.607	**50.331**
	FC	**18.172**	36.063	32.594
Lighthouse.jpg	SPHIT	1.306	69.944	29.683
	CS	7.157	0.397	**52.169**
	FC	**20.079**	34.409	32.798
Nature.jpg	SPHIT	1.265	0.250	54.143
	CS	9.683	0.203	**55.075**
	FC	**28.319**	24.378	34.294
Desert.jpg	SPHIT	1.426	152.039	26.313
	CS	7.129	0.477	**51.379**
	FC	**19.208**	40.509	32.089

Table 2 Compression algorithm simulation results for edge images

Edge images	Compression algorithms and performance metrics			
	Compression algorithm	Compression ratio	MSE	PSNR (dB)
Penguins.jpg	SPHIT	1.265	0.250	34.143
	CS	5.170	0.808	**49.088**
	FC	**10.881**	32.581	35.028
Tulips.jpg	SPHIT	1.324	77.181	29.255
	CS	5.201	0.971	**48.292**
	FC	**9.033**	42.472	31.883
Lena.jpg	SPHIT	1.316	2.941	43.445
	CS	5.439	0.552	**50.745**
	FC	**11.688**	35.152	32.195
Kola.jpg	SPHIT	1.089	47.738	31.342
	CS	4.225	0.898	**48.631**
	FC	**4.782**	31.052	38.6821

3.4   Illustration of Results

PSNR and compression ratio obtained in the analysis are represented in terms of bar diagrams. Figure 5 shows that the images exhibit the maximum compression ratio under fractal coding. The increase in compression ratio for images in fractal coding can be explained in terms of the repetition of similar features at periodic intervals. Figure 6 shows that the images have the highest PSNR value while using the compressive sensing algorithm. It is evident in the figures that the reconstructed image quality is good as the PSNR values for all the three algorithms indicate significantly higher values >28.

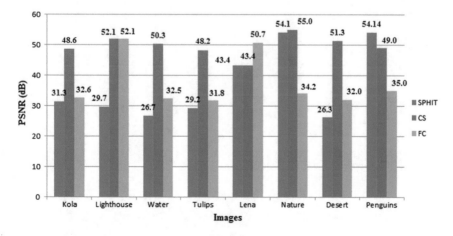

Fig. 5 CR comparison of sample images using three different types of algorithm

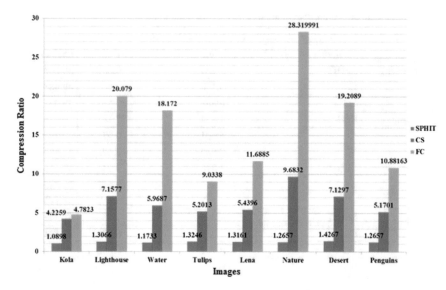

Fig. 6 PSNR comparison of sample images using three different types of algorithm

Figures 7 and 8 represent the reconstructed images obtained from the three compression algorithms. The fractal coding algorithm produces blurred image since the compression ratio is high corroborating the fact that when compression ratio increases image quality decreases. From compressive sensing, high PSNR is obtained for the images with low variations and comparatively a better value than SPHIT is obtained in the case of images with combination of image information. This also substantiates the fact that good quality of image is obtained when the compression ratio is low. Therefore, we can attain a good image quality by adopting compressive sensing technique.

Fig. 7 Reconstructed smooth images

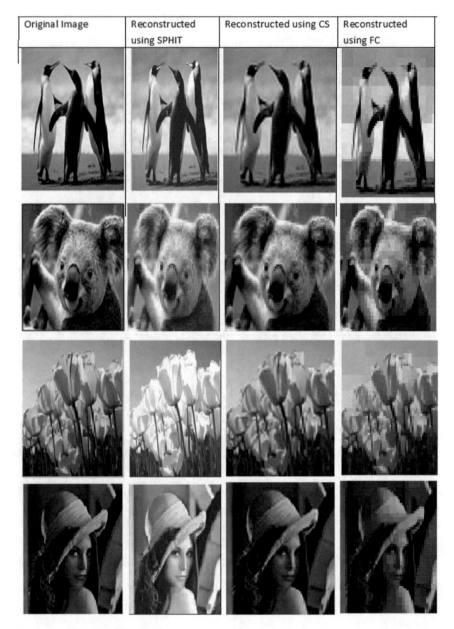

Fig. 8 Reconstructed edge images

4 Conclusion

In this paper, we have analyzed different images based on the image profile by employing the three image compression techniques specifically SPIHT using the Haar wavelets, CS with sym5 wavelet filter, Gaussian measurement matrix, and OMP reconstruction algorithm and Fractal coding by quadtree decomposition and Huffman coding to compare the metrics particularly compression ratio (CR), mean square error (MSE), peak signal-to-noise ratio (PSNR). Therefore from the results obtained by simulation, it is visible that an image with large area of smooth regions can achieve high compression ratio from fractal coding technique and the good quality of reconstructed image can be obtained by compressive sensing technique. The images with edge regions can glean only less compression ratio by fractal coding technique but can bring out a satisfying quality of reconstructed image from compressive sensing technique. SPHIT shows up very low compression ratios and PNSR value in both the cases. Therefore, it is important that we choose the appropriate image compression technique depending upon the nature of the image and the required performance metrics.

References

1. Ebrahim M, Chong CW (2014) A Comprehensive review of distributed coding for algorithms for visual sensor network. Int J Commun Netw Inf Secur (IJCNIS) 6(2):104
2. Arora K, Shukla M (2014) A comprehensive review of image compression techniques. Int J Comput Sci Inf Technol 5(2):1169–1172
3. Pokle PB, Bawane NG (2013) Comparative study of various image compression techniques. Int J Sci Eng Res 4(5)
4. Singh P, Sharma S, Singh B (2012) Comparative analysis of embedded zero tree and fractal image compression techniques. Int J Adv Res Fractal Image Compres Tech 2(2)
5. Nahar PC, Kolte MT (2014) An introduction to compressive sensing and its applications. Int J Sci Res Publ 4(6)
6. Rawat CS, Meher S (2013) A hybrid image compression scheme using DCT and fractal image compression. Int Arab J Inf Technol 10(6):553–562
7. Chourasiya R, Shrivastava A (2012) A study of image compression based transmission algorithm using SPHIT for low bit rate application. Adv Comput Int J (ACIJ) 3(6):117–122
8. Patil BS (2015) Image security in wireless sensor networks using wavelet coding. Int J Emerg Technol 6(2):239 (Special Issue on NCRIET-2015)
9. Chetan E, Deepak Sharma E (2015) Fractal image compression using quad tree decomposition & DWT. Int J Sci Eng Res (IJSER) 3(7)
10. Li X, Bi G (2015) Image reconstruction based on the improved compressive sensing algorithm. In: IEEE international conference on digital signal processing, July-2015
11. Donoho DL (2006) Compressed sensing. IEEE Trans Inf Theory 52(4):1289–1306

A Novel Video Analytics Framework for Microscopic Tracking of Microbes

Devarati Kar and B. Rajesh Kanna

Abstract Micro-organisms or microbes are single- or multi-cellular living organisms viewed under a microscope because they are too tiny to be seen with naked eyes. Tracking them is important as they play a vital role in our lives in terms of breaking down substances, production of medicines, etc., as well as causing several diseases like malaria, tuberculosis, etc., which need to be taken care of. For a pathological study, the images of these microbes are captured from the microscope and image processing is done for further analysis. These operations involved for the analysis requires skilled technicians for error-free results. When the number of images increases, it becomes cumbersome for those technicians as there is a chance of ambiguity in results, which hampers the sensitivity of the study. Further, image processing is a bit challenging and time-consuming as a single image provides only a snapshot of the scene. In this situation, video has come into the picture which works on different frames taken over time making it possible to capture motion in the images keeping track of the changes temporally. Video combines a sequence of images, and the capability of automatically analyzing video to determine temporal events is known as video analytics. The aim of this paper is to develop a new computing paradigm for video analytics which will be helpful for the comprehensive understanding of the microbial data context in the form of video files along with effective management of that data with less human intervention. Since video processing requires more processing speed, a scalable cluster computing framework is also set up to improve the sensitivity and scalability for detecting microbes in a video. The HDP, an open source data processing platform for scalable data management, is used to set up the cluster by combining a group of computers or nodes. Apache Spark, a powerful and fast data processing tool is used for the analysis of these video files along with OpenCV libraries in an efficient manner which is

D. Kar (✉) · B. Rajesh Kanna
School of Computing Science and Engineering,
VIT University, Chennai, Tamil Nadu, India
e-mail: devaratikar2407@gmail.com

B. Rajesh Kanna
e-mail: rajeshkanna.b@vit.ac.in

© Springer Nature Singapore Pte Ltd. 2018
A. K. Nandi et al. (eds.), *Computational Signal Processing
and Analysis*, Lecture Notes in Electrical Engineering 490,
https://doi.org/10.1007/978-981-10-8354-9_11

monitored with a Web UI known as Apache Ambari for keeping in track all the nodes in the cluster.

Keywords Microbes · Pathology · Image and video processing Cluster computing · Hortonworks data platform · Apache spark OpenCV

1 Introduction and Motivation

Necessity is the mother of invention which led to the invention of the computer-aided techniques and gave rise to a new era of research for the medical diagnosis in the field of image and video processing. The most common protocol to study about the microorganisms was invented by Zacharias Janssen, the microscope. Microscopic imaging is used in the area of biomedical analysis, pathological survey, etc., because these concepts help in the multi-dimensional characterization and visualization of biological processes. Microorganisms or microbes are very tiny living creatures which require a microscope to be seen as these cannot be seen with naked eyes. The microbes are required to be seen for the need of studying them as they have many roles in our lives. Microbes benefit us with the breakdown of substances, production of medicines, fermentation, and many such things. At the same time, they are harmful in certain situations like they produce toxins, damage our body tissues and organs, diseases like malaria, tuberculosis happen because of these microbes only and lot more. Diagnosis is a very important part to be played by technicians in the laboratory for the analysis of these microbes, and it is based on the blood and tissue samples of the infected person. This process of diagnosis of the disease using some microbial samples and providing an analysis is known as pathology. It is an age-old and evergreen technique for the analysis and assessment of the microbes but when the number of samples increases, it becomes laborious and tedious for the skilled technicians to provide error-free results of the analysis. Digital pathology has come up and everyone got a sigh of relief. With the help of computerized methodologies, digital pathology has gained popularity as these techniques enhance the task done by the technicians by providing them an extracted information which has emerged as a new field of study and is having a great impact on the whole world.

1.1 Digital Image Processing System—An Existing System for Pathological Diagnosis

Images are collected through an acquisition method, generally from the microscope and stored for image processing. This process is known as *Image Acquisition*.

These images might have some visibility issues like noise and blur. So these images need to be preprocessed with the techniques like noise filtering, histogram equalization, and many more, which is the stage of *Image Preprocessing and Image Enhancement*. Later comes the *Image segmentation* which is the process of dividing a digital image into multiple segments (sets of pixels). The goal of segmentation is to simplify and/or change the representation of an image into something that is more meaningful and easier to analyze. For extracting particular features, we do image segmentation like edge detection, clustering. After segmentation also, some unwanted features still remain in the image which can be removed with the help of classification techniques like Principal Component Analysis (PCA), Bayesian Classifiers. This stage is known as *Feature Selection*. Finally, we do a *Performance evaluation* of the system and find out how much accurate the results of the analysis is.

1.2 Problem Definition

A single digital image has lot of restrictions in digital microscopy, because it covers only the snapshot of the part of the specimen. When more image samples involve and it needs to be processed, the time taken by a single computer is more. Therefore, it reduces the efficiency of the diagnosis. On the other hand, video combines a sequence of images to form a moving picture [1]. Images do not give us so much information and as the microbial images are very small, they need to be zoomed to maximum extent for a better analysis. Moreover, there are number of growing stages of microbes where we can do identification as each stage has its own significance. In this way, we lose quite a lot of information. Video, work in frames, which is the number of images displayed per second and the different frames of a video taken over time fixes the dynamics in the scene, making it possible to capture motion in the video sequence and study changes temporally [2]. This concept is wholly known as video analytics or video content analytics. The ability of automatically analyzing video files to detect and determine temporal events and not based on a single event with the aid of computers and processing techniques is the main notion behind video analytics. Moreover, when we monitor so many stages of growth of the microbes in a single computer, the results are not efficient as the sample images are magnified to a great extent to gather maximum information. Continuous surveillance is also required for monitoring of the samples and therefore we require a cluster of computers to solve our purpose. If we use a cluster setup, we can do scalable data management of the study. As a result, we need a faster processing framework which is accomplished by video analytics along with comprehensive understandings of data content and cluster computing framework. Video analytics is gaining popularity because of its effectiveness in the field of real-time monitoring of events like traffic surveillance systems, healthcare analytics, and many more. Video processing may not be so effective in a single computer system, so we are going for a cluster of computers to do the same.

A single computer cannot process so many video files at a time, and the size of the video files is also huge which requires more space. The video files of the microbes should be processed and analyzed very fast as results are to be given to the decision makers for start of the treatment of the patient. A cluster computing setup is a set of loosely or tightly coupled computers that are connected LAN or WAN and work together and can be viewed as a single system [3, 4]. Video processing is done in a cluster setup consisting of some computers to save time and space.

2 Influence of Digital Image Processing in Existing Digital Pathology

The first and foremost criteria in pathological diagnosis is that we should have a problem definition like tracking of microbes in microbial images and defining the stage of the disease cycle along with the shape of the microbe. This problem requires to be solved with the help of images of the particular microbe. The images are acquired with the help of an electron or compound microscope which is the conventional method, and finally, they are diagnosed by the laboratory technicians to provide results. The future paradigm of pathology is digital where a pathologist will perform a diagnosis through interacting with images on computer screens and performing quantitative analysis [4]. According to World Health Organization (WHO), in the year 2015, reports found 214 million cases of diseases worldwide due to the microbial diseases like malaria, tuberculosis. This resulted in an estimated 438,000 deaths [5]. The most common diseases among these are malaria, tuberculosis, HIV, detection of tumor and cancer. Rate of infection decreased by 37% from 2000 to 2015, but from 2014 it increased to 198 million cases [2]. Even in India, people die due to these diseases, especially malaria, tuberculosis, and the mortality rate is also quite high. So, quick and efficient monitoring in the field of medical science pathology has become the part and parcel of human beings' healthy lives. This is the reason why digital image processing techniques have gained popularity in this regard and are used for microbial analysis not only because it helps us to automate the whole system reducing time and effort of the laboratory technicians but also helping in accurately diagnosing the harmful diseases caused by microbes. Digital image processing diagnosis techniques with algorithms to detect these diseases are saving life everyday with their automated detection systems.

3 Literature Review on the Existing Methods

For this work, many literatures were studied to get a better understanding and knowledge about the existing methods used by researchers for pathological diagnosis. Among them, malaria and tuberculosis are taken for our study because of their harmful impact on people's lives. Malaria is an epidemic disease which is health burden and demands for quick and accurate diagnosis [6]. A review paper on computational microscopic imaging for malaria detection has given an overview on the symptoms of malaria which can be diagnosed either clinically or manually by laboratory technicians. Clinical diagnosis which is based on patients' symptoms gives less specificity, whereas laboratory diagnosis provides accurate results for malaria detection. However, microscopic examination is time-consuming. To supplement this cognitive task, researchers used digital image processing approaches for efficient malaria diagnosis. There are several laboratory methods which are available, viz. microscopic examination of stained thin or thick blood smear, rapid diagnosis test, and molecular diagnosis methods. World health Organization (WHO) recommends microscopic examination of Gimesa-stained blood smear as a golden standard in detecting malaria parasite. Digital image processing diagnosis is pipelined into four stages, namely *image acquisition, preprocessing, segmentation, and feature extraction and classification* [6]. *Image acquisition* is the process of collecting malaria blood smear images from the digital camera attached with microscope, which are viewed in a magnified manner. For computer-aided malaria parasite detection, researchers used thick and thin blood smear images. For identifying stages and types of malaria, we use thin blood smears whereas thick blood smears for quantification of the infected erythrocytes (RBC). For better visualization, we do *preprocessing*. In the existing literature, the possible preprocessing methods used are median filter [7], Gaussian filter [8], Laplacian filter [9], low-pass filter [10], and histogram matching [11]. However, most of the literature suggested that gray world assumption could be the ideal choice, because it is used to eliminate the variation in the brightness of the image [12]. *Segmentation* of infected erythrocyte is the process of isolating RBC (erythrocyte) and eliminating white blood cell, platelets, etc., from preprocessed image which may contain infected and non-infected malarial parasites. From the previous studies, we listed the various segmentation techniques for isolating erythrocytes, namely edge detection algorithm [13], marker-controlled watershed algorithm [14], etc. To extract insights from the RBCs for subsequent human interpretation to detect the infected malaria parasite, we do feature extraction like size and color of the RBCs. This literature describes the feature extraction methods like chain code [7], roundness ratio [7], and area granulometry [15]. Detection and classification of the parasites are done with various classification algorithms used in the literatures. Those are Bayesian classifier [16], k-nearest neighbor classifier [17], AdaBoost algorithm [18], Naïve Bayes tree [19], and support vector machine [20].

4 Proposed System

The images and videos of the microbe specimen are acquired through the microscope by the laboratory technicians and stored in a system. These images and video files are then fed into the cluster computing setup which consists of a set of computers with master and slave nodes and a data analytics framework known as the HortonWorks Data Platform (HDP), which is an open source platform for scalable data management containing the big data ecosystem. Figure 1 shows the data flow diagram of the framework setup to do video analytics on microscopic video footages. For fast processing of image and video, some big data processing tools are used like HDFS and Apache Spark. OpenCV image and video processing libraries which have APIs are used for the purpose of processing the video which will be installed in all the nodes of the cluster. Figure 2 illustrates the flow of the video analytics framework starting from image and video acquisition by laboratory technicians, storing of the files, parallel processing in HDP, and finally results given to patients by doctors which helps in their treatment process benefitting them.

4.1 Experimental Setup

Apache has given an open source, scalable platform known as HortonWorks Data Platform (HDP) for storing, processing, and analyzing large volumes of big data in an efficient and cost-effective manner. The HortonWorks Data Platform consists of the essential set of Apache Hadoop projects like MapReduce, Hadoop Distributed File System (HDFS), HCatalog, Pig, Hive, HBase, Zookeeper, and Ambari. Usefulness of HDP is many which includes data management, data access, data governance and integration, security, operations related to data and finally provisioning the clusters in cloud. Apache Ambari is a tool which helps Hadoop to manage things in an efficient and better way by developing softwares to provision, manage, and monitor Apache Hadoop clusters. It is useful for management of Hadoop as it is intuitive, easy in using and because of its Web UI and it is also backed by its Representational State Transfer (RESTful) APIs. Ambari supports a lot of operating systems such as Redhat Enterprise Linux (RHEL) 6 and 7, CentOS

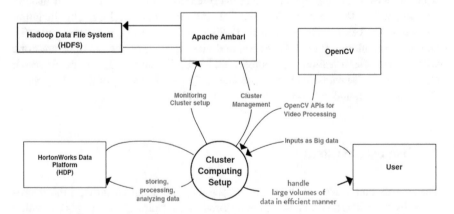

Fig. 1 Data flow diagram for framework setup

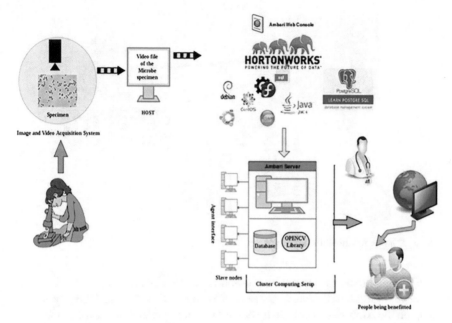

Fig. 2 Proposed video analytics framework

6 and 7, Ubuntu 12 and 14, Debian 7, and so on. Ambari provides an excellent monitoring, managing, and an end-to-end management solution for the HDP clusters for their deployment, operation, configuration, and monitoring services for all nodes in the cluster. Ambari has two components: (i) Ambari-server and (ii) Ambari-agent. *Ambari-server*—This is the master process which communicates with Ambari-agents installed on each node in the cluster. Ambari-server helps to manage all the clusters in the setup. *Ambari-agent*—Ambari-agents are there for periodically sending health status of each of them along with different metrics, installed services status, and many more things. Accordingly, master or server decides the next action of the setup and conveys back to the agent to act. The installation steps will be discussed in the later sections. We should have the minimum system requirements to run the Apache Ambari in our systems. For our setup, we are using five systems as slaves and one system as the master. The master will be configured in such a way that it is able to communicate with all the slaves (Fig. 3).

The system requirements can be described as follows:

Operating Systems Requirements: Linux-based operating system will solve the purpose and it should have 1 HDD free.

Browser Requirements: The Ambari install wizard runs as a browser-based Web application which needs a graphical browser like Google Chrome, Firefox.

Software Requirements: Our hosts should have the yum and rpm installed in them.

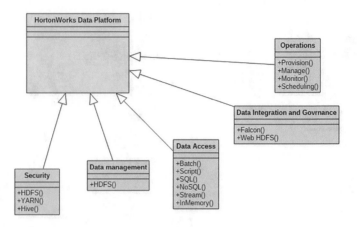

Fig. 3 Primary functions for HortonWorks Data Platform

JDK Requirements: OpenJDK 8 64-bit is needed for the setup.

Database Requirements: An instance of PostgreSQL will be installed by Ambari setup on the Ambari-server host, by default which is a relational database and stores the cluster information.

Memory Requirements: Ambari-server and agent should have at least 1 GB RAM with 500 MB free. Depending on the cluster size, the disk space and the memory will be decided.

4.2 Installation of Tools and Preparation of the Environment

The first and foremost step for the cluster setup is the installation of OS in each computer. We have taken six computers which include one master computer or node and five slave nodes. CentOS 7 is installed in each followed by setting the space directories, setting the network, and so on. Gnome desktop environment was chosen for GUI interface of the OS. The fully qualified domain name (FQDN) is given for each host in our system. *host name–f* can be used to check FQDN. To deploy our Hadoop instance, we need to prepare our deployment environment which is common for the master node as well as the slave nodes and is described as follows:

1. Edit */etc/hosts* file to map the IP address of each machine with the host name. The hosts file is used for the mapping of host names to the IP addresses of the host. So the Ambari-server has this file to determine the IP address that corresponds to a host name.

(a) The text editor is used to open the hosts file in every host using: */etc/hosts*

(b) The IP address and the FQDN should be added in the /etc/hosts file like—

172.16.8.10 master.vitcc.com.

(c) We should not remove the two lines from the host's line—127.0.0.1local-host.localdomainlocalhost::1localhost6.localdomain6 localhost6

2. Enable Network Time Protocol (NTP) on each cluster. NTP is used and needed when the clocks of computer in servers are set manually, and it gets drifted by each passing day, if not closely monitored. It is advisable to enable NTP to synchronize the time of all hosts within the cluster with Ambari-server. The synchronization of all the nodes in the cluster requires NTP upon boot, and the command is—

systemctl enable ntpd
systemctl start ntpd

3. Set Up Passwordless SSH: It is required so that the Ambari-server installed on one host can communicate with other hosts within the cluster to install Ambari-agents. The server uses the key pair to access other hosts.

4. Check DNS: Hosts should be configured with DNS and reverse DNS and for that we need to edit /etc/hosts file in each host. The hosts contain the IP address and fully qualified domain name of each of our hosts.

5. Change and Edit the Network Configuration File: The configuration file of the network is modified in this file—

vi/etc/sysconfig/network

Modification of the HOSTNAME property is to be set into the fully qualified domain name as
Networking=yes
Host name=<fully.qualified.domain.nam>
i.e., <master.vitcc.com>

6. Configuring iptables: Certain ports in the machine must be open to configure the Ambari with the hosts. For that, we can temporarily disable the iptables with the following command—

systemctl disable firewalld
service firewalld stop

7. Disable SELinux and PackageKit: SELinux is a Linux kernel security module that provides security mechanisms to access control policies. The SELinux should be disabled for the Ambari setup to function. Set enforce as '0' and for permanently disabling the SELinux, set *SELinux=disabled* in */etc/selinux/config*

4.3   Installing Ambari-server and Ambari-agent in Master Node

Ambari-server

1. Download the Ambari repository from the HortonWorks for server and copy the files to */etc/yum/repos.d*. The repository from where we download is the link given by HortonWorks itself as—

   public-repo-1.hortonworks.com/ambari/centos7/2.x/updates/2.2.2
   *mv ambari. repo/*etc.*/yum.repos. d*

   On Ambari-server host which has Internet access, the command prompt is used for doing the following steps:

   - We should always log into host as ***root*** user.
   - To check all the repositories in Ambari-server, *yum repolist* helps us. Web server installation in the master node with the command—

     yum install httpd
     systemctl start httpd
     systemctl status httpd

   - After the Web server installation, all the repositories of Ambari are copied to */var/www/html* with the 'cp' command so that it can be accessed in all the nodes.

2. The database of the Ambari-server will be installed with the following command

   yum install ambari-server

3. The immediate thing to do post the installation of Ambari is the configuration of Ambari setup and set it up for provisioning the cluster with the command— *ambari-server setup.* Follow the on-screen instructions and then start the Ambari-server as—*ambari-server start.* We need to set the port for the Ambari-server to start. The default port '8080' may be in use sometimes; we should change it with the following command—

   vi /etc/ambari-server/conf/ambari.properties
   client.api.port=8888 (any port number)

4. For proper running of the setup, we should update the services automatically in all the clusters as—

   systemctl enable httpd
   systemctl start httpd
   systemctl enable ntpd
   systemctl start ntpd
   systemctl disable firewalld.service

5. To give privilege to every user, we should do this with the change mode command which is used to change the permissions of files and directories for all nodes in the cluster. The command for this—

```
vi /etc/rc.d/rc.local
vi chmod a+x /etc/rc.d/rc.local (any user can use the services)
ls –l/etc/(list all the files)
```

Ambari-agent

1. Like the Ambari-server, Ambari-agent also has repositories and they are downloaded as usual kept in */etc/yum/repos.d*
2. Edit */etc/ambari-agent/ambari.ini* and change:

```
host name=master.vitcc.com (localhost)
```

3. After the host name is changed, install the agent, start the agent, and check the status by using the commands

```
yum install ambari-agent –y
ambari-agent start
ambari-agent status
```

4. We have to change the host name of each node in the cluster setup for the Ambari-agent to work in sync or map with the Ambari-server. Here the host is master node and the name is *master.vitcc.com* and the command to change it in every node is—*vi /etc/ambari-agent/conf/ambari-agent.ini*
5. All the repositories required for the Ambari-agent should be there in yum.repos. d directory. To check, we can do
vi */etc/yum.repos.d/ambari.repo*

Ambari Dashboard

1. Apache Ambari Log in Screen: The Ambari service is started now. We need to deploy the cluster using the Ambari Web UI using the browser. We should open a browser and type in http://<your.ambari.server>:(port number):8080, where *<your.ambari.server>* is the Ambari-server's name host. Log into the Ambari-server using the default *user name* and *password: admin/admin* and we should follow the on-screen instructions.
2. In Launch menu, *install Wizard* option should be clicked to start the installation. The following steps need to be done after the login.
3. The '*Cluster*' should be named and noted as we cannot change it again.
4. Select the HDP version; here, it is '*HDP 2.4*'. The host name of the machine is given and provide the list of hosts to be added.
5. Use passwordless *SSH* to connect to all hosts.
6. *Select Stack* and do the confirmation for hosts for Ambari setup. This is done to locate the correct hosts for the cluster, proper directories, processes, packages for the completion of the whole installation.
7. We can choose services based on our need and what is there in Stack during '*Select Stack*'. HDP *Stack* comprises many services.

8. We should *Assign Masters* for all selected services to the hosts properly in the cluster which will display the assignments in *Assign Masters*. Also, *Assign Slaves* and *Clients* like data nodes, node managers, and region servers to appropriate hosts in the cluster.
9. Customize services.
10. Install, start, and test.
11. View of the Ambari Dashboard.
12. Start using the services.

4.4   Integration of Tools

For storage purpose, we have Hadoop Data File System (HDFS). HDFS service is by default given by the HortonWorks Data Platform where we can store all our files be it text files, video files, image files, software packages, etc. Here, we store all our image and video files in HDFS for further processing. For image and video processing, we are using OpenCV which is an open source computer vision library and which is integrated with Apache Spark's PySpark package. To run video files, we also need to setup multimedia software packages in the OS and process the video files in Apache Spark. The integration of OpenCV libraries and Hortonworks data management platform is represented in Fig. 4. Also, the overall configuration and integration of tools can be shown in Fig. 5.

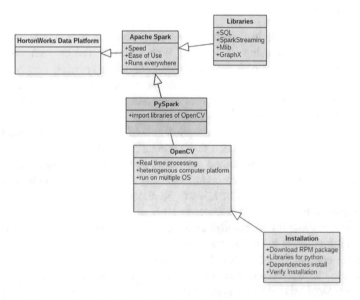

Fig. 4 OpenCV integration with HortonWorks Data Platform

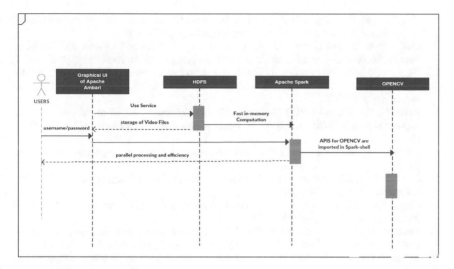

Fig. 5 Sequence diagram for experimental setup

5 Conclusion

The proposed video analytics framework is set up with the motive of helping the healthcare industry in a broader aspect by reducing the fatigue of the laboratory technicians, accelerating the idea of virtual nursing and finally the patients who can get their decisions of treatment in an efficient way. As of now, the setup is completed; sample video files are being processed with the help of tools installed. The future work is that we are progressing with a case study of a tuberculosis or malaria detection system and implement it with the help of the framework.

References

1. Bovik AC (2010) Handbook of image and video processing. Academic Press
2. Jawad FM, Alasadi I (2015) A brief review on Malaria disease (causes, treatment, diagnosis). Ann Pharma Res 3(09):148–150
3. Liu H et al (2013) Performance and energy modeling for live migration of virtual machines. Cluster Comput 16.2:249–264
4. Fahringer T et al (2005) ASKALON: a tool set for cluster and grid computing. Concurrency Comput Pract Experience 17.2–4:143–169
5. Dutta G, Yadav K, Aggarwal JK (2016) A comparative analysis of peripheral blood smear and rapid diagnostic test for diagnosing Malaria. Int J Curr Microbiol App Sci 5.2:802–805
6. Das DKR, Mukherjee R, Chakraborty C (2015) Computational microscopic imaging for malaria parasite detection: a systematic review. J Microsc 260.1:1–19
7. Ross NE et al (2006) Automated image processing method for the diagnosis and classification of malaria on thin blood smears. Med Biol Eng Comput 44.5:427–436

8. Chayadevi M, Raju G (2014) Usage of art for automatic malaria parasite identification based on fractal features. Int J Video Image Process Netw Secur 14:7–15
9. Ghosh P et al (2011) Medical aid for automatic detection of Malaria. In: Computer information systems—analysis and technologies. Springer, Berlin, Heidelberg, pp 170–178
10. Díaz G, González FA, Romero E (2009) A semi-automatic method for quantification and classification of erythrocytes infected with malaria parasites in microscopic images. J Biomed Inform 42(2):296–307
11. Abbas N, Mohamad D (2013) Microscopic RGB color images enhancement for blood cells segmentation in YCBCR color space for k-means clustering. J Theor Appl Inf Technol 55(1):117–125
12. Tek FB, Dempster AG, Kale I (2006) Malaria parasite detection in peripheral blood images. BMVC
13. Suradkar PT (2013) Detection of malarial parasite in blood using image processing. Int J Eng Innovative Technol (IJEIT) 2.10
14. Khiyal MSH, Khan A, Bibi A (2009) Modified watershed algorithm for segmentation of 2D images. Inf Sci Inf Technol 6:877–886
15. Malhi SS, Vera CL, Brandt SA (2013) Relative effectiveness of organic and inorganic nutrient sources in improving yield, seed quality and nutrient uptake of canola. Agric Sci 4(12):1
16. Anggraini D et al (2011) Automated status identification of microscopic images obtained from malaria thin blood smears using Bayes decision: a study case in Plasmodium falciparum. In: 2011 international conference on advanced computer science and information system (ICACSIS). IEEE
17. Khot ST, Prasad RK (2015) Optimal computer based analysis for detecting malarial parasites. In: Proceedings of the 3rd international conference on frontiers of intelligent computing: theory and applications (FICTA) 2014. Springer International Publishing
18. Vink JP et al (2013) An automatic vision-based malaria diagnosis system. J Microsc 250(3):166–178
19. Das DK et al (2013) Machine learning approach for automated screening of malaria parasite using light microscopic images. Micron 45:97–106
20. Sudheer C et al (2014) A support vector machine-firefly algorithm based forecasting model to determine malaria transmission. Neurocomputing 129:279–288

Application of Dynamic Thermogram for Diagnosis of Hypertension

Jayashree Ramesh and Jayanthi Thiruvengadam

Abstract Hypertension (high blood pressure) is blood pressure above than 140 over 90 mmHg (millimeters of mercury). Diagnosis of hypertension is made when one of the above readings is high. By the year 2025, the number of people living with hypertension is about 1.56 billion all over the world. This paper aims at developing a technique to diagnose hypertension noninvasively without using the cuff. In this approach, the dynamic infrared (IR) thermogram of selected body regions like hand (left and right) and neck (left and right) is obtained for about 60 s using IR thermal camera from 50 subjects (normal = 25 and age and sex-matched hypertensive = 25). The average temperature for every millisecond in these selected body regions is measured using ResearchIR software. Correlation is performed between the features extracted from dynamic thermogram and flow rate obtained from carotid Doppler ultrasound scan. The statistical analysis shows that highest correlation is obtained between the rate of temperature change (°C/min) in the neck left side with systolic pressure and hand left side with diastolic pressure (mmHg) (Pearson correlation $r = -0.637$ and 0.668 with $p < 0.01$, respectively). There also exists a linear correlation coefficient between neck right rate of change in temperature (°C/min) and right carotid artery flow rate ($r = -0.358$ with $p < 0.05$). An automated classifier using SVM network was designed with features extracted from dynamic thermogram for diagnosis of hypertension. The accuracy of the classifier was about 80% with sensitivity and specificity values 76.9 and 83.3%, respectively. The accuracy of the classifier when all the correlated features ($n = 17$) were included was 93% (sensitivity 90% and specificity 94%), whereas when highly correlated features ($n = 15$) were alone included, the sensitivity improved to 94% (accuracy 90% and specificity 85%).

Keywords Dynamic thermal imaging · SVM · Hypertension · Mean Energy · Kurtosis · Statistical parameters

J. Ramesh (✉) · J. Thiruvengadam
Department of Biomedical, SRM University, Chennai, India
e-mail: jayashree94music@gmail.com

J. Thiruvengadam
e-mail: Jayanthi.t@ktr.srmuniv.ac.in

© Springer Nature Singapore Pte Ltd. 2018
A. K. Nandi et al. (eds.), *Computational Signal Processing and Analysis*, Lecture Notes in Electrical Engineering 490,
https://doi.org/10.1007/978-981-10-8354-9_12

1   Introduction

Diagnosis of hypertension is made when the systolic reading (the pressure when the heart pumps blood around the body) or diastolic reading (pressure when the heart relaxes and refills with blood) is high. People predicted to be living with hypertension would be 1.56 billion worldwide [1]. In India, hypertension has increased from 20 to 40% in urban areas and 12 to 17% in rural areas [2]. In 2011–2012, about one-third of all people living in low- and middle-income countries had hypertension which was a great economic burden [3]. Hypertension is a disease that paves the way for other cardiovascular diseases if left untreated. This work focuses on the diagnosis of hypertension using dynamic thermal imaging. The current method for hypertension diagnosis varies from operator to operator. There are no standard techniques to predict hypertension with quondam. It needs repeated measurements, to diagnose whether a person is hypertensive. The traditional process of inflation, deflation and examining korotkoff sounds with a stethoscope can be replaced by the technique of continuous dynamic thermal imaging. This setup plays a vital role in ICU and CCU units where almost all patients are critical and do not want any disturbance. Thus, this method can be used for mass and initial screening. Chekmenev developed a model to match thermogram with pulse rate [4]. Hence, our work focuses on the diagnosis of hypertension using features measured from dynamic thermal imaging. Thermal imaging is a noninvasive and non-contact method where the heat patterns in the skin are studied using a thermal camera [5, 6]. Thermal imaging has a wide variety of application in the medical field that includes inflammatory diseases, complex regional pain syndrome and Raynaud's phenomenon, breast cancer, various types of arthritis, injuries, and other traumas [7–9].

2   Materials and Methods

Study Population: A medical camp was conducted in a private hospital where about 50 subjects registered both men and women participated. Normal and known hypertensive subjects in the age-group of 39 ± 5.01 years were included in the study. The following measurements like BP, ultrasound Doppler scan, S_PO_2, sex, age, hip and waist circumference, and height and weight were obtained. The consent form was obtained from all the subjects. And details of the procedure were clearly explained. Institutional ethical clearance was obtained from the institutional ethical clearance committee letter dated 1034/IEC/2016.

Pressure Measurement: The subject was made to relax in sitting position, and his blood pressure both systole and diastole was measured using standard sphygmomanometer. Only a trained person took a measurement for all 50 subjects.

Thermal Imaging: The thermal camera was fixed on the stand, and it was connected to the laptop. The initial setting for taking video was taken care. A standard protocol as proposed by the International Association of Certified

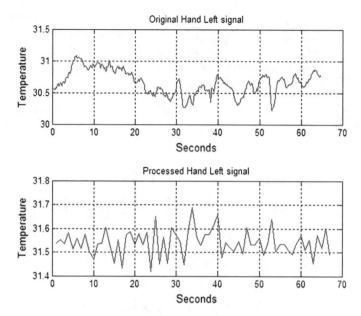

Fig. 1 Original and processed hand left signal

Thermographers was followed during the entire procedure [10]. The process was carried out in a completely dark room which was fully air-conditioned wherein the temperature was maintained at 25 °C. The subjects were made to wait in the room for 15 min. The subject was asked to stand, and his left and right hands were stretched and held in support, while the video signal was taken. Similarly, the subject was made to turn his head left and right sides for the neck images. The data were taken using thermal camera Therma Cam A305sc, FLIR Systems, USA. Then, the average temperature was measured using FLIR tool. The process of removing noise and decomposition of the thermal signal was done using wavelet transform with a level of db4. Figure 1 shows the change in temperature signal for a minute before and after noise removal.

After the noise removal, the rate of temperature change was calculated using equation [11]:

$$C = F_{AT} - I_{AT}/F_{RT} - I_{RT} \; (°C/min) \qquad (1)$$

C Rate of change in temperature,
F_{AT} Final average temperature,
I_{AT} Initial average temperature,
F_{RT} Final relative time,
I_{RT} Initial relative time.

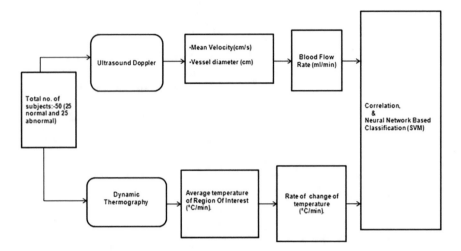

Fig. 2 Overall block diagram

Ultrasound Doppler Scan: The subject was made to lie on the bed, and the ultrasound images and flow patterns are obtained. The ultrasound Doppler scans of right and left common carotid arteries were taken using MINDRAY DC-N3 Doppler machine. The rate of blood flow was calculated from mean velocity and vessel diameter [12]. The formula to calculate flow rate was:

$$Q = \pi * D^2/4 * V_m * 60 \ (\text{ml/min}) \tag{2}$$

Q Flow rate
π 3.14
D Vessel diameter (cm)
V_m Mean velocity (cm/s)

The overall block diagram showing the work flow is as given in Fig. 2.

3   Results

3.1   Correlation Between Rate of Temperature Change with Flow Rate and Blood Pressure

As given in Table 1, the statistical analysis shows that the correlation obtained between rate of temperature change (°C/min) in the neck left side showed negative correlation with systolic pressure (Pearson correlation $r = -0.637$ with $p < 0.01$) and positive correlation was obtained between hand left side with diastolic pressure

Table 1 Pearson coefficient r value obtained between rate of temperature change with blood flow rate and pressure

Rate of change of temperature (°C/min)		Right flow rate (ml/min)	Left flow rate (ml/min)	Systolic pressure (mmHg)	Diastolic pressure (mmHg)
Left	Hand			0.543**	0.668**
	Neck		−0.356*	−0.637**	
Right	Hand			−0.622**	
	Neck	−0.358*		0.319**	0.496**

*Correlation is significant at the level $p < 0.05$
**Correlation is significant at the level $p < 0.01$

Fig. 3 Correlation between neck left rate of temperature change and left carotid artery flow rate

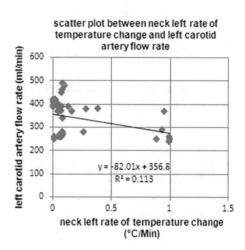

scatter plot between neck left rate of temperature change and left carotid artery flow rate

$y = -82.01x + 356.8$
$R^2 = 0.113$

neck left rate of temperature change (°C/Min)

(mmHg) (Pearson correlation $r = 0.668$ with $p < 0.01$). There exists a negative correlation coefficient between neck left rate of change in temperature (°C/min) and left carotid artery flow rate ($r = -0.356$ with $p < 0.05$), and neck right rate of change in temperature with right carotid artery flow rate ($r = -0.358$ with $p < 0.05$) as shown in Figs. 3 and 4.

3.2 Correlation Between Features from Dynamic Thermogram with Flow Rate and Blood Pressure

As given in Table 2, there exists a linear correlation between hand left mean and neck right median with left and right carotid artery flow rates ($r = -0.365$ ($p < 0.01$) and −0.295 with $p < 0.05$) which is also negative. The analysis also showed the correlation obtained between kurtosis in the hand right side with systolic was positive (Pearson correlation $r = 0.8588$ with $p < 0.01$) as shown in

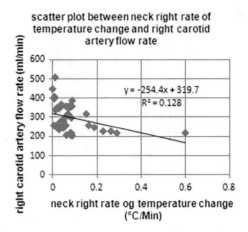

Fig. 4 Correlation between neck right rate of temperature change and right carotid artery flow rate

Fig. 5 and negative correlation between neck right side kurtosis with diastolic pressure (mmHg) (Pearson correlation $r = -0.6647$ with $p < 0.01$) as shown in Figs. 5 and 6.

Table 2 Pearson coefficient r value obtained between features from dynamic thermogram with blood flow rate and pressure

Features obtained from dynamic thermogram			Systolic pressure (mmHg)	Diastolic pressure (mmHg)	Carotid artery flow rate (ml/min)	
					Left	Right
Kurtosis	Left	Hand	0.712**	0.634**		
		Neck	−0.337[a]	−0.344*		
	Right	Hand	0.855**	0.491**		
		Neck	−0.516**	-0.664**		
Mean	Left	Hand	0.533**	0.309*	−0.365**	
		Neck	0.614**	0.459**		
	Right	Hand	0.615**	0.611**		−0.293*
Median	Left	Neck		0.328*	−0.289*	
	Right	Neck		0.461**		−0.295*
Energy	Left	Hand	0.507**			
		Neck	0.648**	0.627**		
	Right	Hand	0.594**	0.365*		
		Neck	0.658**	0.534**		

*Correlation is significant at the level $p < 0.05$
**Correlation is significant at the level $p < 0.01$

Fig. 5 Correlation between hand right kurtosis and systolic pressure

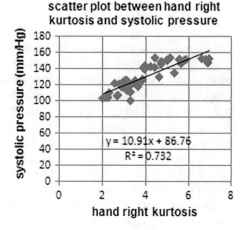

Fig. 6 Correlation between neck right kurtosis and diastolic pressure

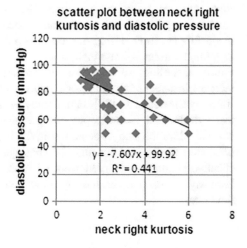

Support Vector Machine: Data-based machine learning technique covers a wide area from pattern recognition to functional regression and density estimation. SVM is a supervised machine learning technique that has been used to classify the rate of change in temperature and its features. It was used in MATLAB version 2012. SVM that produced an accuracy of 80%, the sensitivity of 76.9%, and specificity of 83.3% for all features ($n = 20$). The correlated features ($n = 17$) were classified with an accuracy of 93%, sensitivity of 90%, and specificity of 94% [13]. The highly correlated features ($n = 15$) were classified with an accuracy of 90%, sensitivity of 94%, and specificity of 85.

4   Discussions

Circulation and inspiration are related to each other. During inspiration and expiration, the airflow regulates the temperature of blood by convection. The blood flow depends on heart rate. Hence, change in flow rate and heart rate is reflected in temperature of blood. According to Hirsch, heart rate increases during inspiration and decreases during expiration. According to Jere Mead and James L. Whittenberger, the temperature of the expired gas increases. So as flow rate increases, the temperature decreases [14–16]. Likewise, in our study, there is a correlation between flow rate and rate of temperature change in neck left and neck right. Ziad et al. stated that blood flow in the cerebral region obeys Joule–Thomson effect [17]. According to Joule–Thomson effect, the pressure is inversely proportional to temperature. During systole, the blood gets pumped from the left ventricle to the aorta through a small orifice called the mitral valve. When blood flows through a small orifice (at high pressure), its temperature decreases. Likewise in our study, the rate of change of temperature in the regions neck left and hand right correlates negatively with temperature. According to Eugene Braunwald et al., the blood fills the heart during diastole. According to Starling's law, when the blood gets filled, stroke volume increases which increase in the cardiac output [18, 19]. Further according to Frank–Starling mechanism, the cardiac output increases with increase in temperature and positively correlates with diastolic pressure which is stated by Shields et al. [20]. In our study as shown in the above tables, it is evident that systolic and diastolic pressure correlates with all the regions of the body considered for the study. As the sensitivity increases for highly correlated variables, prediction of hypertension may be consequential. Hence, this work can make an effective diagnosis of hypertension. To the best of our knowledge, there has not been any approach to correlate dynamic thermogram with blood pressure and blood flow rate.

5   Conclusion

The correlation obtained was highest for neck right kurtosis and diastolic pressure and hand right kurtosis and systolic pressure. The flow rates showed the highest correlation between hand left mean with left carotid artery flow rate and right neck rate of temperature change and right carotid artery flow rate. From the above study, it has been concluded that blood pressure could be measured without contact at an accuracy of 80%, the sensitivity of 76.9%, and specificity of 83.3% and accuracy of 93%, sensitivity of 90%, and specificity of 94% with correlated parameters. As an extension, there would be the development of a handheld device with an IR sensor to diagnose hypertension which would be portable and contact-free.

References

1. http://www.world-heart-federation.org/heart-facts/fact-sheets/cardiovascular-dwasease-rwask-factors/quick-facts-on-hypertension-high-blood-pressure/
2. http://timesofindia.indiatimes.com/life-style/health-fitness/health-hypertension/articleshow/35 617449.cmshttps://
3. www.ncbi.nlm.nih.gov/pmc/articles/PMC256060/
4. Chekmenev SY, Farag AA, Essock EA (2007) Thermal imaging of the superficial temporal artery: an arterial pulse recovery model. In: 2007 IEEE conference on computer vision and pattern recognition. IEEE
5. Garbey M (2007) Contact-free measurement of cardiac pulse based on the analysis of thermal imagery. IEEE Trans Biomed Eng 54(8):1418–1426
6. Vollmer M, Klaus-Peter M (2010) Infrared thermal imaging: fundamentals, research and applications. Wiley
7. Ring EFJ, Ammer K (2012) Infrared thermal imaging in medicine. Physiol Meas 33(3):R33
8. Hildebrandt C, Raschner C, Ammer K (2010) An overview of the recent application of medical infrared thermography in sports medicine in Austria. Sensors 10(5):4700–4715
9. Qi H, Diakides NA (2003) Thermal infrared imaging in early breast cancer detection—a survey of recent research. In: Proceedings of the 25th annual international conference of the IEEE on engineering in medicine and biology society, 2003, vol 2. IEEE
10. Ring EFJ, Ammer K (2015) The technique of infrared imaging in medicine. Infrared Imaging. IOP Publishing
11. https://www.physicsforums.com/threads/rate-of-temperature-change.616194/
12. Benetos A (1985) Pulsed Doppler: an evaluation of diameter, blood velocity and blood flow of the common carotid artery in patients with isolated unilateral stenosis of the internal carotid artery. Stroke 16(6):969–972
13. Pytel K (2015) Anthropometric predictors and artificial neural networks in the diagnosis of hypertension. In: 2015 federated conference on computer science and information systems (FedCSIS). IEEE
14. Mead J, Whittenberger JL (1953) Physical properties of human lungs measured during spontaneous respiration. J Appl Physiol 5(12):779–796
15. Mead J, Whittenberger JL (1954) Evaluation of airway interruption technique as a method for measuring pulmonary air-flow resistance. J Appl Physiol 6(7):408–416
16. Hirsch JA, Bishop B (1981) Respiratory sinus arrhythmia in humans: how breathing pattern modulates heart rate. Am J Physiol Heart Circulatory Physiol 241(4):H620–H629
17. Issa Z, Miller JM, Zipes DP (2012) Clinical arrhythmology and electrophysiology: a companion to Braunwald's heart disease. Expert Consult: Online and Print. Elsevier Health Sciences
18. Braunwald E, Frahm CJ (1961) Studies on Starling's law of the heart IV. Observations on the hemodynamic functions of the left atrium in man. Circulation 24(3):633–642
19. Kitzman DW (1991) Exercise intolerance in patients with heart failure and preserved left ventricular systolic function: failure of the Frank-Starling mechanism. J Am Coll Cardiol 17 (5):1065–1072
20. Shiels HA, White E (2008) The Frank–Starling mechanism in vertebrate cardiac myocytes. J Exp Biol 211.13:2005–2013

A Comparative Study of Isolated Word Recognizer Using SVM and WaveNet

John Sahaya Rani Alex, Arka Das, Suhit Atul Kodgule
and Nithya Venkatesan

Abstract In this paper, speaker-independent isolated word recognition system is proposed using the Mel-Frequency Cepstral Coefficients feature extraction method to create the feature vector. Support vector machine, sigmoid neural net, and the novel wavelet neural network are used as classifiers and the results are compared in terms of the maximum accuracy obtained and the number of iterations taken to achieve this. The effect of stretch factor on the accuracy of classification for WaveNets is shown in the results. The number of features is also varied using dimension reduction technique and its effect on the accuracies is studied. The data is prepared using feature scaling and dimensionality reduction before training SVM and NN classifiers.

Keywords Isolated word recogniser · Mel-frequency cepstral coefficients
Support vector machine · Artificial neural network · WaveNet

1 Introduction

Speech recognition system is subdivided into two main branches—isolated word recognition and continuous speech recognition. Isolated words form the basic elements of speech and such recognition systems require a brief pause between two individual words. Previously, a lot of feature extraction and classification and clustering algorithms have been used for this purpose. In this paper, features are extracted by MFCCs and comparison between three algorithms: normal sigmoid neural network, SVM, and the novel WaveNets are done. The effect of stretch factor

J. S. R. Alex (✉)
School of Electronics Engineering, VIT University, Chennai Campus,
Vandalur-Kelambakkam Road, Chennai, India
e-mail: jsranialex@vit.ac.in

A. Das · S. A. Kodgule · N. Venkatesan
School of Electrical Engineering, VIT University, Chennai Campus,
Vandalur-Kelambakkam Road, Chennai, India

© Springer Nature Singapore Pte Ltd. 2018
A. K. Nandi et al. (eds.), *Computational Signal Processing
and Analysis*, Lecture Notes in Electrical Engineering 490,
https://doi.org/10.1007/978-981-10-8354-9_13

and number of features used for classification is also dealt with and brought to an appropriate conclusion.

Speech utterances under stress are taken, advanced acoustic features are extracted, and multi-class SVM is used for recognition [1]. Continuous speech recognition is done using HMM-SVM method to overcome the difficulties faced in the field by just SVM application [2]. HMM-SVM combination is used for audio-visual speech recognition [3]. HMM-NN hybrid is used for solving the speech recognition task known as HNN [4]. Feature extraction is done with MFCC with neural network as classifier [5]. Neural network is superior to Gaussian mixture model when it comes to distant conversational speech recognition and the paper used hybrid neural network and hidden acoustic Markov model for recognition [6]. Different types of activation functions are used in the hidden layer in the wavelet neural network [7]. A mother wavelet function is selected and a variant of wavelet neural network is used for target threat assessment [8].

2   Methodology

For making a speaker-independent isolated speech recognition system, TIDIGIT database was used. Each of the speakers is asked to say the numbers 0–9. The speakers constitute men, women, teenagers, and children. Then, the speech samples are randomly sorted. MFCC technique is used for feature extraction procedure. Different classifiers are applied to the feature vector and the accuracies of each are compared with each other.

2.1   Mel-Frequency Cepstral Coefficients (MFCCs)

Speech recognizer (SR) depends on the robustness of a feature extraction method. Mel-Frequency Cepstral Coefficient method is one of the popular methods used in SR. This method tries to imitate the human auditory perception system. Speech signal is framed for the duration of 25 ms with an overlapping duration of 10 ms. Short-time Fourier Transform (STFT) is performed on the windowed speech signal. The signals are passed through mel-scale triangular filter banks. The filtered output is compressed to imitate the human auditory system using logarithmic block. The cepstral coefficients are de-correlated with discrete cosine transform (DCT). Finally, the first 13 outputs are only considered for each frame as static features. From static features, derivatives and accelerated features are computed as dynamic features.

2.2 Support Vector Machine

Support vector machine is a mode of supervised learning algorithm that is generally used for binary classification of a dataset. SVMs tend to possess high efficiency in higher dimensional data. SVM involves forming a decision boundary of $(n - 1)$ dimensions to classify an n-dimensional dataset. SVM could be considered as an optimization problem where the decision boundary is to be determined such that it maximizes the distances between itself and the support vectors. SVMs tend to be useful due to their kernels. Linear, polynomial, and RBF are few of the kernels that can be used in SVMs. Depending upon the spatial distribution of the datasets, different kernels provide different accuracies and fits for the data. By iteratively optimizing over a data using different kernels, the best possible fit can be achieved.

2.3 Artificial Neural Network (ANN)

Artificial neural networks are self-learning systems which try to replicate a function as closely as possible depending on its training inputs. A neural network consists of multiple layers of interconnected neural units. There can be single or multiple inputs to these units but only one output. Every connection has a certain weight and every single unit has a bias. The sum of multiplication of each weight with outputs of neurons from previous layer forms the input to the neuron. This added to the bias of the particular neuron is used to calculate its output using the activation function.

The inputs of each training set are fed to the input layer and their outputs are fed forward to the next layer. This continues until the effect has propagated through all the hidden layers of the network and the output of the penultimate layer is fed as input to the output layer which in turn generates the final output. This is known as feed-forward network and this final output should closely tally with the outputs of each training dataset. The learning occurs through backpropagation. The errors in output generated after each iteration are differentiated with respect to each weight and bias to find out how changes in weights and biases affect the error term. These error along with a learing rate applied to the weights to get improved wieghts in each layer. The goal is to alter the randomly initialized weights and biases in such a way so as to get a combination which will decrease the error as much as possible.

Too much training of network backfires as it leads to overtraining of the network. This means the accuracy of the network is very high for the training datasets but fails to be accurate in the validation or testing datasets. Care must be taken in choosing the learning rate as too low a learning rate takes a long time to train a network and too high a learning rate leads to overshooting of the minimum point in the convex error curve during the gradient decent.

In this paper, both sigmoid and wavelet activation functions are used. Sigmoid activation function is a very common function but the wavelet function is new. The wavelet function used here is Gaussian. The idea of implementation is to make the wavelet activations follow a similar graph like the sigmoid function. A stretch factor is introduced which stretches the whole function and the calculations are done in a particular domain of the wavelet functions only. Outside this specific domain determined by the stretch factor, the network would fail to train itself and an error is generated.

3 Simulation

3.1 Mel-Frequency Cepstral Coefficients (MFCCs)

TIDIGITS corpus is used for the simulation. This corpus consists of ten English digits 'one' through 'nine' and 'oh.' Each digit is spoken twice by each speaker. The speech of 55 men and 57 women is used to train the models, and this leads to a total of 2464 training digits. The test data consists of speech from 56 men and 57 women for a total of 2486 test digits.

MFCCs are computed for the TIDIGITS data and they are limited to 1500 feature vectors per digits.

3.2 Support Vector Machine

The MFCC features are directly used as input to the SVM model. As the classification of digits requires ten classifications, ten SVM models are used for each sample. The data is modeled such that 80% of the data is used for training and the rest 20% is used for testing. An iteration limit of 10000 iterations is set to avoid irregularities. The SVM is run with a box constraint or the soft margin of $2 * e^7$ to implement proper fitting of the dataset. The kernels used in SVM were

- Linear kernel as shown in (1)

$$K(u,v) = f(u) \cdot f(v) = f1(u)f1(v) + f2(u)f2(v) \tag{1}$$

- Polynomial kernel as shown in (2)

$$K(u, v) = (f(u) \cdot f(v))2 \tag{2}$$

- RBF kernel as shown in (3)

$$K(u, v) = \exp^{-\gamma\left(\|u-v\|^2\right)} \tag{3}$$

- Sigmodal kernel as shown in (4)

$$K(u, v) = \tanh(au'v + r) \tag{4}$$

3.3 Artificial Neural Network (ANN)

The first step is the data preparation step. Two types of data preparations are dealt with in this paper:

- Feature Scaling—Initially, the features are scaled to a range of −1 to 1. This is done so that the gradient decent algorithm works at a faster rate while training the network. First, the length of range of the coefficients across the columns is found and each coefficient in that row is divided with that particular range. This is done for all rows.
- Dimensionality Reduction—Dimensionality reduction is done so that neural net training is faster as there would be less interconnections. This dimension reduction is done using variance method. First, the means of the coefficients across every row are found out. Then variance is calculated across every row using the mean of the respective rows. Now just the rows with the largest variances are chosen manually as features. The intuition behind this is that larger the variances, more is the separation between coefficients and thus it would be easier to separate them and train a network accordingly than a tightly knit set of coefficients. A good substitution of this technique is principal component analysis (PCA).

The total inputs to each neuron are given by (5):

$$x = w_1 * u_1 + w_2 * u_2 + w_3 * u_3 + \cdots + w_n * u_n + b \tag{5}$$

where

x	is the total input.
$w_1, w_2, w_3, \ldots, w_n$	are the weights of the connections connecting the first, second, third, ... nth neuron of the previous layer to the particular neuron.
$u_1, u_2, u_3, \ldots, u_n$	are the outputs of the first, second, third, ... nth neuron of the previous layer.
b	is the bias of that particular neuron

The activation functions used are Sigmoid and Gaussian.

Equation (6) shows the sigmoid activation function:

$$f(x) = \frac{1}{1 + e^{-x}} \tag{6}$$

Figure 1 shows the Sigmoid activation function.Equation (7) shows the Gaussian activation function:

$$f(x) = \frac{x}{\sqrt{2 * \pi}} * e^{\frac{-x^2}{2}} \tag{7}$$

Figure 2 shows the Gaussian wavelet activation function along with different stretch factors and amplitude scaling.

Fig. 1 Sigmoid activation function

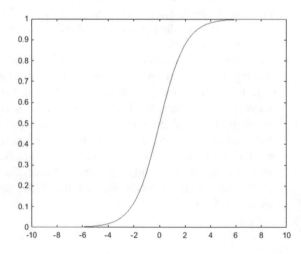

Fig. 2 Gaussian wavelet activation function along with different stretch factors and amplitude scaling

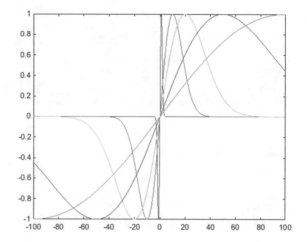

The following changes were done to the above equations before using them as activation functions:

- First, the wavelets' positions along the y-axis are changed so that the modulus of the minimum of a function is equal to its maximum. This is done by adding or subtracting a constant from the function.
- Then, the whole wavelet is divided by the maxima so that the present global maximum becomes 1 and present global minimum become −1. So, the targets while using wavelet functions are changed to bipolar outputs 1 and −1 as opposed to sigmoid function where the targets remain as 1 and 0.
- Lastly, a stretch factor is introduced by replacing x with $\frac{x}{sf}$ where sf is the stretch factor greater than 1. The idea is to replicate a sigmoid function as closely as possible.

The limitation of wavelet is that better to limit the total input in each neuron to the domain $(-sf, sf)$. Outside this domain, the network will fail to train itself.

For the simulation of every ANN, a three-layer network is taken with 360 neurons in the input layer or the number of features taken after dimensionality reduction, one for each coefficient, 50 neurons in the middle hidden layer, and 10 in the output layer for 0–9. The number is determined looking at the output by the one versus all method. The weights and biases are randomly initialized. Outputs are obtained by feed-forward network. Supervised learning was done through backpropagation.

4 Observations

Table 1 shows the dependency of accuracy of classification using ANN on the number of features chosen with the variance method for different activation functions keeping stretch factor constant at 500. The number of iterations taken to reach that accuracy is written below each accuracy.

Table 1 Accuracies versus features

Activation function	Number of features				
	10	50	100	200	360
Sigmoid	76.7 (44)	85.0 (60)	93.3 (62)	95 (72)	93.3 (66)
Gaussian	65 (9)	75 (11)	88.3 (15)	91.7 (13)	93.3 (14)

Note All accuracies are approximated to 1 decimal point

Table 2 Accuracies versus stretch factor

Activation function	Stretch factor				
	5	50	500	5000	5000000
Gaussian	65 (10)	88.3 (12)	91.7 (13)	91.7 (15)	93.3 (13)

Note All accuracies are approximated to 1 decimal point

Table 3 Accuracies versus Kernel function

Kernel function	Accuracies
RBF	8.3
Polynomial	88.3
Linear	10
Sigmoid	45

Note All accuracies are approximated to 1 decimal point

Table 2 shows the dependency of accuracy of classification using ANN on stretch factor different activation functions keeping number of features constant as 500. The number of iterations taken to reach that accuracy is written below each accuracy.

Table 3 shows the dependency of accuracy of classification using SVM on the kernel function used.

5 Results and Conclusion

From the observations, it is seen that in the comparison of ANN classifier with SVM classifier, ANN classifier gives more accuracy with its peak at 95% at 72nd iteration while SVM's maximum accuracy is at 88.3% using polynomial kernel function. But the training of ANN is a tenuous process while SVM does not take much time to be trained.

From Table 1, it is clear that the sigmoid function gives more accuracy than the wavelet function. But the wavelet function achieves their maximum accuracy at a much faster rate. It is observed that the number of iterations required is around 5–6 times less for wavelet functions than the normal sigmoid function. From Table 2, it is observed that as the stretch factor is increased, the accuracy also increases.

Now when the number of iterations is studied, it is found out that generally to achieve a higher accuracy, the number of interactions required is much more. The accuracy versus number of iterations follows a parabolic path where initially the accuracy increases rapidly with increase in accuracy but then the learning rate slows down and eventually it becomes almost parallel to the x-axis. There is no significant change in accuracy with increase in iterations.

The future scope of this research is to increase the accuracies of the wavelet functions without much change in the number of iterations required to converge. The accuracy aimed at is either similar or more than what the sigmoid function is providing. The SVM classifier seems to perform worst in this case but there may be chances to improve their accuracies too by changing the kernel functions.

Overall the novel, WaveNet method proved very effective as the time taken to train these networks is very low as compared to the Sigmoid ANN.

References

1. Besbes S, Lachiri Z (2016) Multi-class SVM for stressed speech recognition. In: 2016 2nd international conference on advanced technologies for signal and image processing (ATSIP), 21–23 March 2016. https://doi.org/10.1109/atsip.2016.7523188
2. Padrell-Sendra J, Martín-Iglesias D, Díaz-de-María F (2006) Support vector machines for continuous speech recognition. In: 2006 14th European on signal processing conference, 4–8 Sept 2006
3. Gurban M, Thiran J-P (2005) Audio-visual speech recognition with a hybrid SVM-HMM system. In: 2005 13th European on signal processing conference, 4–8 Sept 2005
4. Riis SK (1998) Hidden neural networks: application to speech recognition. In: Proceedings of the 1998 IEEE international conference on acoustics, speech and signal processing, 1998, 15–15 May 1998. https://doi.org/10.1109/icassp.1998.675465
5. Barua P, Ahmad K, Khan AAS, Sanaullah M (2014) Neural network based recognition of speech using MFCC features. In: 2014 international conference on informatics, electronics & vision (ICIEV), 23–24 May 2014. https://doi.org/10.1109/iciev.2014.6850680
6. Renals S, Swietojanski P (2014) Neural networks for distant speech recognition. In: 2014 4th joint workshop on hands-free speech communication and microphone arrays (HSCMA), 12–14 May 2014. https://doi.org/10.1109/hscma.2014.6843274
7. Zainuddin Z, Pauline O (2007) Function approximation using artificial neural networks. Int J Syst Appl Eng Dev 1(4)
8. Wang G, Guo L, Duan H (2013) Wavelet neural network using multiple wavelet functions in target threat assessment. Sci World J 2013 (Article ID 632437)

Feature Ranking of Spatial Domain Features for Efficient Characterization of Stroke Lesions

Anish Mukherjee, Abhishek Kanaujia and R. Karthik

Abstract Development of automatic framework for efficient characterization of brain lesions is a significant research concern due to the complex properties exhibited by the brain tissues. This study focuses on observing the properties of such composite structures in order to identify optimal features for characterizing the properties of normal and abnormal brain tissues. This work initially applies Fuzzy C Mean algorithm to identify the region of interest. After segmentation, four different types of features are extracted from the region of interest. These features include first-order parameters, Gray-level Co-occurrence Matrix (GLCM) parameters, Laws texture features, and Gray-Level Run-Length Matrix (GLRLM) parameters. These identification features were ranked in order of pertinence with the help of Mutual Information and Statistical Dependence-based feature ranking algorithms. Based on the inference obtained from the Mutual Information and Statistical Dependence-based feature ranking algorithms, twelve best features are selected for characterizing the properties of the normal and abnormal brain tissues.

Keywords Lesion · Feature ranking · Mutual information · Statistical dependence

1 Introduction

Stroke is an incident which affects the blood flow in the blood vessels, which are responsible for the supply of body fluid to the brain. The World Health Organization (WHO) claims that roughly 15 million individuals worldwide suffer from stroke, out of which almost one-third die and another one-third people are

A. Mukherjee · A. Kanaujia · R. Karthik (✉)
School of Electronics Engineering, VIT University, Chennai, India
e-mail: r.karthik@vit.ac.in

A. Mukherjee
e-mail: anish.mukherjee2014@vit.ac.in

A. Kanaujia
e-mail: abhishek.kanaujia2014@vit.ac.in

© Springer Nature Singapore Pte Ltd. 2018
A. K. Nandi et al. (eds.), *Computational Signal Processing
and Analysis*, Lecture Notes in Electrical Engineering 490,
https://doi.org/10.1007/978-981-10-8354-9_14

rendered permanently incapacitated [1]. Stroke is the third foremost source of death worldwide after cancer and heart diseases. Stroke poses a serious threat because of the huge multitude of people who are at risk. The proposed system aims to detect the lesions by developing an automatic framework for the effective characterization of the lesion affected areas of the brain MRI.

Computed Tomography (CT) and Magnetic Resonance Imaging (MRI) are the two methodologies used in brain imaging. Significant research has been done in the field of computer-aided detection. Generally, the non-enhanced Computed Tomography scan is the radiologist primary point of reference [2] Lee et al. suggested a technique to identify lesion using Adaptive filtering [3]. To improve the quality of detection, the infarct is enhanced to preserve the detail information about the edges. Przeslaskowski et al. recommended a technique to identify the ischemic stroke by using a method involving wavelet transformation [4–6]. Chawla et al. suggested a technique to distinguish a stroke by using a multi-feature-based methodology in which features are extracted from the intensity and the wavelet domain [7]. Usinskas et al. implemented a technique to identify the lesions on the basis of mean, standard deviation, and histogram of the lesion affected area [8]. Maldijan et al. proposed a segmentation-based method [9]. The aforementioned method included identification of the region of hypo-density inside the lentiform nucleus and insula in the patients. Fauzi et al. proposed a segmentation-based approach to extract the abnormality from the brain scan image [10]. It includes a two-phase approach, the first phase involves the segmentation of image with the help of watershed-based algorithm and then by finally using decision tree to segment the abnormality. Tang et al. proposed a mechanism to detect lesions with a method of feature characterization [11]. The method included selecting the region of interest with the help of an adaptive circle. The method involves characterizing the stroke lesions with the help of textural parameters. The method proposed by the authors of this article is based on ranking of the features that are best suited to characterize the lesions and thereby distinguishing the normal MRI scan from the abnormal MRI scan.

2 Proposed Methodology

The phases through which the proposed approach goes through have been illustrated in Fig. 1. The detection system has five stages, acquisition of the MRI scan image of the brain, segmentation of the portion of the brain scan with the help of Fuzzy C Mean algorithm, extracting first-order, second-order, and higher-order features of the segmented image, selecting the best features to represent the segmented image with the help of Mutual Information and Statistical Dependence-based ranking algorithms and finally characterizing the segmented portion of the brain with the help of the selected features.

Fig. 1 Proposed
methodology

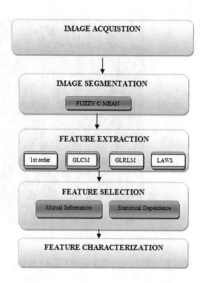

3 Image Acquisition

The input datasets in MRI modality were acquired from multiple offline and online sources. A total of 15 datasets were used in this study. Among 15 datasets, 8 datasets represent normal and the other 7 represents abnormal category. Only axial plane, slices were included for this research.

4 Image Segmentation

Considering an image of number of elements N, which has to be segmented with the help of C number of cluster centers [12], V_i (where $i = 1, 2, 3, 4, ..., C$). The value for the cluster centers V_i is calculated with help of the membership function, μ_{ij} (where $j = 1, 2, 3, ..., N$). The membership function μ_{ij} is defined as,

$$\mu_{ij} = \frac{1}{\sum_{l=1}^{C} \left(\frac{d^2{}_{ij}}{d^2_{ij}}\right)^{\left(\frac{2}{m-1}\right)}} \tag{1}$$

And the number of optimal cluster center V_i is defined as,

$$V_i = \frac{\sum_{j=1}^{N} \mu_{ij}^m x_j}{\sum_{j=1}^{N} \mu_{ij}^m} \tag{2}$$

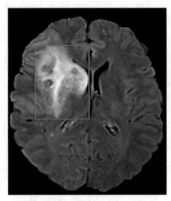

(a) Sample slice with lesion   (b) Segmented lesions portion
                                          from the MRI slice

Fig. 2 **a** Sample slice with lesion, **b** segmented lesions portion from the MRI slice

The Fuzzy C Mean algorithm is iterative in nature and the iteration stops when the value of the objective J_{fcm} is optimized. d_{ij} is defined as the minimum Euclidean distance between the ith element and the ith cluster. The significance of the objective function is optimized when the Euclidean distance d_{ij} is less than the value of the membership function μ_{ij}. The objective function J_{fcm} is defined as,

$$J_{fcm} = \sum_{i=1}^{C} \sum_{j=1}^{N} \mu_{ij}^{m} d_{ij}^{2} \tag{3}$$

Figure 2a shows the input images on which the Fuzzy C Mean algorithm is applied and Fig. 2b shows the segmented output.

5 Feature Extraction

In the proposed system, first-order, second-order, and higher-order statistical features are observed. The first-order features include mean, standard deviation, skewness, and kurtosis. For second-order features, 19 GLCM features and 7 GLRLM features are extracted. Higher-order statistical features comprise of Laws textures energy measures which are six in total. Thus, the total 36 features that are obtained are used to describe the texture in the region with the help of statistical vectors.

5.1 First-Order Features

The first-order statistical features comprise of the first 4 moments of the probability density function [13]. Information up to the fourth moment is extracted because generally the higher-order moments do not contain relevant information required to describe the image. The first moment of the probability density function, the arithmetic mean is usually of limited utility to describe a region of interest accurately enough to distinguish it from the remaining regions. The standard deviation of the region of interest provides information about the roughness of the region. The third moment is known as the skewness of a matrix. The skewness of a matrix provides us information about the difference in degree of illumination of a texture pixel (texel) from the average illumination of the pixel in the matrix. A positive skew indicates the texel is darker than average and negative skew denotes the texel is lighter than the average. The fourth moment or kurtosis of a matrix provides detail about the uniformity of the gray-level distribution in the image.

5.2 Second-Order Features

GLCM features are extracted to differentiate the textural superiority of the region of interest.

5.2.1 Gray-Level Co-Occurrence Matrix Features

The GLCM is a prominent technique developed by Haralick et al. deployed to obtain the textural information [14]. The GLCM features provide information about the change in gray-level intensity as a function of distance and direction. The parameter $p(i,j)$ used in calculating the GLCM feature is the probability density function of change in intensity level from 'i' to 'j' when moving distance 'd' is in the direction 'θ' in the polar form or else the mathematical equivalent of it is in the Cartesian coordinate Δx, Δy. Haralick proposed that the 19 GLCM features provide textural information which is used by the proposed system to serve the purpose of distinguishing the region of interest.

5.3 Higher-Order Features

The higher-order features are Laws Texture Energy Measures feature and GLRLM features.

5.3.1 Gray-Level Run-Length Matrix Comprises

Gray-level run-length matrix (GLRLM) features are higher-order statistical texture descriptors which are often used to obtain the parameters which would distinguish the image from other set of images [15]. The gray-level run-length matrix (GLRLM) computes the span of 'i' intensity levels with the run length of 'j.' It has been observed that coarse features have long runs of a pixel intensity value while a fine detail has a relatively small run of a pixel intensity value. Thereby this feature can be useful for detecting and segregating a fine detail from coarse details in an image [16]. Apart from the traditional run-length matrices, a few modified matrices are also being used to extract potentially useful textural information from the images [17]. The traditional GLRLM features were presented by Galloway et al. [18]. The features were Gray_Level_Non_uniformity (GLN), Short_Runs_Emphasis (SRE), Run_Length_Non_Uniformity (RLN), Long_Runs_Emphasis (LRE), and Run_Percentage (RP). The modified run-length matrices which are considered are short Gray_Level_Run_Emphasis (SGLRE) and High_Gray_Level_Run_Emphasis (HGLRE).

5.3.2 Laws Texture Energy Measures

Measures about the texture of an image can be obtained by spatial domain analysis or frequency domain analysis. In spatial domain, texture is represented by the gray-level change in the entire image plane. To obtain the texture, information matched filters and the outputs from the matched filters are worked upon to compute variance and mean [19]. There are different spatial domain masks which provide information about different aspect of the image texture [20]. Three masks are used by the proposed system, namely Level (L), Edge (E), and Spot (S). These masks are convolved to obtain six outputs.

6 Feature Selection

For selecting the best features to represent the lesions, the proposed system makes use of two feature selection algorithms. Mutual Information-based Selection algorithm and Statistical Dependence-based Selection algorithm are the two algorithms. Features' selection technique is for the most part used to diminish elements space dimensionality. The algorithms used by the proposed system instead of reducing the dimensionality and selecting a particular subset of feature data score the features on the basis of their usefulness in classifying the features using predetermined distribution in the image.

6.1   Mutual Information-Based Selection Algorithm

Mutual Information is feature scoring algorithm which significantly reduces the cost of computation and measures whether the co-occurrence is dependent upon the class labels and a classification system can be made or not. The higher value of the function MI suggests that the feature values are heavily dependent upon the class label and a mathematical model for classification can be built using the feature as the basis of classification [21].

$$\text{MI} = \sum_{y=Y} \sum_{z=Z} p(y,z) \log \left(\frac{p(y,z)}{p(y)p(z)} \right) \tag{4}$$

where $p(x, y)$ is joint probability distribution function of X and Y distribution in the image.

6.2   Statistical Dependence-Based Selection Algorithm

Statistical Dependence is also a feature scoring algorithm used to find the dependence of the features on their respective class labels. The feature values are quantized into Q levels. The quantization bins are chosen in a adaptively so that all the bins have approximately the same number of values and the bins are checked for co-occurrence and higher value of co-occurrence suggests that there is dependence of the feature value on the class labels and the features are thereby scored on the basis of the higher co-occurrence [22]. The mathematical equation governing this algorithm is stated below

$$\text{SD} = \sum_{y=Y} \sum_{z=Z} p(y,z) \frac{p(y,z)}{p(y)p(z)} \tag{5}$$

7   Result and Discussion

The proposed experiments were carried on 'Intel core i5 4210U' with clock speed of 1.70 GHz. The implementation environment is MATLAB 14. The MI and SD algorithms rank the extracted features on the basis of their ability to distinguish between a normal and abnormal brain tissue. The best 12 features are selected for the effective characterization of the lesions on the basis of the value of the weight matrix generated by the MI and SD algorithms. To verify the effectiveness of the selected features, the selected features are used to identify the abnormal and normal brain images using the K-Nearest Neighbor algorithm [23]. Initially, the training set

Table 1 Observation of feature ranking

Feature	Rank
Information_Measure_Of_Correlation1	1
Inverse_Difference	2
Information_Measure_Of_Correlation2	3
Auto_Correlation	4
Difference_Entropy	5
Dissimilarity	6
Energy	7
Entropy	8
Homogeneity	9
Maximum_Probability	10
Sum_Average	11
Sum_Entropy	12
Difference_Variance	13
Contrast	14
Correlation	15
Cluster_Shade	16
Cluster_Prominence	17
Sum_Of_Squares_Variance	18
Sum_Variance	19
RLN	20
LGRLE	21
HGRLE	22
Skewness	23
SRE	24
LRE	25
GLN	26
RP	27
Mean	28
Standard_Deviation	29
Kurtosis	30
LL	31
EE	32
SS	33
LE	34
ES	35
LS	36

comprises of all the 36 extracted features and it is tested against a training set comprising of the best 12 selected features. The test set contained 10 elements the first five being normal MRI scan and the remaining being MRI with tumor lesions. The normal brain images are denoted by the label class '1' and the abnormal brain

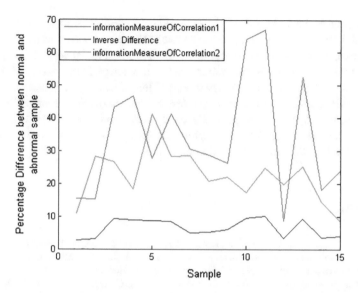

Fig. 3 Observation obtained for percentage deviation with respect to feature statistics

images are represented by the label class '2.' When the training set comprises of the selected 12 features, all the prediction by the K-Nearest Neighbor algorithm is made accurately which signifies accurate characterization of the stroke lesions. The features are ranked in Table 1.

The training set with selected feature is more accurate than the training set comprising of all the features. Therefore, this validates the effective characterization of the stroke lesions. The difference in feature value between normal and abnormal brain tissues expressed in percentage is shown in Fig. 3.

The average percentage variation between the abnormal and normal brain tissue for the best three characterization parameters namely Information of Correlation1, Inverse Difference, Information of Correlation2 is 37.05%.

8 Conclusion

Thus, an efficient approach for the characterization of the stroke lesion has been discussed in this work. The framework is quite promising in characterizing the stroke lesions with high accuracy and can assist in computer-aided detection of stroke. Amongst all the textural features that were extracted, the features which had the highest scores when MI and SD algorithms were applied are observed to be the best distinguishing features. Features that obtained the highest scores were used to distinguish the normal and abnormal images by using a dataset of 30 images as training set and the test set had 10 images. The test set had 10 images in which 5 were normal and 5 were abnormal. By applying K-nearest neighbor algorithm, the selected 12 features have a score of 0.99982712 and 1.000024 from MI and SD algorithms, respectively. They accurately distinguished the abnormal images from the normal images while the training set comprising all the extracted features distinguished them with an accuracy of 80%. Thus, it can be concluded that ranking of the spatial domain features leads to effective characterization of the brain lesions.

The proposed work focuses solely on characterization of stroke lesion using different feature using different feature primitives. This work can further be extended using different classification algorithms.

References

1. Thom T, Haase N, Rosamond W, Howard VJ, Rumsfeld J, Manolio T (2006) Heart disease and stroke statistics—2006 update: are port from the American Heart Association Statistics Committee and Stroke Statistics Subcommittee. Circulation 113:e85–e151
2. Adams H, Adams R, Del Zoppo G, Goldstein LB (2005) Guidelines for the early management of patients with ischemic stroke, 2005 guidelines update, a scientific statement from the Stroke Council of the American Heart Association/American Stroke Association. Stroke 36:916–921
3. Lee Y, Takahashi N, Tsai DY, Fujita H (2006) Detectability improvement of early sign of acute stroke on brain CT images using an adaptive partial smoothing filter. Proc Soc Photo Opt Instrum Eng Med Imaging 6144:2138–2145
4. Karthik R, Menaka R (2016) Statistical characterization of ischemic stroke lesion from MRI using discrete wavelet transformations. ECTI Trans Electr Eng Electron Commun 14(2): 57–64
5. Karthik R, Menaka R (2016) A critical appraisal on wavelet based features from brain MR images for characterization if ischemic stroke injuries. Electron Lett Comput Vis Image Anal 15(3):1–6
6. Przelaskowski A, Sklinda KP, Bargie PJ, Walecki J, Biesiadko-Matus-zewska M, Kazubek M (2007) Improved early stroke detection: wavelet-based perception enhancement of computerized tomography exams. Comput Biol Med 37:524–533
7. Chawla M, Sharma S, Sivaswamy J, Kishore LT (2009) A method for automatic detection and classification of stroke from brain CT images. In: EMBC 2009, Annual International Conference of the IEEE, 2009, pp 3581–3584
8. Usinskas A, Dobrovolskis RA, Tomandl BF (2004) Ischemic stroke segmentation on CT images using joint features. Informatica 15(2):283–290
9. Maldjian JA, Chalela J (2001) Automated CT segmentation and analysis for acute middle cerebral artery stroke. Am J Neuroradiol 22(6):1050–1055
10. Fauzi MFA, Komiya R, Haw S-C (2008) Unsupervised abnormalities extraction and brain segmentation. In: International conference on intelligent system and knowledge engineering, vol 1, pp 1185–1190
11. Tang F, Ng DKS, Chow DHK (2011) An image feature approach for computer-aided detection of ischemic stroke. Comput Biol Med 41:529–536
12. Benaichouche AN, Oulhadj H, Siarry P (2013) Improved spatial fuzzy c-means clustering for image segmentation using PSO initialization, Mahalanobis distance and post-segmentation correction. Digital Signal Process 23:1390–1400
13. Schilling C (2012) Analysis of atrial electrograms, vol 17. KIT Scientific Publishing
14. Haralick RM (1979) Statistical and structural approaches to texture. Proc IEEE 67(5):786–804

15. Chu A, Sehgal CM, Greenleaf JF (1990) Use of gray value distribution of run lengths for texture analysis. Pattern Recognit Lett 11:415–420
16. Dasarathyand BR, Holder EB (1991) Image characterizations based on joint gray-level run-length distributions. Patt Recogn Lett 12:497–502
17. He D-C, Wang L (1990) Texture unit, texture spectrum, and texture analysis. Trans Geosci Remote Sens IEEE 28(4):509–512
18. Galloway MM (1975) Texture analysis using gray level run lengths. Comput Graph Image Process 4:172–179
19. Liua Y, Zhanga D, Lua G, Mab W-Y (2007) A survey of content-based image retrieval with high-level semantics. Patt Recogn 40:262–282
20. Laws KI (1979) Texture energy measures. In: Proceedings of the image understanding workshop, pp 47–51, Nov 1979
21. Pohjalainen J (2013) Feature selection methods and their combinations in high dimensional classification of speaker likability, intelligibility and personality traits. Comput Speech Lang
22. Karp RM (1972) Reducibility among combinatorial problems. Complex Comput Comput 85–103
23. Li S, Harner J, Adjeroh D (2011) Random kNN feature selection—a fast and stable alternative to random forests. BMC Bioinf 12

Active Vibration Control Based on LQR Technique for Two Degrees of Freedom System

Behrouz Kheiri Sarabi, Manu Sharma and Damanjeet Kaur

Abstract Recently, researchers have proposed active vibration control technique to control undesired vibrations in structures. In this work, procedure of active vibration control is discussed in a simple way. For that, the mathematical model of structure, the optimal placement of sensor and actuator, and control laws of active vibration control are discussed. Finally, vibration control using LQR technique has been applied on two degrees of freedom system to illustrate the active vibration control.

Keywords Active vibration control · Mathematical model · Optimal placement
Control law · Two degrees of freedom system

1 Introduction

Vibrations generated by an IC engine or by a bumpy road profile, if not isolated sufficiently can result in acute inconvenience of the passengers traveling in an automobile. Compressor of a household refrigerator can be major source of noise if shell of the compressor is poorly designed. There have been instances when mighty bridges have been damaged by vibrations generated by march of soldiers or by a specific flow of wind. Mechanical machines having rotating parts such as pumps, turbines, fans, compressors have to be meticulously designed, properly aligned, and sufficiently balanced so as to keep low vibration levels while in operation. Structure of an aeroplane particularly its wings are prone to flow-induced vibrations and therefore special materials having high structural damping and high strength are

B. Kheiri Sarabi (✉) · M. Sharma · D. Kaur
UIET, Panjab University, Chandigarh 160014, India
e-mail: behrooz.kheiri@gmail.com

M. Sharma
e-mail: manu@pu.ac.in

D. Kaur
e-mail: damanee@pu.ac.in

© Springer Nature Singapore Pte Ltd. 2018
A. K. Nandi et al. (eds.), *Computational Signal Processing
and Analysis*, Lecture Notes in Electrical Engineering 490,
https://doi.org/10.1007/978-981-10-8354-9_15

needed. Material for construction of an aeroplane should also have low density to save fuel cost. Sometimes vibrations are required but mostly vibrations are cause of discomfort, unwanted noise, and wastage of energy. Vibrations may occur due to external excitation, unbalanced force, friction, etc. There are three types of vibrations, viz. free vibration, forced vibration, and self-excited vibration. Vibrations generated in a structure due to some initial displacement or/and velocity or/and acceleration are called free vibrations. Forced vibrations occur when the structure is continuously excited by some harmonic or random force. In the case of self-excited vibration, exciting force is a function of motion of the vibrating body. Since time immemorial man has been trying to dissipate undesirable vibrations occurring in structures and machines. Several ways have been developed to control vibrations and newer techniques are being developed. Passive vibration control (PVC), active vibration control (AVC), and semi-active vibration control are main ways to control vibrations. In passive control, mass and/or stiffness and/or damping of the structure are changed so as to control structural vibrations, this may increase overall mass of the system. On the other hand, in active control, an external source of energy is used to control structural vibrations. As shown in Fig. 1, an actively controlled structure essentially consists of sensors to sense structural vibrations, a controller to manipulate sensed vibrations, and actuators to deform the structure as per orders of the controller. Such a structure is also called 'intelligent structure' because it exhibits desired dynamic characteristics even in the presence of an external load and disturbances in the environment. In semi-active vibration control technique, passive as well as active techniques are simultaneously used. Active vibration control may fail if an external source of energy gets exhausted or sensing mechanism ill performs or actuating mechanism ill performs or controller malfunctions. Therefore, semi-active control has found importance as in this technique passive technique can still control the vibrations if active technique fails. AVC is suited for applications where stringent weight restrictions are present, e.g., aerospace, nanotechnology, robotics. In situations where low-frequency vibrations are present, AVC is more effective than PVC. In AVC, different type of sensors can be used, e.g., strain gauge [1], piezoelectric accelerometer [2], piezoelectric patch [3], piezoelectric fiber

Fig. 1 Schematic diagram of an intelligent structure

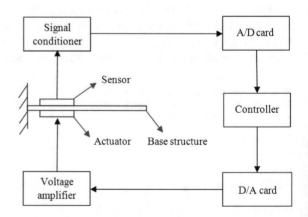

reinforced composite (PFRC) [4], polyvinylidene fluoride (PVDF) [5]. Similarly, different type of actuators can be used, e.g., magneto-rheological damper [6], electro-rheological damper [7], piezoelectric patch [8], piezoelectric stack [9]. Piezoelectric patches have been extensively used in AVC both as sensors as well as actuators. Piezoelectric materials have coupled electromechanical properties. Piezoelectric material layers can be pasted on the base structure [10] or segmented piezoelectric patches can be pasted on the surface [11], piezoelectric layer can be sandwiched between two layers of structure [12], segmented piezoelectric patches can be embedded in the composite structure [13], wires of piezoelectric material can be embedded in the composite structure [14] etc.

In Fig. 1, one piezoelectric sensor and one piezoelectric actuator are instrumented on a cantilevered plate. Signal sensed by the sensor is conditioned by signal conditioner and then fed to host PC through A/D card, control voltage generated by host PC is converted into an analog voltage, suitably amplified and then applied on the piezoelectric actuator. Following steps have to be followed for implementing active vibration control on a typical structure:

- Create mathematical model of structure instrumented with sensors and actuators
- Find optimal locations of sensors and actuators
- Design a suitable controller

Numerous simulations have to be performed using the mathematical model of the structure so as to evaluate performance of an AVC scheme under various loads. Following sections briefly discuss typical steps that need to be followed while actively controlling a structure.

2 Preliminary

2.1 Mathematical Model of a Structure

The first step in AVC is capturing the physics of the system in mathematical form. For creating mathematical model of an intelligent structure, governing equations of motion of the base structure, and electro-mechanics of sensors and actuators are required [15]. Governing equations of motion of the structure can be written using experimental tests [16], finite element techniques [17], and Hamilton's principle [18]. Finite element technique is powerful and widely accepted technique for analyzing an intelligent structure. Generally, to derive equations of motion of a structure, following steps can be followed:

- Define the structure using an appropriate coordinate system and draw its schematic diagram
- Draw free-body diagrams of the structure
- Write equilibrium relations using free-body diagrams.

2.2 Optimal Placement of Sensors and Actuators

Placement of sensors and actuators at appropriate locations over a base structure using an optimization technique is called an optimal placement. One important issue in active vibration control is to find the optimal position and size of sensors/actuators. Limited number of sensors and actuators can be placed over a structure in many ways. Effective optimal placement of sensors/actuators over a structure increases the performance of an AVC scheme. Usually, to find optimal location of sensors/actuators, a criterion is extremized which is called an optimization criterion. Then a suitable optimization technique is employed to search optimal location of sensors and actuators over the base structure.

Optimization Criteria The process of optimal placement of sensors/actuators over a structure aims at maximizing the efficiency of an AVC scheme. Some criterion is fixed based upon end application and is subsequently extremized so as to obtain locations of sensors/actuators. Some of the possible optimization criteria which can be used in AVC are:

Remark 1 Maximizing force applied by actuators: In AVC, actuators are desired to be placed over the structure in such a way that forces applied by actuators on the base structure are large. For instance, in case of a cantilevered plate, force applied by actuators can be maximized when actuators are placed near the root of the cantilevered plate. Hence, output force by an actuator can be considered as a criterion for optimal placement of actuators.

Remark 2 Maximizing deflection of the base structure: In AVC, actuators are desired to be placed over the structure in such a way that maximum deflection of the base structure is obtained. Therefore, deflection of base structure can be considered as a criterion for optimal placement of actuators.

Remark 3 Maximizing degree of controllability/observability: One of the necessary conditions in any control process is controllability and observability. Effective and stable AVC depends on the degree of controllability/observability of the system. Controllability and observability can be checked by using rank test. Optimal placement of actuators/sensors can be determined by using degree of controllability/observability as optimization criterion.

Remark 4 Minimizing the control effort: With ever-increasing cost of energy, a very natural criterion for optimal placement of sensor and actuators over a structure is amount of control effort. Also, it is to be appreciated that limited supply of external energy is usually available to suppress the structural vibrations. Therefore, minimizing the control energy can be considered as a criterion for optimal placement of sensors/actuators.

Remark 5 Minimizing the spillover effects; A continuous structure has infinite natural frequencies or modes. Usually, first few modes of vibration have most of vibrational energy. Therefore in AVC, controller is designed to control first few modes only and not all the modes. Uncontrolled residual modes can make the system unstable and reduce the control effectiveness. This phenomenon is called as spillover effect. Therefore, placement of sensors/actuators can be selected that minimize the spillover effects.

Optimization Techniques A large number of optimization techniques are available and still new techniques are continuously coming [19]. Optimization is an art of finding the best convergent mathematical solution that extremizes an objective function. Best mathematical solution is calculated by maximizing the efficiency function or/and minimizing the cost function of the system. Optimization techniques can be classified as classical techniques (single-variable function, multi-variable function with no constraints), numerical methods for optimization (linear programming [20], nonlinear programming [21], integer programming [22] and quadratic programming [23]) and advanced optimization methods (univariate search method [24], swarm intelligence algorithm [25], simulated annealing algorithm [26], genetic algorithms [27], Tabu search [28]).

2.3   Control Law

Finally, a control technique has to be used to generate a suitable control signal in an AVC application. Control strategies, which have been applied in AVC so far, can be classified as: classical control, modern control, and intelligent control.

Classical Control Classical control techniques are described using system transfer functions. Feedback controller [29], feedforward controller [30], and proportional–integral–derivative (PID) controller [31] have been frequently used in active vibration control. In feedback control, manipulated signal is calculated using error between setpoint and dynamic output signal. In feedforward controller, controller signal is based on signal and disturbance signal. The velocity feedback controller is one of the very famous control techniques which have been used practically in AVC [32].

Modern Control The classical control methods are limited to single-input–single-output (SISO) control configurations and being used for linear time-invariant systems only. To solve multi-input and multi-output (MIMO) systems, modern control techniques are used. Governing equations of the plant are converted into a state-space format. In optimal control, control gains are derived by extremizing a performance index. In eigenvalue assignment method, those control gains are used which give desired eigenvalue of the plant in closed loop. Control gains can also be calculated by satisfying Lyapunov stability criterion. Many times an observer is used to estimate states of the plant, and these estimated states are used in the control law. Adaptive controller [33], optimal controller [34], robust controller [35], sliding mode control [36], μ-controller [37], etc., have been frequently used in active vibration control.

Intelligent Control Classical and modern control theories find it difficult to control uncertain and nonlinear systems effectively. Therefore, most of nonlinear systems are stabilized using controllers based on intelligent control. Intelligent control exhibits intelligent behavior, rather than using purely mathematical method to keep the system under control. Intelligent control is most suited for applications where mathematical model of a plant is not available or it is difficult to develop mathematical model of a plant. It is based on qualitative expressions and experiences of people working with the process. Neural network-based control techniques require a set of inputs and outputs for training the neural network by a training algorithm. Once neural network has been trained for specific purpose, the network gives useful outputs even for unknown/unforeseen inputs. No mathematical model of the plant is required for neural network-based control. Fuzzy logic is based on simple human reasoning. Fuzzy logic involves: fuzzification, rule base generation, and defuzzification. Simple if-then rules specify the control law. Input variables are fuzzified using fuzzy sets in step called fuzzification. Crisp output is obtained by defuzzifying output variables in step called defuzzification. Mamdani fuzzy controllers [38] and Takagi–Sugeno fuzzy controllers [39] are extensively used techniques in fuzzy logic.

3   Illustration of Principle of Active Vibration Control

Let us understand principle of active vibration control through an example of a two degrees of freedom spring–mass–damper system as shown in Fig. 2.

Mass 'M_1' is connected to mass 'M_2' through spring of stiffness 'K_2' and damper with damping coefficient 'C_2.' Mass 'M_1' is connected to a boundary through a spring of stiffness 'K_1' and a damper with damping coefficient 'C_1.' Mass 'M_2' is connected to a boundary through a spring of stiffness 'K_3' and a damper with damping coefficient 'C_3.' Actuator 'f' is capable of exerting force 'f' on mass 'M_2.' Free-body diagrams of the two masses are drawn in Fig. 3.

Equations of motion are written from free-body diagrams as:

$$M_1\ddot{x}_1 + K_2(x_1 - x_2) + C_2(\dot{x}_1 - \dot{x}_2) + K_1x_1 + C_1\dot{x}_1 = 0$$
$$M_2\ddot{x}_2 + C_3\dot{x}_2 + K_3x_2 + f + K_2(x_2 - x_1) + C_2(\dot{x}_2 - \dot{x}_1) = 0 \tag{1}$$

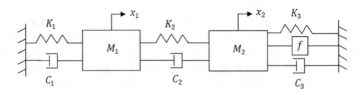

Fig. 2 Schematic diagram of two degrees of freedom spring–mass–damper system

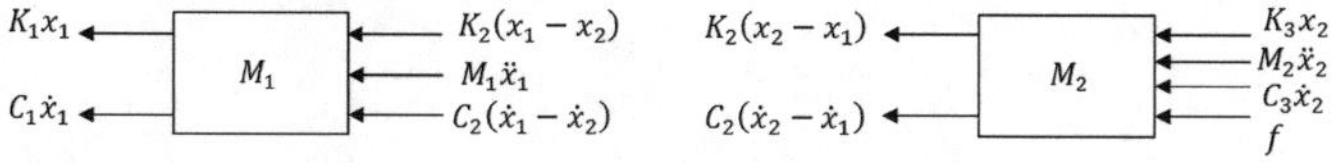

K_1x_1 ◄───── | M_1 | ◄───── $K_2(x_1 - x_2)$ $K_2(x_2 - x_1)$ ◄───── | M_2 | ◄───── K_3x_2
◄───── $M_1\ddot{x}_1$ ◄───── $M_2\ddot{x}_2$
$C_1\dot{x}_1$ ◄───── ◄───── $C_2(\dot{x}_1 - \dot{x}_2)$ $C_2(\dot{x}_2 - \dot{x}_1)$ ◄───── ◄───── $C_3\dot{x}_2$
                                                                              ◄───── f

Fig. 3 Free-body diagrams of the two masses

So, we have a system of two second-order ordinary differential equations which are coupled with each other. These equations can be converted into a state-space format by taking:

$$x_3 = \dot{x}_1 \quad \text{and} \quad x_4 = \dot{x}_2$$

now (1) can be rewritten as:

$$M_1\dot{x}_3 + K_2(x_1 - x_2) + C_2(x_3 - x_4) + K_1x_1 + C_1x_3 = 0$$
$$M_2\dot{x}_2 + C_3x_4 + K_3x_2 + f + K_2(x_2 - x_1) + C_2(x_3 - x_4) = 0 \tag{2}$$

These equations can be expressed as matrix equation of motion as:

$$\{\dot{x}\}_{4\times1} = [A]_{4\times4}\{x\}_{4\times1} + [B]_{4\times1}\{f\}_{1\times1} \tag{3}$$

where

$$\{x\} = \{x_1 \quad x_2 \quad x_3 \quad x_4\}^{\mathrm{T}}$$

$$[A] = \begin{bmatrix} 0 & 0 & 1 & 0 \\ 0 & 0 & 0 & 1 \\ \frac{-K_1 - K_2}{M_1} & \frac{K_2}{M_1} & \frac{-C_1 - C_2}{M_1} & \frac{-C_2}{M_1} \\ \frac{-K_2}{M_2} & \frac{-K_2 - K_3}{M_2} & \frac{C_2}{M_2} & \frac{-C_2 - C_3}{M_2} \end{bmatrix}$$

$$[B] = \{0 \quad 0 \quad 0 \quad -1\}^{\mathrm{T}}$$

Control law for this system can be expressed as:

$$f_{1\times1} = -\{k\}_{1\times4}\{x\}_{4\times1} \tag{4}$$

where $\{k\}_{1\times4}$ is vector of control gains. This vector of control gains can be easily obtained using optimal control, pole-placement technique, Lyapunov control, etc. State-space equations can be solved using following algebraic equation:

$$x_{4\times1}(n+1) = \bar{A}_{4\times4}x_{4\times1}(n) + \bar{B}_{4\times1}f(n) \tag{5}$$

where $\bar{A}_{4\times4}$ and $\bar{B}_{4\times1}$ are discretized forms of A and B matrices discretized using a small sampling time interval. Alternatively, second-order coupled differential equations of motion of the system in closed loop can be solved using suitable numerical technique like Newmark-β method. Simulink software of MATLAB can

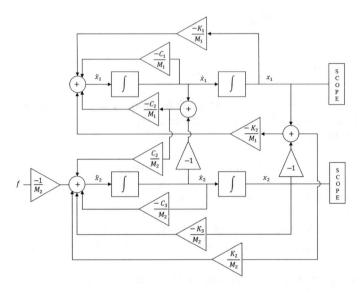

Fig. 4 Simulink model of two degrees of freedom system

also be employed to solve this system. Simulink model for this system is produced in Fig. 4.

For $M_1 = 2M_2 = 10$ kg, $K_1 = 2K_2 = 3K_3 = 1000$ N/m, $C_1 = C_2 = C_3 = 0.8$ Ns/m optimal gains can be obtained using LQR command in MATLAB as:

$$K = [\,1.2347 \quad -0.3189 \quad -0.5925 \quad -1.7024\,]$$

If at time = 0 s, $x^T = [0.1000]$ then controlled and uncontrolled responses of both the masses are as plotted in Fig. 5.

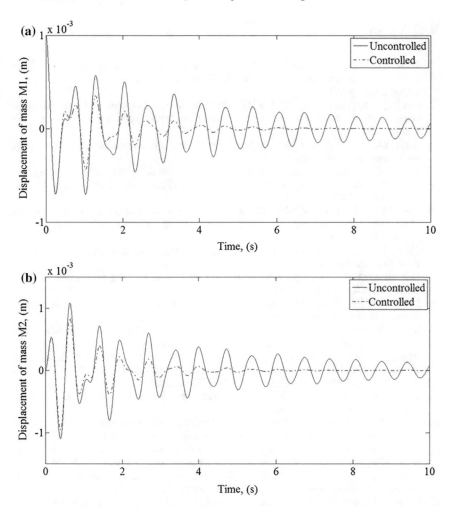

Fig. 5 Controlled/uncontrolled time responses of a two degrees of freedom system

4 Conclusions

In this work, a comprehensive introduction has been given on active vibration control through an illustrative example. This work can help students and researchers who want to start doing research on active vibration control. For active vibration control, one needs to derive mathematical model of structure, then find the optimal placement of sensors and actuators over structure, and finally employ appropriate control law. To show the procedure of active vibration control, an example of two degrees of freedom has been presented in this paper.

References

1. Pisoni AC, Santolini C, Hauf DE, Dubowsky S (1995) Displacements in a vibrating body by strain gage measurements. In: Proceedings-SPIE the International Society for Optical Engineering, pp 119–119
2. Qiu ZC, Zhang XM, Wang YC, Wu ZW (2009) Active vibration control of a flexible beam using a non-collocated acceleration sensor and piezoelectric patch actuator. J Sound Vib 326:438–455
3. Sarabi Kheiri B, Sharma M, Kaur D (2016) Simulation of a new technique for vibration tests, based upon active vibration control. IETE J Res 63:1–9
4. Kapuria S, Yasin MY (2010) Active vibration suppression of multilayered plates integrated with piezoelectric fiber reinforced composites using an efficient finite element model. J Sound Vib 329:3247–3265
5. Clark YG, Fuller RLCR, Zander AC (1994) Experiments on active control of plate vibration using piezoelectric actuators and polyvinylidene fluoride (PVDF) modal sensors. J Vib Acoust 116:303–308
6. Shinya S (2013) Vibration control of a structure using magneto-rheological grease damper. Front Mech Eng 8:261–267
7. Li J, Gruver WA (1998) An electrorheological fluid damper for vibration control. In: IEEE conference ICRA, pp 2476–2481.
8. Kheiri Sarabi B, Sharma M, Kaur D, Kumar N (2016) A Novel technique for generating desired vibrations in structure. Integr Ferroelectr 176:236–250
9. Simões RC, Steffen V, Der Hagopian J, Mahfoud J (2007) Modal active vibration control of a rotor using piezoelectric stack actuators. J Vib Control 13:45–64
10. Baillargeon BP, Vel SS (2005) Exact solution for the vibration and active damping of composite plates with piezoelectric shear actuators. J Sound Vib 282:781–804
11. Inman DJ (2006) Vibration with control. Wiley, West Sessex
12. Crawley E, de Luis J (1987) Use of piezoelectric actuators as elements of intelligent structures. AIAA J 25:1373–1385
13. Raja S, Prathap G, Sinha PK (2002) Active vibration control of composite sandwich beams with piezoelectric extension-bending and shear actuators. Smart Mater Struct 11:63
14. Chandrashekhara K, Tanneti R (1995) Thermally induced vibration suppression of laminated plates with piezoelectric sensors and actuators. Smart Mater Struct 4(1995):281–290
15. Kheiri Sarabi B, Sharma M, Kaur D (2014) Techniques for creating mathematical model of structures for active vibration control. In: IEEE conference (RAECS)
16. Bailey T, Ubbard JE (1995) Distributed piezoelectric-polymer active vibration control of a cantilever beam. J Guid Control Dyn 8:605–611
17. Kheiri Sarabi B, Sharma M, Kaur D (2017) An optimal control based technique for generating desired vibrations in a structure. Iran J Sci Technol 40:219–228
18. Baillargeon BP, Senthil Vel S (2005) Exact solution for the vibration and active damping of composite plates with piezoelectric shear actuators. J Sound Vib 282:781–804
19. Kheiri Sarabi B, Maghade DK, Malwatkar GM (2012) An empirical data driven based control loop performance assessment of multi-variate systems. In: IEEE conference (INDICON), pp 70–74
20. Gou B (2008) Optimal placement of PMUs by integer linear programming. IEEE Trans Power Syst 23:1525–1526
21. Theodorakatos N, Manousakis N, Korres G (2015) Optimal placement of phasor measurement units with linear and non-linear models. Electr Power Compon Syst 43:357–373
22. Gou B (2008) Generalized integer linear programming formulation for optimal PMU placement. IEEE Trans Power Syst 23
23. Powell M (1978) The convergence of variable metric methods for nonlinearly constrained optimization calculations. In: Nonlinear programming 3. Academic Press

24. Vanderplaats GN (1984) An efficient feasible directions algorithm for design synthesis. AIAA J 22:1633–1640
25. Rajdeep D, Ganguli R, Mani V (2011) Swarm intelligence algorithms for integrated optimization of piezo-electric actuator and sensor placement and feedback gains. Smart Mater Struct 20:1–14
26. Jose M, Simoes M (2006) Optimal design in vibration control of adaptive structure using a simulating annealing algorithm. Compos Struct 75:79–87
27. Han JH, Lee I (1999) Optimal placement of piezoelectric sensors and actuators for vibration control of a composite plate using genetic algorithms. Smart Mater Struct 8:257–267
28. Kincaid RK, Keith EL (1998) Reactive Tabu search and sensor selection in active structural acoustic control problems. J Heuristics 4:199–220
29. Chen L, Lin C, Wang C (2002) Dynamic stability analysis and control of a composite beam with piezoelectric layers. Compos Struct 56:97–109
30. Beijen MA, van Dijk J, Hakvoort WBJ, Heertjes MF (2014) Self-tuning feedforward control for active vibration isolation of precision machines. In: Preprints of the 19th World Congress The International Federation of Automatic Control Cape Town, South Africa
31. Jovanovic AM, Simonovic AM, Zoric ND, Luki NS, Stupar SN, Ilic SS (2013) Experimental studies on active vibration control of a smart composite beam using a PID controller. Smart Mater Struct 22
32. Lam KY, Peng XQ, Liu GR, Reddy JN (1997) A finite-element model for piezo-electric composite laminates. Smart Mater Struct 6:583–591
33. Xu SX, Koko TS (2004) Finite element analysis and design of actively controlled piezoelectric smart structures. Finite Elem Anal Des 40:241–262
34. Dong XJ, Meng G, Peng JC (2006) Vibration control of piezoelectric smart structures based on system identification technique: numerical simulation and experimental study. J Sound Vib 297:680–693
35. Hu YR, Ng A (2005) Active robust vibration control of flexible structures. J Sound Vib 288:43–56
36. Gu H, Song G (2007) Active vibration suppression of a smart flexible beam using a sliding mode based controller. J Vib Control 13:1095–1107
37. Lei L, Benli W (2008) Multi objective robust active vibration control for flexure jointed struts of Stewart platforms via H∞ and μ synthesis. Chin J Aeronaut 21:125–133
38. Marinaki M, Marinakis Y, Stavroulakis GE (2010) Fuzzy control optimized by PSO for vibration suppression of beams. Control Eng Pract 18:618–629
39. Chen CW, Yeh K, Chiang WL, Chen CY, Wu DJ (2007) Modeling, H∞ control and stability analysis for structural systems using Takagi-Sugeno fuzzy model. J Vib Control 13:1519–1534

Mathematical Model of Cantilever Plate Using Finite Element Technique Based on Hamilton's Principle

Behrouz Kheiri Sarabi, Manu Sharma and Damanjeet Kaur

Abstract In this work, the finite element model of a cantilevered plate is derived using Hamilton's principle. A cantilevered plate structure instrumented with one piezoelectric sensor patch and one piezoelectric actuator patch is taken as a case study. Quadrilateral plate finite element having three degrees of freedom at each node is employed to divide the plate into finite elements. Thereafter, Hamilton's principle is used to derive equations of motion of the smart plate. The finite element model is reduced to the first three modes using orthonormal modal truncation, and subsequently, the reduced finite element model is converted into a state-space model.

Keywords Mathematical model · Cantilevered plate · Finite element
Hamilton's principal · State-space

1 Introduction

Mathematical model of an active structure is required for implementing an active vibration control (AVC) scheme [1]. Finite element techniques [2] and experimental modal analysis [3] have been used frequently in AVC applications. Equations of motion of beam can be derived using Lagrangian theory [4], Kirchhoff theory [5], Euler–Bernoulli beam theory [6], etc. In active vibration control, mathematical model of composite structure, hybrid structure, sandwich structure, and multilayer structures can be carried out using classical theory (CLT) [7], first-order shear deformation theory (FSDT) [8], third-order theory (TOT) [9], high-order shear deformation theory [10], layerwise displacement theory [11], Reedy theory [12], etc. Using these methods, displacement variables are calculated along the thickness

B. Kheiri Sarabi (✉) · M. Sharma · D. Kaur
UIET, Panjab University, Chandigarh 160014, India
e-mail: behrooz.kheiri@gmail.com

M. Sharma
e-mail: manu@pu.ac.in

© Springer Nature Singapore Pte Ltd. 2018
A. K. Nandi et al. (eds.), *Computational Signal Processing and Analysis*, Lecture Notes in Electrical Engineering 490,
https://doi.org/10.1007/978-981-10-8354-9_16

through all the layers, with different material properties of the layers. Finite element techniques can be used to create mathematical models of structures [13, 14]. Finite element technique can be based on Hamilton's principle [15, 16], Galerkin's approach [17], Rayleigh–Ritz theory [18], Hook's law [19], etc. After extracting the equations of motion of structures in terms of differential equations, model can be reduced to smaller order using modal truncation.

Generally, mathematical model of structures is created using physical specifications of structure (mass, stiffness, and damping). It has been shown that environmental effects also can change the equations of motion. Thermal effect [20], electric-field effect [21], hydro-effect [22], and magnetic-field effect [23] can be incorporated in mathematical model of structures. In this paper, mathematical model of a cantilevered plate instrumented with piezoelectric patches is developed. Many real-life structures are in the shape of a cantilevered plate such as satellite structures, aircraft wings, and turbine blades. Piezoelectric patches are used extensively in active vibration control as sensors and actuators. Finite element techniques have emerged as confident techniques to model dynamics of mechanical structures. Finite element method based on Hamilton's principle is used in this work to find the dynamic equations of structure. In Sect. 2, a cantilevered plate is considered and an expression is derived for Lagrangian of the system. In Sect. 3, equations of motion of the smart plate are derived using Hamilton's principle. Equations of motion are decoupled using modal analysis in Sect. 4, and in Sect. 5, decoupled equations of motion are used to create a state-space model of the system. Finally, in Sect. 6, conclusions are drawn.

2   Finite Element Model of a Smart Cantilevered Plate

Consider a thin cantilevered mild-steel plate of size 16 cm × 16 cm, as shown in Fig. 1. This plate is discretized into '64' equal elements of size 2 cm × 2 cm. One

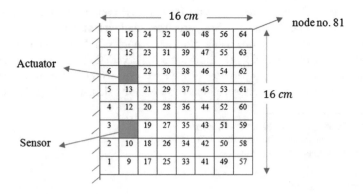

Fig. 1 Cantilevered plate instrumented with piezoelectric

Fig. 2 Quadrilateral plate
element

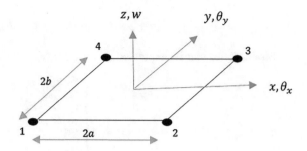

piezoelectric sensor patch is pasted at 11th element, and one piezoelectric patch is pasted at 14th element.

Quadrilateral plate element, as shown in Fig. 2, with four node points is used for finite element modeling. Six degrees of freedom are possible at each node, but here for sake of need and simplicity, three degrees of freedom have been considered at each node as:

w — displacement normal to the plane of the plate

$\theta_y = -\frac{\partial w}{\partial y}$ — rotation about x-axis

$\theta_x = \frac{\partial w}{\partial x}$ — rotation about y-axis

At the piezoelectric patch location, the structure is composite with one layer of mild steel and one piezoelectric layer. Constitutive equations of piezoelectricity can be written as:

$$\{D\} = [e]\{\varepsilon\} + [\zeta]\{E\}$$
$$\{\sigma\} = [D_p]\{\varepsilon\} - [e]^T\{E\} \tag{1}$$

where D, E, ε, and σ are the electric displacement, electric field, strain, and stress vectors, respectively. D_p, e, and ζ are the elasticity, piezoelectric constant, and dielectric constant matrices, respectively. For elastic material of the base structure, the constitutive equation is:

$$\{\sigma\} = [D_s]\{\varepsilon\} \tag{2}$$

where D_s is the elasticity constant matrix of the main structure.

The normal displacement w can be expressed in terms of nodal displacements as:

$$w = [N_1 N_2 N_3 N_4]\{u^e\} = [N]\{u^e\} \tag{3}$$

where N_1, N_2, N_3, and N_4 are shape functions and $\{u^e\}$ is vector of elemental degrees of freedom which is given by:

$$\{u^e\}^T = \left[w_1 \theta_{x1} \theta_{y1}, \ldots, w_4 \theta_{x4} \theta_{y4} \right] \tag{4}$$

Strain vector can be expressed as function of nodal displacements as:

$$\{\varepsilon\} = z \left\{ \begin{array}{c} -\dfrac{\partial^2 w}{\partial x^2} \\[2mm] -\dfrac{\partial^2 w}{\partial y^2} \\[2mm] -2\dfrac{\partial^2 w}{\partial x \partial y} \end{array} \right\} \tag{5}$$

Substituting (3) in (5), we have:

$$\{\varepsilon\} = z \left\{ \begin{array}{c} -\dfrac{\partial^2}{\partial x^2} \\[2mm] -\dfrac{\partial^2}{\partial y^2} \\[2mm] -2\dfrac{\partial^2}{\partial x \partial y} \end{array} \right\} [N]\{u^e\} \tag{6}$$

Taking

$$[B_u] = \left\{ \begin{array}{c} -\dfrac{\partial^2}{\partial x^2} \\[2mm] -\dfrac{\partial^2}{\partial y^2} \\[2mm] -2\dfrac{\partial^2}{\partial x \partial y} \end{array} \right\} [N]$$

we have,

$$\{\varepsilon\} = z[B_u]\{u^e\} \tag{7}$$

In the present case, electric voltage is applied only perpendicular to the plane of the piezoelectric patch; therefore, the electric field vector becomes:

$$\{E\} = - \left\{ \begin{array}{c} 0 \\ 0 \\ 1 \\ \dfrac{1}{h_p} \end{array} \right\} v = -\{B_v\}v \tag{8}$$

where v is voltage applied across the piezoelectric patch and h_p is thickness of piezoelectric patch. Total kinetic energy of one finite element can be written as:

$$T_e = \frac{1}{2} \int_{Q_s} \rho_s \dot{w}^2 dQ + \int_{Q_p} \rho_p \dot{w}^2 dQ \tag{9}$$

where ρ is density and subscripts s and p refer to main structure and piezoelectric structure, respectively. Total potential energy of one finite element can be written as:

$$U_e = \frac{1}{2} \int_{Q_s} \{\varepsilon\}^T \{\sigma\} dQ + \frac{1}{2} \int_{Q_p} \{\varepsilon\}^T \{\sigma\} dQ \tag{10}$$

The electrostatic potential energy stored in one element is:

$$W_{\text{elec}} = \frac{1}{2} \int_{Q_p} \{E\}^T \{D\} dQ \tag{11}$$

Sum of energy stored by the surface force and the energy required to apply surface electrical charge on piezoelectric is:

$$W_{\text{ext}} = \int_{s1} \{w\}^T \{f_s^e\} ds - \int_{s2} vq ds \tag{12}$$

where $\{f_s^e\}$ is the surface force vector, q is applied surface electrical charge density, and s_1 and s_2 are surface area of plate and surface area of piezoelectric patch, respectively. Lagrangian for one finite element of plate is:

$$L = T_e - U_e + (W_{\text{elec}} + W_{\text{ext}}) \tag{13}$$

After substituting the values of kinetic energy, potential energy, and work done by one element, the Lagrangian becomes:

$$L = \left(\frac{1}{2} \int_{Q_s} \rho_s \dot{w}^2 dQ + \int_{Q_p} \rho_p \dot{w}^2 dQ \right) - \left(\frac{1}{2} \int_{Q_s} \{\varepsilon\}^T \{\sigma\} dQ + \frac{1}{2} \int_{Q_p} \{\varepsilon\}^T \{\sigma\} dQ \right)$$
$$+ \left(\frac{1}{2} \int_{Q_p} \{E\}^T \{D\} dQ + \int_{s1} \{w\}^T \{f_s^e\} ds - \int_{s2} vq ds \right) \tag{14}$$

Substituting w, $\{\varepsilon\}$, and $\{E\}$ in (14), we have:

$$
L = \left(\frac{1}{2}\rho_s \int_{Q_s} \{\dot{u}^e\}^T [N]^T [N]\{\dot{u}^e\}\mathrm{d}Q + \frac{1}{2}\rho_p \int_{Q_p} \{\dot{u}^e\}^T [N]^T [N]\{\dot{u}^e\}\mathrm{d}Q \right)
$$
$$
- \left(\frac{1}{2}\int_{Q_s} (z[B_u]\{u^e\})^T \{\sigma\}\mathrm{d}Q + \frac{1}{2}\int_{Q_p} (z[B_u]\{u^e\})^T \{\sigma\}\mathrm{d}Q \right) \tag{15}
$$
$$
+ \left(\frac{1}{2}\int_{Q_p} (-[B_v]v)^T \{D\}\mathrm{d}Q + \int_{s1} ([N]\{u^e\})^T \{f_s^e\}\mathrm{d}s - \int_{s2} qv\mathrm{d}s \right)
$$

or,

$$
L = \left(\frac{1}{2}\rho_s \int_{Q_s} \{\dot{u}^e\}^T [N]^T [N]\{\dot{u}^e\}\mathrm{d}Q + \frac{1}{2}\rho_p \int_{Q_p} \{\dot{u}^e\}^T [N]^T [N]\{\dot{u}^e\}\mathrm{d}Q \right)
$$
$$
- \left(\frac{1}{2}\int_{Q_s} z\{u^e\}^T [B_u]^T [\sigma]\mathrm{d}Q + \frac{1}{2}\int_{Q_p} z\{u^e\}^T [B_u]^T [\sigma]\mathrm{d}Q \right) \tag{16}
$$
$$
+ \left(-\frac{1}{2}\int_{Q_p} v[B_v]^T \{D\}\mathrm{d}Q + \int_{s1} \{u^e\}^T [N]^T \{f_s^e\}\mathrm{d}s - \int_{s2} qv\mathrm{d}s \right)
$$

Substituting values of $\{D\}$ and $[\sigma]$ in (31), we have:

$$
L = \left(\frac{1}{2}\rho_s \int_{Q_s} \{\dot{u}^e\}^T [N]^T [N]\{\dot{u}^e\}\mathrm{d}Q + \frac{1}{2}\rho_p \int_{Q_p} \{\dot{u}^e\}^T [N]^T [N]\{\dot{u}^e\}\mathrm{d}Q \right)
$$
$$
- \left(\frac{1}{2}\int_{Q_s} z\{u^e\}^T [B_u]^T [D_s]\{\varepsilon\}\mathrm{d}Q \right) - \left(\frac{1}{2}\int_{Q_p} z\{u^e\}^T [B_u]^T ([D_p]\{\varepsilon\} - [e]^T [E])\mathrm{d}Q \right)
$$
$$
- \left(\frac{1}{2}\int_{Q_p} v[B_v]^T ([e]\{\varepsilon\} - [\zeta][E])\mathrm{d}Q \right) + \left(\int_{s1} \{u^e\}^T [N]^T \{f_s^e\}\mathrm{d}s - \int_{s2} qv\mathrm{d}s \right)
$$
$$
\tag{17}
$$

Substituting $\{\varepsilon\}$ and $[E]$ from (22) and (23) in above equation, we have:

$$
L = \left(\frac{1}{2}\rho_s \int_{Q_s} \{\dot{u}^e\}^T [N]^T [N]\{\dot{u}^e\} \mathrm{d}Q + \frac{1}{2}\rho_p \int_{Q_p} \{\dot{u}^e\}^T [N]^T [N]\{\dot{u}^e\} \mathrm{d}Q \right)
$$
$$
- \left(\frac{1}{2} \int_{Q_s} z\{u^e\}^T [B_u]^T \left(z[D_s][B_u]\{u^e\} \right) \mathrm{d}Q \right)
$$
$$
- \left(\frac{1}{2} \int_{Q_p} z\{u^e\}^T [B_u]^T \left(z[D_p][B_u]\{u^e\} - [e]^T [B_v]v \right) \mathrm{d}Q \right) \qquad (18)
$$
$$
- \left(\frac{1}{2} \int_{Q_p} v[B_v]^T \left(z[e][B_v]\{u^e\} - [\zeta][B_v]v \right) \mathrm{d}Q \right)
$$
$$
+ \left(\int_{s1} \{u^e\}^T [N]^T \{f_s^e\} \mathrm{d}s - \int_{s2} qv \mathrm{d}s \right)
$$

3 Deriving Equations of Motion of Smart Plate Using Hamilton's Principle

According to Hamilton's principle, integration of Lagrangian between any two arbitrarily selected time intervals t_0 and t_1 must satisfy the following equation:

$$
\delta \int_{t_0}^{t_1} L \mathrm{d}t = 0 \qquad (19)
$$

Substituting expression of Lagrangian from (18), we have:

$$\delta \int_{t_0}^{t_1} L dt = \delta \int_{t_0}^{t_1} \left(\frac{1}{2}\rho_s \int_{Q_s} \{\dot{u}^e\}^T [N]^T [N]\{\dot{u}^e\} dQ + \frac{1}{2}\rho_p \int_{Q_p} \{\dot{u}^e\}^T [N]^T [N]\{\dot{u}^e\} dQ \right)$$

$$- \left(\frac{1}{2} \int_{Q_s} z^2 \{u^e\}^T [B_u]^T [D_s][B_u]\{u^e\} dQ \right)$$

$$- \left(\frac{1}{2} \int_{Q_p} z^2 \{u^e\}^T [B_u]^T [D_p][B_u]\{u^e\} + zv\{u^e\}^T [B_u]^T [e^T]^T [B_v] dQ \right)$$

$$- \left(\frac{1}{2} \int_{Q_p} vz[B_v]^T [e^T][B_u]\{u^e\} - v^2[B_v]^T [\zeta][B_v] dQ \right)$$

$$+ \left(\int_{s1} \{u^e\}^T [N]^T \{f_s^e\} ds - \int_{s2} qv ds \right) dt = 0$$

$$(20)$$

In (20), we can take variation with respect to both variables $\{u^e\}$ and v one by one, separately. Taking variation with respect to $\{u^e\}$, we have:

$$\int_{t_0}^{t_1} \left(\frac{1}{2}\rho_s \int_{Q_s} \{\delta\dot{u}^e\}^T [N]^T [N]\{\dot{u}^e\} dQ + \frac{1}{2}\rho_s \int_{Q_s} \{\dot{u}^e\}^T [N]^T [N]\{\delta\dot{u}^e\} dQ \right.$$

$$+ \frac{1}{2}\rho_p \int_{Q_p} \{\delta\dot{u}^e\}^T [N]^T [N]\{\dot{u}^e\} dQ + \frac{1}{2}\rho_p \int_{Q_p} \{\dot{u}^e\}^T [N]^T [N]\{\delta\dot{u}^e\} dQ$$

$$- \frac{1}{2} \int_{Q_s} z^2 \{\delta u^e\}^T [B_u]^T [D_s][B_u]\{u^e\} dQ - \frac{1}{2} \int_{Q_s} z^2 \{u^e\}^T [B_u]^T [D_s][B_u]\{\delta u^e\} dQ$$

$$+ \frac{1}{2} \int_{Q_p} z^2 \{\delta u^e\}^T [B_u]^T [D_p][B_u]\{u^e\} dQ - \frac{1}{2} \int_{Q_p} zv[B_v]^T [e^T][B_u]\{\delta u^e\} dQ$$

$$\left. + \int_{s1} \{\delta u^e\}^T [N]^T \{f_s\} ds \right) = 0$$

$$(21)$$

Since $[D_p]$ and $[D_s]$ are symmetric matrices, the equation becomes

$$
\int_{t_0}^{t_1} \left(\frac{1}{2}\rho_s \int_{Q_s} \{\delta\dot{u}^e\}^T [N]^T [N]\{\dot{u}^e\}dQ \right.
$$

$$
+ \frac{1}{2}\rho_s \int_{Q_s} \{\delta\dot{u}^e\}^T [N]^T [N]\{\dot{u}^e\}dQ + \frac{1}{2}\rho_p \int_{Q_p} \{\delta\dot{u}^e\}^T [N]^T [N]\{\dot{u}^e\}dQ
$$

$$
+ \frac{1}{2}\rho_p \int_{Q_p} \{\delta\dot{u}^e\}^T [N]^T [N]\{\dot{u}^e\}dQ - \frac{1}{2}\int_{Q_s} z^2\{\delta u^e\}^T [B_u]^T [D_s][B_u]\{u^e\}dQ
$$

$$
- \frac{1}{2}\int_{Q_s} z^2\{\delta u^e\}^T [B_u]^T [D_s][B_u]\{u^e\}dQ - \frac{1}{2}\int_{Q_p} z^2\{\delta u^e\}^T [B_u]^T [D_p][B_u]\{u^e\}dQ
$$

$$
- \frac{1}{2}\int_{Q_p} z^2\{\delta u^e\}^T [B_u]^T [D_p][B_u]\{u^e\}dQ - \frac{1}{2}\int_{Q_p} zv\{\delta u^e\}^T [B_u]^T [e^T]^T [B_v]dQ
$$

$$
- \frac{1}{2}\int_{Q_p} zv\{\delta u^e\}^T [B_u]^T [e^T][B_v]dQ + \int_{s1} \{\delta u^e\}^T [N]^T \{f_s\}ds \right) = 0
$$

$$(22)$$

or,

$$
\int_{t_0}^{t_1} \left(\rho_s \int_{Q_s} \{\delta\dot{u}^e\}^T [N]^T [N]\{\dot{u}^e\}dQ \right.
$$

$$
+ \rho_p \int_{Q_p} \{\delta\dot{u}^e\}^T [N]^T [N]\{\dot{u}^e\}dQ - \int_{Q_s} z^2\{\delta u^e\}^T [B_u]^T [D_s][B_u]\{u^e\}dQ
$$

$$
- \int_{Q_p} z^2\{\delta u^e\}^T [B_u]^T [D_p][B_u]\{u^e\}dQ - \int_{Q_p} zv\{\delta u^e\}^T [B_u]^T [e^T][B_v]dQ
$$

$$
+ \int_{s1} \{\delta u^e\}^T [N]^T \{f_s\}ds \right) dt = 0
$$

$$(23)$$

Performing integration by parts of first two terms of (23), we get:

$$
\rho_s \int_{Q_s} \left(\{\delta u^e\}^T [N]^T [N] \{\dot{u}^e\} \right) \big|_{t_0}^{t_1} dQ
$$

$$
- \rho_s \int_{t_0}^{t_1} \int_{Q_s} \left(\{\delta u^e\}^T [N]^T [N] \{\ddot{u}^e\} \right) dQ dt + \rho_p \int_{Q_p} \left(\{\delta u^e\}^T [N]^T [N] \{\dot{u}^e\} \right) \big|_{t_0}^{t_1} dQ
$$

$$
- \rho_p \int_{t_0}^{t_1} \int_{Q_p} \left(\{\delta u^e\}^T [N]^T [N] \{\ddot{u}^e\} \right) dQ dt + \int_{t_0}^{t_1} \left[- \int_{Q_s} \left(z^2 \{\delta u^e\}^T [B_u]^T [D_s][B_u]\{u^e\} \right) dQ \right.
$$

$$
- \int_{Q_p} \left(z^2 \{\delta u^e\}^T [B_u]^T [D_p][B_u]\{u^e\} \right) dQ - \int_{Q_p} \left(zv \{\delta u^e\}^T [B_u]^T [e^T]^T [B_v] \right) dQ
$$

$$
\left. + \int_{s1} \left(\{\delta u^e\}^T [N]^T \{f_s\} \right) ds \right] = 0
$$

$$
\tag{24}
$$

or,

$$
- \rho_s \int_{t_0}^{t_1} \int_{Q_s} \left(\{\delta u^e\}^T [N]^T [N] \{\ddot{u}^e\} \right) dQ dt
$$

$$
- \rho_p \int_{t_0}^{t_1} \int_{Q_p} \left(\{\delta u^e\}^T [N]^T [N] \{\ddot{u}^e\} \right) dQ dt + \int_{t_0}^{t_1} \int_{Q_s} \left(z^2 \{\delta u^e\}^T [B_u]^T [D_s][B_u]\{u^e\} \right) dQ dt
$$

$$
- \int_{t_0}^{t_1} \int_{Q_p} \left(z^2 \{\delta u^e\}^T [B_u]^T [D_p][B_u]\{u^e\} \right) dQ dt - \int_{t_0}^{t_1} \int_{Q_p} \left(zv \{\delta u^e\}^T [B_u]^T [e^T]^T [B_v] \right) dQ dt
$$

$$
+ \int_{t_0}^{t_1} \int_{s1} \left(\{\delta u^e\}^T [N]^T \{f_s\} \right) ds dt = 0
$$

$$
\tag{25}
$$

By taking out common term $\{\delta u^e\}^T$, the equation becomes:

$$\{\delta u^e\}^T \int_{t_0}^{t_1} \left[-\rho_s \int_{Q_s} ([N]^T[N]\{\ddot{u}^e\})\mathrm{d}Q - \rho_p \int_{Q_p} ([N]^T[N]\{\ddot{u}^e\})\mathrm{d}Q \right.$$
$$- \int_{Q_s} (z^2[B_u]^T[D_s][B_u]\{u^e\})\mathrm{d}Q - \int_{Q_p} (z^2[B_u]^T[D_p][B_u]\{u^e\})\mathrm{d}Q \qquad (26)$$
$$\left. - \int_{Q_p} \left(zv[B_u]^T[e^T]^T[B_v]\right)\mathrm{d}Q + \int_{s1} ([N]^T\{f_s\})\mathrm{d}s \right] = 0$$

To satisfy the above equation, the terms inside the bracket must be equal to zero as:

$$- \rho_s \int_{Q_s} ([N]^T[N]\{\ddot{u}^e\})\mathrm{d}Q - \rho_p \int_{Q_p} ([N]^T[N]\{\ddot{u}^e\})\mathrm{d}Q$$
$$- \int_{Q_s} (z^2[B_u]^T[D_s][B_u]\{u^e\})\mathrm{d}Q + \int_{Q_p} (z^2[B_u]^T[D_p][B_u]\{u^e\})\mathrm{d}Q \qquad (27)$$
$$- \int_{Q_p} \left(zv[B_u]^T[e^T]^T[B_v]\right)\mathrm{d}Q + \int_{s1} ([N]^T\{f_s\})\mathrm{d}s = 0$$

After rearranging, we get:

$$\left[\rho_s \int_{Q_s} ([N]^T[N])\mathrm{d}Q + \rho_p \int_{Q_p} ([N]^T[N])\mathrm{d}Q \right] \{\ddot{u}^e\}$$
$$+ \left[\int_{Q_s} (z^2[B_u]^T[D_s][B_u])\mathrm{d}Q + \int_{Q_p} (z^2[B_u]^T[D_p][B_u])\mathrm{d}Q \right] \{u^e\} \qquad (28)$$
$$+ \left[\int_{Q_p} \left(z[B_u]^T[e^T]^T[B_v]\right)\mathrm{d}Q \right]$$
$$v = \int_{s1} ([N]^T\{f_s\})\mathrm{d}s$$

Equation (28) can be written as follows and is called as matrix equation of motion of one element:

$$\left(\left[m_p\right] + \left[m_s\right]\right)\{\ddot{u}^e\} + \left(\left[k_p\right] + \left[k_s\right]\right)\{u^e\} + \left[k_{uv}\right]v = \left\{F_s^e\right\} \tag{29}$$

where

$\left[m_p\right] = \rho_p \int\limits_{Q_p} [N]^T[N]dQ$ is elemental mass matrix for piezoelectric

$\left[m_s\right] = \rho_s \int\limits_{Q_s} [N]^T[N]dQ$ is elemental mass matrix for plate structure

$\left[k_s\right] = \int\limits_{Q_s} z^2[B_u]^T[D_s][B_u]dQ$ is elemental stiffness matrix for plate structure

$\left[k_p\right] = \int\limits_{Q_p} z^2[B_u]^T[D_p][B_u]dQ$ is elemental stiffness matrix for piezoelectric

$\left[k_{uv}\right] = \int\limits_{Q_p} z[B_u]^T[e]^T[B_v]dQ$ is elemental electromechanical interaction matrix

$\left\{F_s^e\right\} = \int\limits_{s1} [N]^T\{f_s\}ds$ is elemental external force acting on the structure

Similarly taking variation of (20) with respect to 'v', the equation becomes:

$$\delta \int\limits_{t_0}^{t_1} Ldt = \int\limits_{t_0}^{t_1} \left[-\frac{1}{2}\int\limits_{Q_p} z\,\delta v[B_u]^T\left[e^T\right]^T[B_v]\{u^e\}dQ \right.$$

$$-\frac{1}{2}\int\limits_{Q_p} \delta v[B_v]^T\left[e^T\right]z[B_u]\{u^e\}dQ \tag{30}$$

$$\left. +\int\limits_{Q_p} \delta v[B_v]^T\left[\zeta^T\right][B_v]vdQ - \int\limits_{s2} \delta v\,qds \right]dt = 0$$

Again taking out common term 'δv', the above equation becomes:

$$\delta v\int\limits_{t_0}^{t_1} \left[-\frac{1}{2}\int\limits_{Q_p} z[B_v]^T\left[e^T\right][B_u]\{u^e\}dQ - \frac{1}{2}\int\limits_{Q_p} z[B_v]^T\left[e^T\right][B_u]\{u^e\}dQ \right.$$

$$\left. +\int\limits_{Q_p} v[B_v]^T\left[\zeta^T\right][B_v]dQ - \int\limits_{s2} qds \right]dt = 0 \tag{31}$$

After simplification, we have:

$$\delta v \int_{t_0}^{t_1} \left[-\int_{Q_p} z[B_v]^T [e^T] [B_u] \{u^e\} \, dQ \right.$$

$$\left. + \int_{Q_p} v[B_v]^T [\zeta^T] [B_v] \, dQ - \int_{s2} q \, ds \right] = 0 \tag{32}$$

The above equation will be satisfied if terms inside the bracket are equal to zero, i.e.:

$$-\int_{Q_p} z[B_v]^T [e^T] [B_u] \{u^e\} dQ$$

$$+ \int_{Q_p} v[B_v]^T [\zeta^T] [B_v] dQ \tag{33}$$

$$- \int_{s2} q \, ds = 0$$

On further simplification, we get:

$$-\int_{Q_p} \left(z[B_v]^T [e^T] [B_u] + v[B_v]^T [\zeta^T] [B_v] \right) \{u^e\} dQ = \int_{s2} q \, ds \tag{34}$$

Equation (34) can be expressed in simple form as:

$$-[k_{vu}]\{u^e\} + [k_{vv}]v = \bar{q}$$

The total voltage generated across piezoelectric patch is due to structural vibrations and externally applied charge. The voltage developed across piezoelectric patch can be expressed as:

$$v = \frac{\bar{q} + [k_{uv}]\{u^e\}}{[k_{vv}]} \tag{35}$$

where

$\bar{q} = \int_{s2} q \, ds$ is external charge applied on piezoelectric surface

$[k_{vv}] = \int_{Q_p} [B_v]^T [\zeta] [B_v] v \, dQ$ is the capacitance of piezoelectric sensor

Substituting Eq. (35) in Eq. (29) gives equation of motion as:

$$
\left([m_p] + [m_s]\right)\{\ddot{u}^e\} + \left([k_s] + [k_p]\right)\{u^e\}
+ [k_{uv}][k_{vv}]^{-1}(\bar{q} + [k_{vu}]\{u^e\}) = \{F_s^e\}
\tag{36}
$$

Following assembly procedure, mass matrices, stiffness matrices, and force vectors of individual elements can be assembled to produce global matrix equation of motion of the two-dimensional smart cantilevered plate instrumented with piezoelectric sensor and actuator as:

$$
[M]\{\ddot{u}\} + [K]\{u\} = \{F\}
\tag{37}
$$

where $[M]$ and $[K]$ are global mass and stiffness matrices, respectively, of the system and $\{F\}$ is vector of excitation forces. In Fig. 3, the finite element mesh of smart plate is shown, having 64 elements and 81 nodes. Each node has three degrees of freedom.

Therefore, global matrices of final equation of motion have order of 243×243 as:

$$
[M]_{243 \times 243}\{\ddot{u}\}_{243 \times 1} + [K]_{243 \times 243}\{u\}_{243 \times 1} = \{F\}_{243 \times 1}
\tag{38}
$$

For cantilevered plate, whose one side is fixed in mechanical clamp, first 27 nodes are fixed to boundary. Therefore, the order of equation of motion of cantilevered plate is reduced as:

$$
[M]_{216 \times 216}\{\ddot{u}\}_{216 \times 1} + [K]_{216 \times 216}\{u\}_{216 \times 1} = \{F\}_{216 \times 1}
\tag{39}
$$

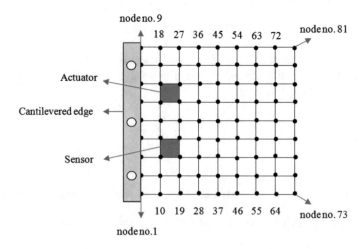

Fig. 3 Nodes of cantilevered plate

$\{F\}$ is vector of excitation forces plus actuation forces applied by actuator to control the mechanical vibrations.

$$\{F\}_{216\times1}= [d]_{1\times216}\{f_a(t)\}_{216\times1} + \{f_d(t)\}_{216\times1} \tag{40}$$

where $[d]$ is matrix of actuator location, $\{f_d(t)\}$ is vector of disturbance force, and $\{f_a(t)\}$ is $m \times 1$ dimensional actuator force vector. The elasticity matrix for intelligent structure (cantilevered plate instrument with piezoelectric patches) is expressed as:

$$[D_s] = \frac{Y}{(1-v)^2} \begin{bmatrix} 1 & \mu & 0 \\ \mu & 1 & 0 \\ 0 & 0 & \frac{1-\mu}{2} \end{bmatrix} \tag{41}$$

where Y and μ are Young's modulus and Poisson's ratio of the material of the base structure.

4 Modal Analysis

To analyze a multidegree of freedom system, it can be converted to a single degree-of-freedom system using modal technique. Orthonormal modal transformation is given by

$$\{d\}_{3\times1}= [U]_{3\times216}\{\eta(t)\}_{216\times3} \tag{42}$$

where $[U]$ and η are orthonormal modal matrix and orthonormal modal transformation variable, respectively. A damping matrix is considered according to the following relation in equation of motion:

$$[C]_{216\times216}= 2 \times \xi \times \omega_n \times [M] \tag{43}$$

where ξ is damping ratio and such that $0.001 < \xi < 0.007$ and $\omega_n = 2\pi$. Equation (39) can be written in modal domain as:

$$[M][U]\{\ddot{\eta}\} + [C][U]\{\dot{\eta}\} + [K][U]\{\eta\} = [D]\{f_a\} + \{f_e\} \tag{44}$$

where[U] is orthonormal modal matrix. Pre-multiplying (3.60) by $[U]^T$ becomes:

$$[U]^T[M][U]\{\ddot{\eta}\} + [U]^T[C][U]\{\dot{\eta}\} + [U]^T[K][U]\{\eta\}$$
$$= [U]^T[D]\{f_a\} + [U]^T\{f_e\} \tag{45}$$

After simplification, we have:

$$[I]\{\ddot{\eta}\} + [c]\{\dot{\eta}\} + [\lambda^2]\{\eta\} = [U]^T[D]\{f_a\} + [U]^T\{f_e\} \tag{46}$$

The equation of motion of rth mode is:

$$\ddot{\eta}_r + c_r\dot{\eta}_r + \lambda_r^2\eta_r = q_r(t) + y_r(t) \quad r = 1, \ldots, n \tag{47}$$

η_r, q_r, y_r represent the modal displacement, modal control force, and modal excitation force, respectively, of rth mode, and λ_r is rth mode natural frequency. For first three modes, we can write:

$$\begin{aligned} \ddot{\eta}_1 + c_1\dot{\eta}_1 + \lambda_1^2\eta_1 &= q_1 \\ \ddot{\eta}_2 + c_2\dot{\eta}_2 + \lambda_2^2\eta_2 &= q_2 \\ \ddot{\eta}_3 + c_3\dot{\eta}_3 + \lambda_3^2\eta_3 &= q_3 \end{aligned} \tag{48}$$

The total modal external force acting on intelligent structure is:

$$\{F\} = \{f_a(t)\} + \{f_e(t)\} \tag{49}$$

where $\{f_e(t)\}$ is a $n \times 1$ dimensional vector of excitation force/disturbance force and $\{f_a(t)\}$ is $m \times 1$ dimensional actuator force vector. Actuator force vector can be expressed as:

$$\{f_a(t)\} = [U]^T[k_{uv}][k_{vv}]^{-1}(\varepsilon_{33} \times a)/h_{\text{piezo}} \tag{50}$$

where 'ε_{33}' is permittivity of piezoelectric patch, 'a' is area of piezoelectric patch, and 'h_{piezo}' is the thickness of piezoelectric patch.

5 State-Space Model

A mathematical model which is required for optimal controller must be in state-space format as:

$$\{\dot{x}\} = [A]\{x\} + [B]\{u\} \tag{51}$$

where $\{x\}$ and $\{u\}$ are system state and control vector, and $[A]$ and $[B]$ are system state matrix and control matrix, respectively. The output relation is given as:

$$\{y\} = [C]\{x\} \tag{52}$$

where[C] is output matrix. In order to convert equation of motion to state-space, let us take:

$$\dot{\eta}_1 = x_1$$
$$\dot{\eta}_2 = x_2$$
$$\dot{\eta}_3 = x_3$$

Modal equations of first three modes can be rewritten as:

$$\dot{x}_1 + c_1 x_1 + \lambda_1^2 \eta_1 = q_1$$
$$\dot{x}_2 + c_2 x_2 + \lambda_2^2 \eta_2 = q_2 \tag{53}$$
$$\dot{x}_3 + c_3 x_3 + \lambda_3^2 \eta_3 = q_3$$

These equations can be expressed in matrix form as:

$$\begin{bmatrix} 0 & 0 & 0 & 1 & 0 & 0 \\ 0 & 0 & 0 & 0 & 1 & 0 \\ 0 & 0 & 0 & 0 & 0 & 1 \\ 1 & 0 & 0 & 0 & 0 & 0 \\ 0 & 1 & 0 & 0 & 0 & 0 \\ 0 & 0 & 1 & 0 & 0 & 0 \end{bmatrix} \begin{Bmatrix} \dot{x}_1 \\ \dot{x}_2 \\ \dot{x}_3 \\ \dot{\eta}_1 \\ \dot{\eta}_2 \\ \dot{\eta}_3 \end{Bmatrix}$$
$$+ \begin{bmatrix} 0 & 0 & 0 & 1 & 0 & 0 \\ 0 & 0 & 0 & 0 & 1 & 0 \\ 0 & 0 & 0 & 0 & 0 & 1 \\ -\lambda_1^2 & 0 & 0 & c_1 & 0 & 0 \\ 0 & -\lambda_2^2 & 0 & 0 & c_2 & 0 \\ 0 & 0 & -\lambda_3^2 & 0 & 0 & c_3 \end{bmatrix} \begin{Bmatrix} x_1 \\ x_2 \\ x_3 \\ \eta_1 \\ \eta_2 \\ \eta_3 \end{Bmatrix} = \begin{Bmatrix} 0 \\ 0 \\ 0 \\ q_1 \\ q_1 \\ q_3 \end{Bmatrix} \tag{54}$$

Let

$$
[m_{ss}] = \begin{bmatrix} 0 & 0 & 0 & 1 & 0 & 0 \\ 0 & 0 & 0 & 0 & 1 & 0 \\ 0 & 0 & 0 & 0 & 0 & 1 \\ 1 & 0 & 0 & 0 & 0 & 0 \\ 0 & 1 & 0 & 0 & 0 & 0 \\ 0 & 0 & 1 & 0 & 0 & 0 \end{bmatrix}, \quad \{x\} = \begin{Bmatrix} x_1 \\ x_2 \\ x_3 \\ \eta_1 \\ \eta_2 \\ \eta_3 \end{Bmatrix}
$$

$$
+ [k_{ss}] = \begin{bmatrix} 0 & 0 & 0 & 1 & 0 & 0 \\ 0 & 0 & 0 & 0 & 1 & 0 \\ 0 & 0 & 0 & 0 & 0 & 1 \\ -\lambda_1^2 & 0 & 0 & c_1 & 0 & 0 \\ 0 & -\lambda_2^2 & 0 & 0 & c_2 & 0 \\ 0 & 0 & -\lambda_3^2 & 0 & 0 & c_3 \end{bmatrix}, \quad f = \begin{Bmatrix} 0 \\ 0 \\ 0 \\ q_1 \\ q_2 \\ q_3 \end{Bmatrix}
$$

(55)

Equation (55) and (56) can be manipulated to write as:

$$
\{\dot{x}\} = -[m_{ss}]^{-1}[k_{ss}]\{x\} + [m_{ss}]^{-1}f \tag{56}
$$

Let $[A] = -[m_{ss}]^{-1}[k_{ss}]$, then we have:

$$
\{\dot{x}\} = [A]\{x\} + [m_{ss}]^{-1}f \tag{57}
$$

Substituting (51), we have:

$$
\{\dot{x}\} = [A]\{x\} + [m_{ss}]^{-1}[k_{uv}][k_{vv}]^{-1}\frac{\varepsilon_{33} \times a}{h_{piezo}} \tag{58}
$$

Let

$$
[B] = [m_{ss}]^{-1}[k_{uv}][k_{vv}]^{-1}\frac{\varepsilon_{33} \times a}{h_{piezo}} \tag{59}
$$

Therefore, state-space model of cantilevered plate becomes:

$$
\begin{aligned} \{\dot{x}\} &= [A]\{x\} + [B]\{u\} \\ \{y\} &= [C]v_{sen} \end{aligned} \tag{60}
$$

Fig. 4 First three natural
frequencies of plate

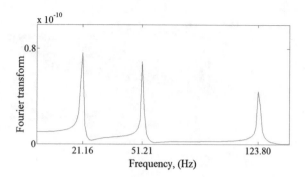

Where [C] is output matrix and v_sen is voltage generated by piezoelectric
sensor as:

$$v_{sen} = [k_{vu}][k_{vv}]^{-1}\{u\} \tag{61}$$

To understand modal analysis, (16) is converted into an eigenvalue problem and
the first three natural frequencies are obtained as 21.16, 51.21, and 123.80 Hz.
Figure 4 gives typical frequency spectrum of time response of the plate in open
loop when edge opposite to the cantilevered edge is displaced by 2 mm. Mode
shapes of first three modes are shown in Fig. 5.

Transient response of plate contains several frequencies. To generate transient
response for individual modes, equation of motion (45) has been transformed into
modal equation of motion by performing orthonormal modal transformation. Upon
simplification, decoupled equations of motion of individual modes are obtained in
(49). Equation (49) is solved in time domain to generate transient response of
individual modes.

Sensor voltage response when plate is disturbed by 2 mm is shown in Fig. 6,
where theoretical sensor voltage and experimental sensor voltage are shown.

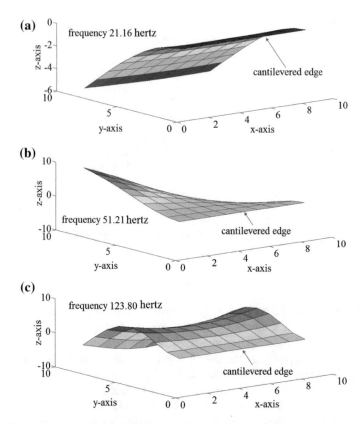

Fig. 5 Mode shapes of **a** first mode, **b** second mode, and **c** third mode

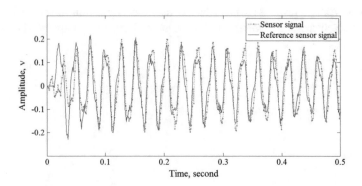

Fig. 6 Experimental time response of sensor voltage

6 Conclusions

In this work, mathematical model of a cantilevered plate instrumented with piezoelectric sensor and actuator patch is derived in detail. The finite element technique based on Hamilton's principle is employed to do so. The cantilevered plate is discretized into 64 elements and 81 nodes. The finite element has four nodes in the corner, and each node has three degrees of freedom. The equation of motion of a cantilevered plate is expressed by using a global mass matrix, stiffness matrix, and external force vector. The equations of motions are truncated to first three modes using modal truncation. Thereafter, the model is converted into a state-space model. This work can help researcher to understand finite element technique, modal truncation, and state-space model of structure effectively.

References

1. Kheiri Sarabi B, Sharma M, Kaur D (2014) Techniques for creating mathematical model of structures for active vibration control. IEEE, Recent Adv Eng Comput Sci (RAECS)
2. Benjeddou A (2000) Advances in piezoelectric finite element modeling of adaptive structural elements: a survey. Comput Struct 76:347–363
3. Allemang RJ (1983) Experimental modal analysis for vibrating structures
4. Moita JMS, Soares CMM, Soares CAM (2002) Geometrically non-linear analysis of composite structures with integrated piezoelectric sensors and actuators. Compos Struct 57:253–261
5. Reddy JN (2007) Theory and analysis of elastic plates and shells, 2nd edn. CRC Press, Boca Raton
6. Park SK, Gao XL (2006) Bernoulli-Euler beam model based on a modified couple stress theory. J Micromech Microeng 16:2355
7. Umesh K, Ganguli R (2009) Shape and vibration control of a smart composite plate with matrix cracks. Smart Mater Struct 18:1–13
8. Heidary F, Eslami MR (2004) Dynamic analysis of distributed piezothermoelastic composite plate using first-order shear deformation theory. J Therm Stresses 27:587–605
9. Peng XQ, Lam KY, Liu GR (1998) Active vibration control of composite beams with piezoelectrics: a finite element model with third order theory. J Sound Vib 209:635–650
10. Kulkarni SA, Bajoria KM (2003) Finite element modeling of smart plates/shells using higher order shear deformation theory. Compos Struct 62:41–50
11. Robbins DH, Reddy JN (1991) Analysis of piezoelecrically actuated beams using a layer-wise displacement theory. Comput Struct 41:265–279
12. Reddy JN (1999) On laminated composite plate with integrated sensors and actuators. Eng Struct 21:568–593
13. Kheiri Sarabi B, Sharma M, Kaur D (2016) Simulation of a new technique for vibration tests, based upon active vibration control. IETE J Res 63:1–9
14. Kheiri Sarabi B, Sharma M, Kaur D, Kumar N (2016) A novel technique for generating desired vibrations in structure. Integr Ferroelectr 176:236–250
15. Kheiri Sarabi B, Sharma M, Kaur D, Kumar N (2017) An optimal control based technique for generating desired vibrations in a structure. Iran J Sci Technol Trans Electr Eng 41:219–228
16. Petyt M (1990) Introduction to finite element vibration analysis, 2nd edn. Cambridge University Press, New York

17. Acharjee S, Zabaras N (2007) A non-intrusive stochastic Galerkin approach for modeling uncertainty propagation in deformation processes. Comput Struct 85:244–254
18. Pradhan KK, Chakraverty S (2013) Free vibration of Euler and Timoshenko functionally graded beams by Rayleigh-Ritz method. Compos B Eng 51:175–184
19. Dovstam K (1995) Augmented Hooke's law in frequency domain. A three dimensional, material damping formulation. Int J Solids Struct 32:2835–2852
20. Tzou HS, Howard RV (1994) A piezothermoelastic thin shell theory applied to active structures. J Vibr Acoust 116:295–302
21. Sharma S, Vig R, Kumar N (2015) Active vibration control: considering effect of electric field on coefficients of PZT patches. Smart Struct Syst 16:1091–1105
22. Smittakorn W, Heyliger PR (2000) A discrete-layer model of laminated hygrothermopiezo-electric plates. Mech Compos Mater Struct 7:79–104
23. Dyke SJ, Spencer BF, Sain MK, Carlson JD (1996) Modeling and control of magnetorhe-ological dampers for seismic response reduction. Smart Mater Struct 5:565

Evaluation of Cepstral Features of Speech for Person Identification System Under Noisy Environment

Puja Ramesh Chaudhari and John Sahaya Rani Alex

Abstract Robust feature extraction techniques play an important role in speaker recognition system. Four speech feature extraction techniques such as Mel-Frequency Cepstral Coefficient (MFCC), Linear Prediction Cepstrum Coefficient (LPCC), Perceptual Linear Predictive (PLP), and Wavelet Cepstral Coefficient (WCC) techniques are analyzed for extracting speaker-specific information. The design of WCC method is done for this work. Hidden Markov Model (HMM) is used to model each speaker from the speaker-specific speech features. The conventional Person Identification System (PIS) is normally employed in an environment where the background noise is unavoidable. To simulate such environment, an additive white Gaussian noise of different SNRs is added with a studio quality speech data. Evaluation of PIS is performed using the Hidden Markov Toolkit (HTK). Multiple experiments are performed. Acoustic modeling of speaker and evaluation is done for clean and noisy environment. The experiment results indicate that 100% accuracy for text-independent PIS in a clean environment. Furthermore, it is observed that MFCC is proven to be better noise robust than PLP and LPC. It is also noted that dynamic features such as delta and acceleration features are combined with static features improve the performance of the PIS in noisy environment.

Keywords Biometric · MFCC · LPCC · PLP · WCC · HMM
Text-independent · Speaker recognition

P. R. Chaudhari · J. S. R. Alex (✉)
School of Electronics Engineering, VIT University, Chennai Campus,
Vandalur-Kelambakkam Road, Chennai, India
e-mail: jsranialex@vit.ac.in

P. R. Chaudhari
e-mail: chaudharipuja7@gmail.com

© Springer Nature Singapore Pte Ltd. 2018
A. K. Nandi et al. (eds.), *Computational Signal Processing and Analysis*, Lecture Notes in Electrical Engineering 490,
https://doi.org/10.1007/978-981-10-8354-9_17

1 Introduction

In today's digital world, each customer indeed needs robust and secure environment for online transactions and communications. Access point to systems whether into a computer, communications applications, building, campus, or country must be protected. Voice is a behavioral phenomenon which depends on natural parameters to identify a person. Biometric technology is ubiquitous in-person identification system. Various types of physiological parameters are used to achieve high accuracies such as voice, fingerprint, face, and IRIS. In this work, voice is used to identify a person in noisy environment. The person's voice plays an important role for security purpose. In this case, the PIS needs robust feature extraction techniques to extract speech features. The system also needs lot of data samples to calculate the accuracy of the system. For the speaker recognition, feature extraction techniques are very significant prerequisite. Speaker recognition system is used in multiple applications such as any authentication system which is related to the various biometric terms and conditions. Discrete Cosine Transform (DCT) is used in feature extraction methods but affected by noise [1]. Nowadays, speaker recognition system is not consistent or not giving efficient accuracy. To improve the speaker recognition system, Mel-Frequency Cepstral Coefficient (MFCC) and Vector Quantization (VQ) methods are used [2–5]. The VQ and Hidden Markov Model (HMM) provide better performance in the speaker recognition system. VQ generates the same number of vector coefficient as an MFCC for improving performance in the noisy environment. Also, the HMM-based speaker recognition (SR) resulted in recognition rate of 100% accuracy. For classification, Hidden Markov Model (HMM) and neural network are employed [6–8]. In the speaker recognition system, widely used feature extraction techniques are MFCC, Perceptual Linear Predictive (PLP), and Linear Prediction Cepstrum Coefficient (LPCC). The voice signal is nonlinear in nature, so for that the MFCC and the PLP methods are reliable for speaker recognition than the LPC because LPC is linear in nature which affects the non-stationary signal parameters. The MFCC vectors were affected by nearby noisy environment. For better accuracy, effective feature extraction parameters required to generate a feasible recognition rate in speaker recognition [9]. MFCC, PLP, and LPC feature extraction methods also used as parameters of the speaker for the emotion recognition from speech [10]. Hidden Markov Toolkit (HTK) plays most important role in the speaker recognition. Also, it is flexible with any embedded board for implementation [11–13]. MFCC, PLP, and LPC are Fourier transform based and give only frequency-related information. Recently, wavelet transform is used for speech enhancement applications [14, 15]. So in this work, wavelet transform-based feature extraction methods are designed and analyzed for Person Identification System (PIS) in noise environment.

The rest of the paper is divided as follows: Sect. 2 gives a description of system design of PIS and feature extraction methods. Section 3 gives detail about dataset collection and practical implementation. Conclusion is given in Sect. 4.

2 System Design

Generalized PIS is shown in Fig. 1. In this system, speaker-specific features are extracted by feature extraction methods from the speech signal of each speaker. Each speaker acoustic model is trained in the training phase using the speaker-specific features. Each speaker HMM represents the particular speaker. The unknown speaker's feature is extracted from the speech signal and compared with the known set of HMM model of speaker using Viterbi decoder in the testing phase.

A. *Feature Extraction*

From the speech signal, speaker-specific parameter is to be extracted. Speech is a non-stationary signal. Short-time spectral analysis used as a basis for feature extraction methods. MFCC, PLP, and LPC employ Fourier transform to do STFT.

(1) *Mel-Frequency Cepstral Coefficient (MFCC)*

MFCC method is based on human auditory perception system. The framing and windowing for speech signal are done with 25 ms. The overlapping period of 10 ms is taken which should give result without missing any signal pattern of voice sample which provides the continuous data frames. Mel scale filter banks of 24 triangular filters used for separating different auditory bands. To imitate the ear's perception, logarithm of power spectrum is done. The Discrete Cosine Transform (DCT) is used to de-correlate the overlapping spectrum. The first thirteen vector output of DCT block is considered as static MFCC. The static MFCC is used to calculate the dynamic and accelerated vectors of MFCC.

(2) *Linear Predictive Coding (LPC)*

The linear predictive coding method is useful for the linear time-varying signal. In this method, LPC vector is calculated based on the current speech sample that can be nearly related to the linear arrangement of past samples. For calculating LPC, vector first speech signal generates the train of impulse and estimating random noise. A white Gaussian noise is used to model framed signal. This framed speech signal passed through speech analysis filter to create frame blocking window and

Fig. 1 Person identification system

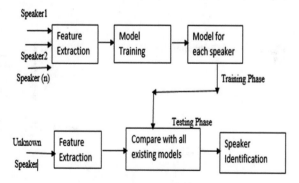

compared with the binary form of speech signal. An autocorrelation method removing error from the signal and calculating the pitch period and time from the autocorrelation sample time. LPC encoded good quality of a speech at a low bit rate and provide extremely efficient speech vectors.

(3) *Perceptual Linear Predictive (PLP)*

PLP is an all-pole model. This method calculates feature extraction of the human auditory system very efficiently. The length of framing window is about 20 ms. The framed window of speech signal applied through the Discrete Fourier Transform (DFT) transform to convert speech segment into the frequency domain. The pre-emphasis signal simulated equal-loudness which reduces largeness of every signal. DCT is used to de-correlate overlapping spectrum.

(4) *Wavelet Cepstral Coefficient (WCC)*

Discrete Wavelet Transform (DWT) offers a flexible time–frequency analysis of non-stationary speech signal. It also localizes time–frequency information. The implementation of DWT is done with successive digital low-pass and high pass-filtering [16–18]. Like the conventional method, in this method also speech signal is framed with 25 ms and windowed with overlapping of 10 ms. Thirteen filter banks are designed with wavelet packet decomposition method.

Each frame passes through the 13 filter banks. This filter bank is compressed by logarithmic to imitate auditory path. DCT is used to convert the frequency domain signal into time domain and calculated the WCC vectors. The block diagram is shown in Fig. 2.

Figure 3 shows the proposed optimal sub-band filter banks, having 8 sub-band of 125 Hz, 4 sub-band of 500 Hz, and 1 sub-band of 1 kHz. Totally, 13 vectors of wavelet cepstral coefficients are calculated in MATLAB. These 13 vectors are used as static wavelet cepstral coefficient (WCC).

B. *Hidden Markov Model (HMM)*

The Hidden Markov Model is used for each speaker to train HMM in the training phase. In this, HMM states are invisible, but the output is visible. HTK toolkit built the HMM for each speaker using 11 parameters of the toolkit. This toolkit is very useful for the acoustic modeling.

Fig. 2 Feature extraction of WCC

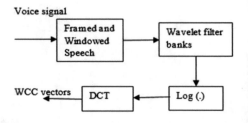

Fig. 3 Tree diagram of WCC

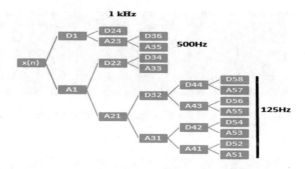

C. *Viterbi Decoding*

It is used to decode feature vector of unknown speech from the most probable path for the HMM created for each speaker output sequence. Using Viterbi decoding, the highest probability path is computed for the unknown speaker among the trained speaker HMMs. The Viterbi algorithm does a maximization step at every time point for the speaker recognition.

3   Implementation and Results

A. *Database of Speakers*

The data have been selected with the 16,000 Hz sampling frequency with specific duration in *.wav* format files. Four different sample narrations are spoken by one person (6 females) in the clear environment. The conventional PIS is normally employed in an environment where the background noise is inevitable. To simulate such environment, the additive white Gaussian noise (AWGN) of different SNRs is added to a studio quality speech data.

B. *Implementation*

In this paper, four different sets of experiment are performed. In this experiment, first set is a clean environment dataset (without noise) for four different extraction methods MFCC, LPC, PLP, and WCC. In this set, model trained with the six persons' voices (each four samples). Second set is with mixed condition in which training is done with clean data and testing is done with noisy data. The third set is a noisy dataset for same feature extraction methods for 39 vectors. Also follows the same procedure for training and testing model. In the fourth experiment, we used wavelet cepstral coefficient dataset with the 13 vectors of each sample in clean environment.

The experiment results are shown in Figs. 4, 5, 6, and 7 which indicate that 100% accuracy for text-independent PIS in a clean environment for MFCC, PLP, LPC, and WCC feature extraction methods with static features. Evaluation using

```
hp-pc@hppc-HP-Pavilion-15-Notebook-PC:~/project/exp1/MFCC1$ ./scripts1/testing.s
cr
./scripts1/testing.scr: line 1: !/bin/tcsh: No such file or directory
Read 6 physical / 6 logical HMMs
Read lattice with 9 nodes / 13 arcs
Created network with 17 nodes / 21 links
File: data/shri4.mfcc
shri   == [4999 frames] -39.3902 [Ac=-196911.6 LM=0.0] (Act=15.0)
```

Fig. 4 Experimental Result of PIS with 13 MFCC features in clean environment

```
hp-pc@hppc-HP-Pavilion-15-Notebook-PC:~/project/exp2-added noise in samples/0db/
MFCC$ ./scripts1/testing.scr
./scripts1/testing.scr: line 1: !/bin/tcsh: No such file or directory
Read 6 physical / 6 logical HMMs
Read lattice with 9 nodes / 13 arcs
Created network with 17 nodes / 21 links
File: data/tanu4_0_snr.mfcc
tanu   == [4999 frames] -32.2380 [Ac=-161157.9 LM=0.0] (Act=15.0)
```

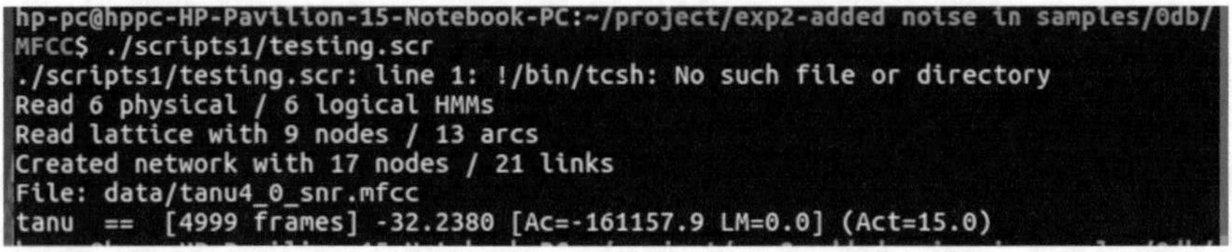

Fig. 5 Experimental Result of PIS with 13 MFCC features in noisy environment

```
hp-pc@hppc-HP-Pavilion-15-Notebook-PC:~/project/exp2-added noise in samples/39ve
ctor/MFCC$ ./scripts1/testing.scr
./scripts1/testing.scr: line 1: !/bin/tcsh: No such file or directory
Read 6 physical / 6 logical HMMs
Read lattice with 9 nodes / 13 arcs
Created network with 17 nodes / 21 links
File: data/neha4_0_snr.mfcc
neha   == [4999 frames] -67.4691 [Ac=-337278.2 LM=0.0] (Act=15.0)
```

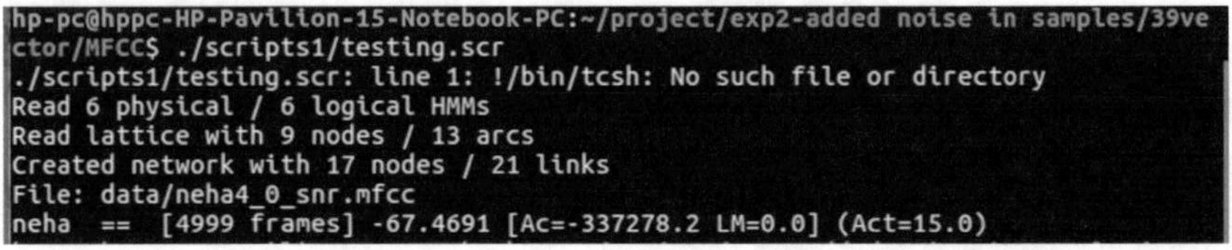

Fig. 6 Experimental Result of PIS with 39 MFCC features in noisy environment

```
hp-pc@hppc-HP-Pavilion-15-Notebook-PC:~/project/speaker_wcc_low$ ./scripts1/te
ing.scr
./scripts1/testing.scr: line 1: !/bin/tcsh: No such file or directory
Read 6 physical / 6 logical HMMs
Read lattice with 9 nodes / 13 arcs
Created network with 17 nodes / 21 links
File: data/shri4_0_snr.mfc
shri   == [9999 frames] -0.4824 [Ac=-4823.7 LM=0.0] (Act=15.0)
```

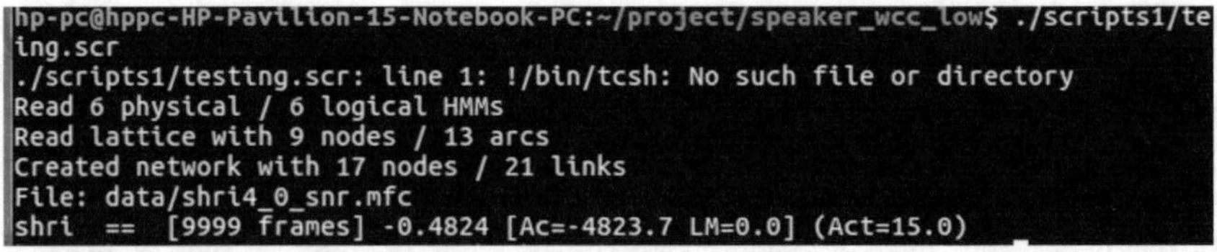

Fig. 7 Experimental Result of PIS with 13 WCC features in clean environment

static features of MFCC and PLP reported 83.33%, static features of WCC reported 66.6%, and static features of LPC reported 50% for 0 dB SNR of AWGN. When dynamic features such as derivatives and acceleration features are combined with the static features resulted in 100% accuracy for MFCC, 83.33% for PLP, and 50% for LPC in a noisy environment for 0 dB SNR of AWGN. Table 1 presents the recognition accuracy of all the methods discussed in this paper in the noisy environment and also in the mixed environment.

Table 1 Recognition Accuracy of Each Method

Exp. No	Dimension of feature vectors	Environment	Methods	Recognition accuracy in %
1	**13**	**Clean**	**MFCC**	**100**
			PLP	**100**
			LPC	**100**
			WCC	**100**
2	13	Mixed environment	MFCC	33.33
			PLP	33.33
			LPC	33.33
3	13	Noise	MFCC	83.33
			PLP	83.33
			WCC	66.6
			LPC	50
4	**39**	**Noise**	**MFCC**	**100**
			PLP	83.33
			LPC	50

4 Conclusion

Person Identification System is much needed in the digital era to access the cyber-physical systems. These systems are employed inevitably in noisy environment. This research work aims at analyzing robust feature extraction methods of speech. Four methods MFCC, PLP, LPC, and WCC are analyzed. The experiment results indicate that 100% accuracy for text-independent PIS in a clean environment for MFCC, PLP, LPC, and WCC feature extraction methods with static vectors. It is also observed from the results that when noisy speech data are used for acoustic modeling of the speaker and for testing, the static MFCC and static PLP features are more robust than static LPC features. Moreover, when dynamic features such as derivatives and acceleration features are combined with the static features in noisy environment, MFCC and PLP features are more robust than LPC. Furthermore, it is observed that MFCC is proven to be better noise robust than PLP and LPC. It is also noted that dynamic features improve the performance of the PIS in noisy environment. The proposed WCC method resulted in less robust than MFCC in noisy environment. This is because of the less filter banks used composed to MFCC method. In future, different wavelet filter banks have to be analyzed.

References

1. Kartik R, Prasanna P, Vara Prasad S (2008) Multimodal biometric person authentication system using speech and signature features. In: TENCON 2008—2008 IEEE region 10 conference 2008, pp 1–6; Maxwell JC (1892) A treatise on electricity and magnetism, 3rd edn. vol 2, Oxford: Clarendon, pp 68–73
2. Martinez J, Perez H, Escamilla E, Suzuki (2012) Speaker recognition using Mel frequency cepstral coefficients (MFCC) and vector quantization (VQ) techniques. In: 22nd international conference on electrical communication and computers (CONIELECOMP), 2012 pp 248–251.K
3. Malode SL, Sahare AA (2012) An improved speaker recognition by using VQ & HMM. In: 3rd international conference on sustain energy international system (SEISCON 2012), IET Chennai pp 1–7; Yorozu Y, Hirano M, Oka K, Tagawa Y (1987) Electron spectroscopy studies on magneto-optical media and plastic substrate interface. IEEE Transl J Magn Japan 2:740–741 (Digests 9th Annu)
4. Tripathi S, Bhatnagar S (2012) Speaker recognition. In: 2012 third international conference on computer and communication technology (ICCCT), pp 283–287
5. Tiwari V (2010) MFCC and its applications in speaker recognition. Int J Emerg Technol 1(1): 19–22
6. Dave N (2013) Feature extraction methods LPC, PLP and MFCC in speech recognition. Int J Adv Res Eng Technol 1(VI)
7. Mukhedhkar AS, Alex JSR (2014) Robust feature extraction methods for speech recognition in noisy environments. In: 2014 first international conference on networks & soft computing 978-1-4799-3486-7/14/$31.00 _c 2014 IEEE
8. Sumithra MG, Thanuskodi K, Archana AHJ (2011) A new speaker recognition system with combined feature extraction techniques, vol 7, issue 4. Department of electronics and communication engineering, pp 459–465
9. Dhonde SB, Jagade SM Feature extraction techniques in speaker recognition: a review. Int J Recent Technol Mech Electr Eng (IJRMEE) 2(5):104–106 ISSN: 2349-7947
10. Chaudhari PR, Alex JSR (2016) Selection of features for emotion recognition from speech. Indian J Sci Technol 9(39). https://doi.org/10.17485/ijst/2016/v9i39/95585
11. Study of speaker recognition systems. [Online]. Available: http://ethesis.nitrkl.ac.in/2450/1/Project_Report.pdf. Accessed 17 Feb 2015
12. www.ee.columbia.edu/ln/LabROSA/doc/HTKBook21/node3.htmlorr, speech.ee.ntu.edu.tw/homework/DSP_HW2-1/htkbook.pdf
13. Rabiner LR (1989) A tutorial on hidden Markov models and selected application in speech recognition. Proc IEEE 77(2)
14. Allabakash S et al (2015) Wavelet transform-based methods for removal of ground clutter and denoising the radar wind profiler data. 9:440–448.
15. Cho J, Park H (2016) Independent vector analysis followed by HMM-based feature enhancement for robust speech recognition. Sig Process 120:200–208. Available at: http://dx.doi.org/10.1016/j.sigpro.2015.09.002
16. Daubechies I (1997) Ten lectures on wavelets. SIAM, Philadelphia, USA
17. Starang G, Nguyen T (1997) Wavelets and filter banks. Wellesley-Cambridge press, Wellesley MA, USA
18. Mallat S (1998) A wavelet tour of signal processing. Academic, New York, 1998

Segmentation of Cochlear Nerve Based on Particle Swarm Optimization Method

S. Jeevakala and A. Brintha Therese

Abstract Sensorineural hearing loss is a hearing impairment happens when there is damage to the inner ear or to the nerve pathways from the internal ear to the brain. Cochlear implants have been developed to help the patients with congenital or acquired hearing loss. The size of the cochlear nerve is a prerequisite for the successful outcome of cochlear implant surgery. Hence, an accurate segmentation of cochlear nerve is a critical assignment in computer-aided diagnosis and surgery planning of cochlear implants. This paper aims at developing a cochlear nerve segmentation approach based on modified particle swarm optimization (PSO). In the proposed approach, a constant adaptive inertia weight based on the kernel density estimation of the image histogram is estimated for fine-tuning the current search space to segment the cochlear nerve. The segmentation results are analyzed both qualitatively and quantitatively based on the performance measures, namely Jaccard index, Dice coefficient, sensitivity, specificity, and accuracy as well. These results indicate that the proposed algorithm performs better compared to standard PSO algorithm in preserving edge details and boundary shape. The proposed method is tested on different slices of eight patients undergone for magnetic resonance imaging in the assessment of giddiness/vertigo or fitness for the cochlear implant. The significance of this work is to segment the cochlear nerve from magnetic resonance (MR) images to assist the radiologists in their diagnosis and for successful cochlear implantation with the scope of developing speech and language, especially in children.

Keywords Cochlear nerve · Cochlear implant · Improved PSO segmentation
Magnetic resonance (MR) image

S. Jeevakala (✉) · A. Brintha Therese
School of Electronics Engineering, VIT University, Chennai Campus, Chennai, India
e-mail: jeevakala.s2013@vit.ac.in

A. Brintha Therese
e-mail: abrinthatherese@vit.ac.in

© Springer Nature Singapore Pte Ltd. 2018
A. K. Nandi et al. (eds.), *Computational Signal Processing
and Analysis*, Lecture Notes in Electrical Engineering 490,
https://doi.org/10.1007/978-981-10-8354-9_18

1 Introduction

Sensorineural hearing loss (SNHL) is the hearing loss causing disability in children or adults [1]. Severe hearing loss is often caused by the inner ear abnormalities such as reduction in the caliber of the cochlear nerve. Cochlear implantation (CI) has turned out to be progressively common in pediatric clinical practice for the treatment of congenital or profound SNHL [2]. Cochlear nerve segmentation plays a vital role in early implantation, especially in children for speech and language acquisition. Since manual segmentation by radiologists is tedious and subjective with high error rate, automatic segmentation has been proposed using modified particle swarm optimization (PSO) technique.

In spite of the fact that there are a few division strategies, for example, [3], region growing [4] and clustering [5], the PSO algorithm is chosen to segment the cochlear nerve. As an effective swarm intelligence algorithm, PSO [6] has attracted much attention from a researcher for solving image segmentation. The PSO algorithm accomplishes the segmentation task because of its efficiency to provide approximation solution to complex optimization at a reasonable computational cost [7]. However, PSO algorithm itself still has some limitations. It is a challenging task to find appropriate constant inertia weight for PSO. PSO's inertia weight shows the inheritance of the current velocity of the particle. It also decides the search capability of the particle. Improper inertia weight may bring out the algorithm to produce oscillation in the region of the optimal solution to be trapped in neighborhood optima. In this proposed algorithm, an appropriate constant inertia weight is chosen through the kernel density estimate using the intensity of pixels from the region of interest (ROI).

Most of the algorithms produce segmentation results with less accuracy which is undesirable for clinical usage. Hence, it is necessary to develop a suitable technique which segments cochlear nerve with good accuracy, reduced computation time, and less segmentation error rate per image

The paper is organized into the following: Preprocessing of MR images and modified PSO segmentation algorithm are briefly given in Sect. 2. Section 3 provides the results and discussions, and Sect. 4 gives the concluding remarks.

2 Methods

The cochlear nerve segmentation of the inner ear MR image is preprocessed by non-local means filter-based algorithm. Then, for simplicity of segmentation, ROI interest is cropped from the MR image. The resolution of the ROI is enhanced using Lanczos interpolation technique. Finally, the interpolated ROI is segmented by modified PSO. The obtained segmented output is evaluated for segmentation accuracy.

2.1 Preprocessing

The preprocessing step aims at reducing the noise and increasing the contrast level between cochlear nerve and other background tissues. The MR images are often corrupted by signal-dependent Rician noise, which deteriorates the image quality and segmentation of subtle tissues/organs. Since Rician noise is signal-dependent, it creates a random fluctuation and brings out signal-dependent bias to the data, which in turn reduces the image contrast [8]. In this present study, a non-local means-based denoising technique [9] is employed to reduce the effects of noise while preserving subtle nerves edges of tissues/organs. Then, the resolution of ROI image is enhanced using Lanczos-2 kernel [10] to smooth the nerve edges adequately.

2.2 Modified Particle Swarm Optimization

Based on the work inspired by Li et al. [11], the proposed method segments the cochlear nerve using constant inertia weight and has the advantage of easy implementation, speed, less segmentation error, and accuracy.

In PSO-based image segmentation, the pixel set relates to search space and pixel intensity correlates to the optimal solutions of global best position. At time t, each particle k moves in a search space with position xtk and velocity vtk which are reliant on nearby best position neighborhood and global best information as in Eq. (1):

$$v_{t+1}^k = w v_t^k + C_1 r_1 (\tilde{g}_t^k - x_t^k) + C_2 r_2 (\tilde{x}_t^k - \tilde{x}_t^k) \tag{1}$$

The position of a particle is calculated as in Eq. (2):

$$x_{t+1}^k = x_t^k + v_{t+1}^k \tag{2}$$

where the parameter w is the inertia weight of PSO, and C_1 and C_2 are constants and their values are between 0 and 2. In general, r_1 and r_2 are both random numbers uniformly distributed between 0 and 1.

Moreover, in this proposed approach the calculation of constant adaptive inertia weight is improved utilizing the kernel density estimation (KDE) of the pixel intensity. To augment the convergence speed and increase optimization quality effectively in modified PSO algorithm, the constant adaptive inertia weight is calculated as in Eq. (3):

$$w = \max(KDE) + 1 \tag{3}$$

where KDE is the kernel density estimate of the ROI.

The inertia weight with large values performs weak exploration, and low values tend to trap in local optima. To overcome this drawback, the maximum pixel intensity of image histogram is chosen as inertia weight. This constant adaptive inertia weight improves the global search capability of the algorithm and thus increases the segmentation efficiency and computational speed.

3 Results and Discussion

A collection of 8 sagittal, T2-weighted, MR inner images of 256 × 256 size are obtained from the Sri Ramachandra Medical Center for validation of the proposed system. The MATLAB© software is used for image processing and segmentation. All PSO-based segmentations are performed with the PSO segmentation toolbox. The process of cochlear nerve segmentation from the inner ear MR image of a patient is shown in Fig. 1. Figure 1a shows the original inner ear MR image, and Fig. 1b shows the Lanczos interpolated ROI of the image.

Figure 2 shows an example of inner ear MRI and its segmentation results by PSO and modified PSO. Evaluation of segmentation performance is verified using Jaccard index, Dice coefficients, sensitivity, specificity, and accuracy. These measures are computed individually for each image using Eqs. (4)–(8).

$$\text{Jaccard index} = \frac{I_{\text{GT}} \cap I_{\text{PSO}}}{I_{\text{GT}} \cup I_{\text{PSO}}} \tag{4}$$

$$\text{Dice Coefficient} = 2 \times \frac{I_{\text{GT}} \cap I_{\text{PSO}}}{I_{\text{GT}} + I_{\text{PSO}}} \tag{5}$$

$$\text{Sensitivity} = \text{TP}/(\text{TP} + \text{FN}) \tag{6}$$

(a)   **(b)**

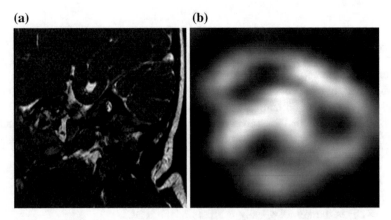

Fig. 1 **a** Original MR image of inner ear. **b** Interpolated image

(a)   **(b)**   **(c)**

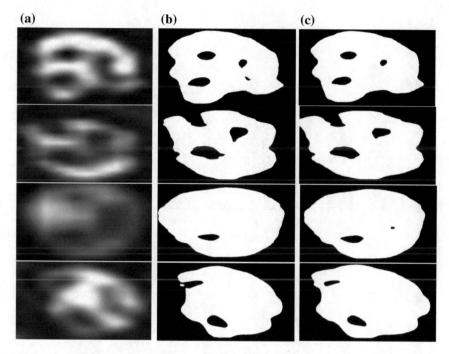

Fig. 2 Segmentation results. **a** ROI. **b** PSO. **c** Modified PSO

$$\text{Specificity} = TN/(TN + FP) \tag{7}$$

$$\text{Accuracy} = (TP + TN)/(TP + FN + TN + FP) \tag{8}$$

where IGT represents manually segmented ground truth cochlear nerve area and IPSO represents cochlear nerve area detected by modified PSO method. TP or true positive means a pixel appears in ground truth cochlear nerve and region detected by PSO method. TN or true negative means a pixel absent in both manually segmented cochlear nerve and cochlear nerve detected by PSO method. FP or false positive means a pixel absent in manually cochlear nerve, but it appears in cochlear nerve detected by PSO method. FN or false negative means a pixel appears in manually segmented cochlear nerve, but it is absent in cochlear nerve detected by PSO method.

Jaccard index shows the similarity between the segmented and ground truth images. Dice coefficients are the measurement of spatial properties of segmented image. Figure 3 shows the comparison of Jaccard index and the Dice coefficients between the PSO and modified PSO method. Sensitivity is the proportion of effectively classified cochlear nerve region, while specificity is the ratio of correctly

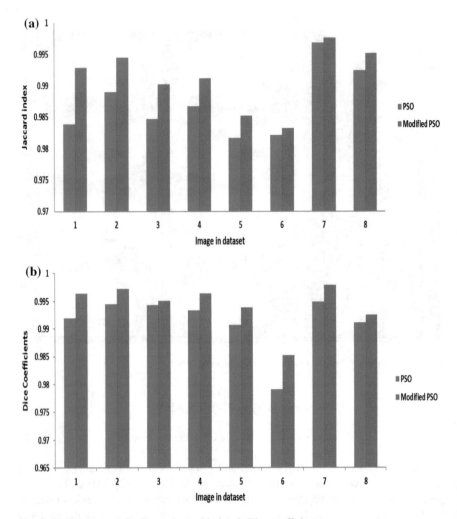

Fig. 3 Performance evaluation. **a** Jaccard index. **b** Dice coefficients

classified background tissues and accuracy is the ratio of correctly characterized both the cochlear nerve and background pixel. In the present study, the observed accuracy results (PSO and modified PSO) are 95.32 and 97.22%, respectively, as shown in Fig. 4. It is found that the modified PSO algorithm shows better performance when compared to standard PSO algorithm. Table 1 shows the computation speed of standard PSO and modified PSO. The adaptive inertia weight increases the global search and computation speed of the algorithm.

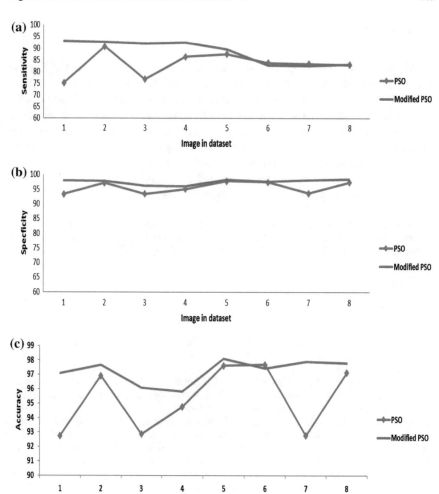

Fig. 4 Comparison analysis between PSO and modified PSO method. **a** Sensitivity. **b** Specificity. **c** Accuracy

Table 1 Computation time in seconds

Image	PSO	Modified PSO
1	4.00	3.01
2	4.54	3.63
3	4.32	4.29
4	5.81	4.62
5	5.24	5.26
6	6.80	5.66
7	6.25	5.77
8	8.10	7.56

4 Conclusion

In this paper, an image segmentation method in the view of modified PSO algorithm is presented. The proposed method uses kernel density estimation of the pixel to overcome the problem of local optima in image segmentation. The cochlear nerve with sub-millimeter size appears distorted and hence cannot be segmented to its correct size. The proposed algorithm successfully segments the cochlear nerve from the inner ear MRI with constant adaptive inertia weight. The experimental results demonstrate that the newly proposed algorithm could segment cochlear nerve exactly from its background tissues and performs better than traditional PSO in accuracy and computation speed of segmentation.

Acknowledgements We express our gratitude to Dr. R. Rajeshwaran for his helpful data on demonstrative detail of internal ear MR images. Likewise, authors might want to thank the Sri Ramachandra Medical Center, Porur, Chennai, India, for giving the MR images.

References

1. Adunka OF, Roush PA, Teagle HF, Brown CJ, Zdanski CJ, Jewells V, Buchman CA (2006) Internal auditory canal morphology in children with cochlear nerve deficiency. Otol Neurotology 27(6):793–801
2. Kim BG, Chung HJ, Park JJ, Park S, Kim SH, Choi JY (2013) Correlation of cochlear nerve size and auditory performance after cochlear implantation in postlingually deaf patients. JAMA Otolaryngol Head Neck Surg 139(6):604–609
3. Lemieux G, Krakow K, Woermann F (1999) Fast, accurate, and reproducible automatic segmentation of the brain in weighted volume MRI data. Magn Reson Med 42:127–135
4. Tang H, Wu E, Ma Q, Gallagher D, Perera G, Zhuang T (2000) MRI brain image segmentation by multi-resolution edge detection and region selection. Comput Med Imag Graph 24:349–357
5. Liew AWC, Yan H (2003) An adaptive spatial fuzzy clustering algorithm for 3-D MR image segmentation. IEEE Trans Med Imag 22:1063–1075
6. Kennedy J, Eberhart R (1995) Particle swarm optimization. In: Proceedings of the IEEE international conference on neural networks, Perth, Australia, vol 4, no 2, pp 1942–1948
7. Lee C-Y, Leou J-J, Hsiao H-H (2012) Saliency-directed color image segmentation using modified particle swarm optimization. Signal Process 92:1–18
8. Wood JC, Johnson KM (1999) Wavelet packet denoising of magnetic resonance images: importance of Rician noise at low SNR. Magn Reson Med 41(3):631–635
9. Jeevakala S, Brintha Therese A (2016) Non local means filter based Rician noise removal of MR images. Int J Pure Appl Math 109(5):133–139
10. Meijering EHW, Niessen WJ, Viergever MA (2001) Quantitative evaluation of convolution-based methods for medical image interpolation. Med Image Anal 5:111–126
11. Li H, He H, Wen Y (2015) Dynamic particle swarm optimization and K-means clustering algorithm for image segmentation. Opt-Int J Light Electron Opt 126(24):4817–4848

Evaluating the Induced Emotions on Physiological Response

Shraddha Menon, B. Geethanjali, N. P. Guhan Seshadri, S. Muthumeenakshi and Sneha Nair

Abstract Emotions are complex occurrence that is felt at each instant and are entrenched in the underlying state of physiological processes. The cardiac activity reproduces direct quantitative analysis of the induced emotion. The objective of this study is to determine the effect of cardiac response in order to assess the intensity of induced emotion. Ten adult participants are participated in this study. The stimulus to provoke emotion was selected from the International Affective Picture System (IAPS). The positive valence low/high arousal images were selected. The perceived emotion was measured using Self-Assessment Manikin Scale (SAM scale) and physiological response to the stimulus was evaluated using heart rate. The heart rate while viewing positive valence and low arousal pictures was significantly low ($p = 0.05$) for when compared with non-image appearance period, whereas it was significantly high ($p = 0.05$) for positive valence and high arousal images. The two-dimensional model of emotion study concludes that the subjective perceived emotion correlates with the induced emotion for the selected categories with arousal playing the most significant role in inducing the changes in cardiac reactions.

Keywords Emotions · IAPS · Heart rate · Arousal · ECG

1 Introduction

Emotions significantly contribute in processes such as cognition, learning and decision making, motivation, and creativity [1]. The emotion induced due to external stimulus or merely by means of thought processes can trigger the brain activation. This brain activation will be manifested in the autonomous nervous system which is responsible for sympathetic and parasympathetic activity.

S. Menon · B. Geethanjali · N. P. Guhan Seshadri (✉) · S. Muthumeenakshi · S. Nair
Department of Biomedical Engineering, SSN College of Engineering, Chennai, India
e-mail: guhan131192@gmail.com

B. Geethanjali
e-mail: geethanjalib@ssn.edu.in

© Springer Nature Singapore Pte Ltd. 2018
A. K. Nandi et al. (eds.), *Computational Signal Processing and Analysis*, Lecture Notes in Electrical Engineering 490,
https://doi.org/10.1007/978-981-10-8354-9_19

Eventually, the changes are reflected as increase or decrease in periphery signals like heart rate and respiration rate. The life's predictable stressor can be dealt by changing the routine and indulge in positive valence stimulus. Emotion is subjective and can be either induced or felt through personal experience in response to stimuli.

The valence-arousal model, a two-dimensional emotional space was presented by [2] having valence ratings on the horizontal axis and the arousal ratings on the vertical axis. The most commonly studied models have been in 2D affective space, assessing valence and arousal [3]. SAM is a nine-pointer Likert scale, a non-verbal pictorial assessment technique that directly measures the pleasure, arousal and dominance associated with a person's affective reaction to wide variety of stimuli [4].

The induced emotions cause an alteration in heart rate and pulse rate, respiration rate. The cardiac activity reproduces direct quantitative analysis of the induced emotion. The current study focuses on cardiac response (ECG) to assess the intensity of emotion and establish a correlation measure between perceived and induced emotions. Hockenbury and Hockenbury [5] stated emotion as a composite psychological state that varies across the globe. The current work focuses on evaluating the emotion provoked on south Indian participants' while viewing various categories of images from IAPS database. The perceived and induced emotions were analyzed using SAM scale (for perceived) and heat rate (for induced) respectively.

2   Methods and Materials

The overall methodology in Fig. 1 has been explained below.

2.1   Selection of Subjects and Stimuli

Ten healthy subjects in the 18–25 age groups without any visual impairment were selected based on their own interest for the study. No training or special dietary instruction was given before the experiments. The participants for this study was selected based their interest and supported by the guidelines formulated Institutional Ethics Committee of SSN College guidelines for human volunteer research. The experiment protocols were explained to them and have signed the informed consent. In the current work, two categories were selected and in each category ten pictures that were selected from the International Affective Picture System [4] (emotionally positive and high arousal pictures and emotionally negative and high arousal pictures).

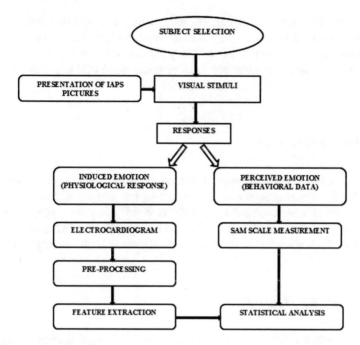

Fig. 1 Methodology flow chart

2.2 Subject Ratings

The rating is done according to the subjective feelings. The legendary researchers like Bradley and Lang and Juslin [4, 6] used Self-Assessment Manikin (SAM) scale to evaluate the pleasure, arousal and dominance in one scale. The perceived emotion was assessed using SAM scale to measure the valence; arousal and the dominance of the selected pictures [4]. The subjects were asked to rate the pictures using SAM scale, on a nine-point scale to assess for the valence, arousal and dominance [4]. Experimental Protocol is given in Table 1.

2.3 Experimental Setup

An array of 20 pictures has been allocated for each of the participants in a shuffling manner. Each picture was presented for 20 s on a 15.6-in. (39.6 cm) monitor. The

Table 1 Experimental protocol

Baseline	Warning	Viewing category A	Resting	Viewing category D	Resting
120 s	5 s	100 s	120 s	100 s	120 s

single lead ECG was recorded for all the participants by using Recorders and Medicare systems which has a port of single lead ECG, India. The recorded signal was filtered to the cutoff of 70 Hz by using Butterworth low pass filter of order 5, and power-line interference was removed by using Notch filer where the cutoff was set at 50 Hz.

2.4 Data Processing

The ECG signals were recorded for each participant during the experiment. The ECG signal was acquired and the sampling rate was fixed at of 256 Hz. The acquired signals were analyzed using LabVIEW 14, using the digital filter toolkit and was applied for each data point for processing using the Pan-Tompkins algorithm. The block diagram for Pan-Tompkins algorithm is shown in Fig. 2.

2.5 Detecting Heart Rate

A series of filters and methods perform low pass, high pass, derivative, squaring, integration, and extracting the RR interval. The bandpass filter reduces the influence of muscle noise, 50 Hz interference, baseline wander, and T-wave interference. 5–15 Hz passband was used to extract the QRS complex. The squaring operation makes the result positive and emphasizes large differences resulting from QRS complex. The high-frequency components are further enhanced. The output of derivative operation will exhibit multiple peaks within the duration of a single QRS complex followed by smoothing of the output through a moving-window integration filter. The thresholding method adapts to alter in the ECG signal by computing running estimates of the signal and noise peaks. Feature extraction using is shown in Fig. 3.

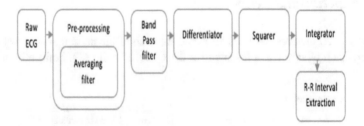

Fig. 2 Block diagram of heart rate detection

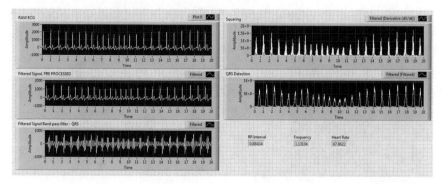

Fig. 3 Features extraction—raw ECG, filtered signal—pre-processed, band-pass filtered, squared signal, QRS detection

2.6 Statistical Analysis

The ratings for the pictures were not normally distributed, so a nonparametric analysis was performed. The Friedman test is the nonparametric equivalent to one-way ANOVA with repeated measures. The independent variables were positive valence high arousal/negative valence high arousal picture stimuli. The measured heart rate, valance, arousal, and dominance were dependent variables. The overall significance was evaluated using Friedman test for both categories, $\gamma^2(31) = 1507.550; p = 0.05$. There existed a significant difference; since the same subjects viewed both the categories, Wilcoxon signed-rank test is carried out for the special grouping of related groups. The significant level was set at $p = 0.05$, and analysis was performed using IBM SPSS Statistics for Windows, Version 20.0.

3 Results

3.1 Behavioral Data Analysis

The SAM scale ratings reviled the emotionally positive valence was perceived as positive and negative category of viewed picture as negative valence. On the other hand, the arousal rate is evidently seen to be higher in the high arousal category than low arousal category. Hence, arousal plays the foremost role while rating the category as high arousal and low arousal.

The valence, arousal and dominance were plotted using SAM scale. And for positive valence high arousal (Fig. 4), the valence ratings are: mean = 6.98 ± 1.122; the arousal ratings are: mean = 5.32 ± 1.402; the dominance ratings are: mean = 6.5 ± 1.59. The negative valence high arousal (Fig. 4), the valence ratings are: mean = 2.22 ± 1.2; the arousal ratings are: mean = 4.94 ± 2.63; the dominance

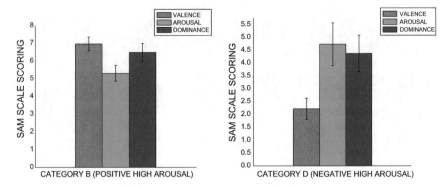

Fig. 4 SAM scale for positive valence low arousal (left) SAM scale for positive valence high arousal (right)

ratings are: mean = 4.38 ± 2.2. The participants' ratings show that the arousal and dominance do not vary greatly when you compare the two categories. But the valence is perceived as positive and is evidently seen to be higher for positive valence category in when compared to negative valence category. Hence, valence/arousal plays the foremost role while rating the category as positive and negative valence.

3.2 Heart Rate

In each category (positive valence high arousal and negative valence high arousal), 10 pictures were used. And total epoch of 100 s were considered and in that the RR interval, and heart rate for each five seconds epoch was estimated shown in Fig. 5. For negative valence high arousal, the variation in RR interval, as seen in Fig. 7a, is less and this is reflected in the heart rate (Fig. 6). The selected IAPS pictures are able to maintain the stable arousal levels throughout the experiment (during viewing).

3.3 Visualizing the Heart Rate Variation

The variation in heart rate for the total epoch of 100 s was divided into 20 epochs and each epoch holds five seconds information about the changes in heart rate. As the data was not normally distributed spearman correlation was applied to locate the correlation between each epoch. In Fig. 7, the heart rate at 5th, 10th, and 15th seconds was highly correlated with 40th, 45th, 50th, 85th to 95th seconds indicates

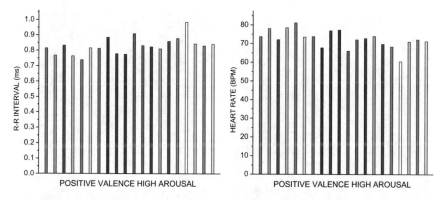

Fig. 5 RR interval while viewing the positive valence high arousal pictures (left) heart rate variation (right)

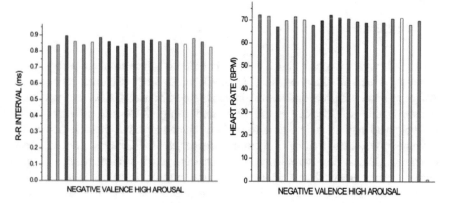

Fig. 6 RR interval while viewing the negative valence high arousal pictures (left) heart rate variation (right)

no variation in heart rate, whereas the rest of the epoch were not correlated with any other epoch indicates high variation in heart rate.

The beat-to-beat variation was high which indicates the participants' experienced low levels of stress and greater resiliency while viewing positive valence high arousal pictures. Whereas in Fig. 7, the heart rate at 40th to 100th second was highly correlated which means there is no variation in heart rate thereby low variation in heart rate was noted while viewing negative valence high arousal pictures. The beat-to-beat variation was low which indicates the participants' experienced high levels of stress and lesser resiliency while viewing negative valence high arousal pictures.

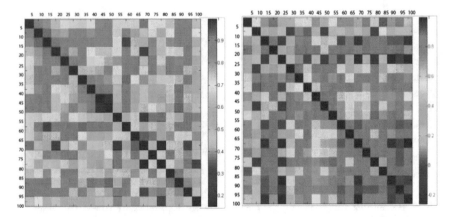

Fig. 7 Variation in heart rate (correlogram) while viewing positive valence high arousal pictures (left) and the variation while viewing negative valence high arousal pictures (right)

4 Discussion

Emotion is varied across different culture. In order to induce emotion, a very strong stimulus is needed. Lang et al. [7], Bradley and Lang [8] used International Affective Picture System and International Digitized Sounds (IADS) as stimulus to induce emotions. Courtney et al. [9] stated using IAPS images were highly authenticated and are widely-used to study of human affective effect. Based on the previous researchers' findings the current work used IAPS images grouped in two categories, namely positive valence low arousal and positive valence high arousal to evaluate the perceived and induced emotion on physiological response. The changes that arise in the brain while viewing negative valence high arousal pictures will be reflected in the periphery signals. The high beta component energy at Brodmann area 47 [10] leads to arousal and that is reflected in the heart rate. The beat-to-beat variation in the heart rate is used to assess the heart rate variation. This variation will vary extensively for individuals when they encounter low level of stress and ability to recover quickly from the exposed stress stimulus. Whereas the heart rate variation is briefly for the individuals when they cannot recover rapidly. Bradley and Lang [4], Justin and Laukka [6], and Guhan Seshadri [11] used the Self-Assessment Manikin (SAM) scale to assess the affective dimensions of valence, arousal and dominance [4]. Based on their conclusion the nine-pointer SAM scale was used for evaluating the perceived emotion of the subjects. The perceived emotion results suggest that for positive valence low arousal category the measured mean valence 7.24, mean arousal 3.6, and mean dominance 6.58. And for positive valence and high arousal category, the reported mean valence 6.98, mean arousal 5.32, and mean dominance 6.5. Hence, arousal plays a significant role in inducing emotion.

Each individual reacts differently to diverse types of emotion stimuli, and this stimulus induces changes in the autonomic nerve systems which are governed by sympathetic and parasympathetic activation. The sympathetic activation enhances the arterial vasoconstriction thereby the total peripheral resistance and cardiac output increases which in turn increase the heart rate and blood pressure [12]. But parasympathetic activation leads the reverse effects. Courtney [9] in their work used skin conductance responses as an index of arousal, the reduction in skin conductance measures the sympathetic activity whereas startle eye blink was used as an index of valence. The various psychological states of individuals have been linked to the respiration the fast rate was noted through excitement and irregular during emotional distress [13]. In this work, the variation in RR interval and heart rate was considered as index of arousal which measures the sympathetic and parasympathetic activation.

The activation and suppression of sympathetic and parasympathetic nervous system were reflected in heart rate and heart rate variability [14]. The researchers [15] monitored the ECG signals of adults during a session of high arousal and a session of low arousal by using combinations of music and images. They found that heartbeat evoked potential (HEP) changes based on the induced arousal. Our result supports the findings of [14, 15] while viewing the positive valence and high arousal pictures the RR interval did not change to a great extent when compared to rest (nonviewing period), and these changes were reflected in the heart rate. The parasympathetic activation was more evident when viewing positive valence high arousal pictures. This indicates the participants' experienced low levels of stress and greater resiliency while viewing positive valence high arousal pictures. Whereas while viewing the negative valence and high arousal pictures, the RR interval changes were more consistent when compared to rest (Nonviewing period) and these changes were reflected in the heart rate. The results indicate the sympathetic activation during negative valence and high arousal viewing, and this indicates the participants' experienced high levels of stress and lesser resiliency.

5 Conclusion

The present study considers only two categories of images, positive valence high arousal, and negative valence high arousal. Ratings showed that the arousal and dominance ratings do not vary greatly for positive and negative arousal. But the rating for valence varies significantly among the two categories which are significantly reflected in RR intervals and heart rate. Thus, the study concludes that the subjective perceived emotion correlates with the induced emotion for the selected categories with arousal playing the most significant role in inducing the changes in cardiac responses. The constant exposure to negative valence high arousal pictures will affect the heart rate and induces stress. In order to support this statement, sample size should be increased and other categories pictures that are negative valence low arousal, positive valence low arousal and neutral, and the correlation between them could be considered.

References

1. Hosseini SA, Naghibi-Sistani MB (2011) Classification of emotional stress using brain activity. INTECH Open Access Publisher
2. Russell JA (1980) A circumplex model of affect. J Pers Soc Psychol 39(6):1161
3. Chanel G, Kronegg J, Grandjean D, Pun T (2006) Emotion assessment: arousal evaluation using EEG's and peripheral physiological signals. In: Multimedia content representation, classification and security, pp 530–537
4. Bradley MM, Lang PJ (1994) Measuring emotion: the self-assessment manikin and the semantic differential. J Behav Ther Exp Psychiatry 25(1):49–59
5. Hockenbury DH, Hockenbury SE (2010) Discovering psychology. Macmillan
6. Juslin PN, Laukka P (2004) Expression, perception, and induction of musical emotions: a review and a questionnaire study of everyday listening. J New Music Res 33(3):217–238
7. Lang PJ, Bradley MM, Cuthbert BN (1999) International affective picture system (IAPS): instruction manual and affective ratings. The center for research in psychophysiology, University of Florida
8. Bradley MM, Lang PJ (2000) Emotion and motivation. In: Handbook of psychophysiology, vol 2, pp 602–642
9. Courtney CG, Dawson ME, Schell AM, Parsons TD (2009) Affective computer-generated stimulus exposure: psychophysiological support for increased elicitation of negative emotions in high and low fear subjects. In: Foundations of augmented cognition. neuroergonomics and operational neuroscience. Springer, Berlin, Heidelberg, pp 459–468
10. Guhan Seshadri NP, Geethanjali B and Muthumeenakshi S (2016). Visualizing the brain connectivity during negative emotion processing—an EEG study. Front Hum Neurosci (Conference abstract: SAN2016 meeting) https://doi.org/10.3389/conf.fnhum.2016.220.00005
11. Guhan Seshadri NP, Geethanjali B, Pravinkumar S, Adalarasu K (2015) Wavelet based EEG analysis of induced emotion on South Indians. Aust J Basic Appl Sci 9(33):156–161
12. Vorobiof G, Blaxall BC, Bisognano JD (2006) The future of endothelin-receptor antagonism as treatment for systemic hypertension. Curr Hypertens Rep 8(1):35–44
13. Boiten F, Frijda N, Wientjes C (1994) Emotions and respiratory patterns: review and critical analysis. Int J Psychophysiol 17:103–128
14. O'Kelly J, James L, Palaniappan R, Taborin J, Fachner J, Magee WL (2013) Neurophysiological and behavioral responses to music therapy in vegetative and minimally conscious states. Front Hum Neurosci 7(884)
15. Luft CDB, Bhattacharya J (2015) Aroused with heart: modulation of heartbeat evoked potential by arousal induction and its oscillatory correlates. Sci Rep 5

Accuracy Enhancement of Action Recognition Using Parallel Processing

C. M. Vidhyapathi, B. V. Vishak and Alex Noel Joseph Raj

Abstract Implementation of action recognition for embedded applications is one of the prime research areas in the fields of both computer vision and embedded systems. In this paper, we propose a novel algorithm to improve the accuracy of human action recognition by implementing parallel processing and incorporating multiple neural networks working in coherence for action classification and recognition. A feature set known as Eigen joints is used to model the actions in the database. The algorithm proposes an efficient method to reduce the feature set required to recognize an action accurately based on the concept of accumulated motion energy. The paper talks about the use of Robot Operating System and its advantages for implementing parallel processing. The paper also presents a comparative study in the accuracies of action recognition between support vector machine (SVM) and Gaussian Naïve Bayes (GNB) classifiers for recognizing the actions for which the networks are trained. In this paper, we also talk about how multiple supervised learning neural networks working in coherence can detect an action whose model is not present in the database.

Keywords Accumulated motion energy · Action recognition · Supervised learning · Support vector machine · Gaussian Naïve Bayes · Eigen joints

C. M. Vidhyapathi (✉) · B. V. Vishak · A. N. Joseph Raj
School of Electronics Engineering, VIT University, Vellore, India
e-mail: vidhyapathi.cm@vit.ac.in

B. V. Vishak
e-mail: vishakhbv@gmail.com

A. N. Joseph Raj
e-mail: alexnoel@vit.ac.in

© Springer Nature Singapore Pte Ltd. 2018
A. K. Nandi et al. (eds.), *Computational Signal Processing and Analysis*, Lecture Notes in Electrical Engineering 490,
https://doi.org/10.1007/978-981-10-8354-9_20

1 Introduction

Action recognition has a wide range of real-time applications like human–computer interaction (HCI), video surveillance, home automation, health care, mobile applications, sign language recognition, etc. [1, 2]. The main problem in implementing action recognition in most of the real-time applications is the lack of accuracy achieved by the system. Embedded systems need to implement action recognition with high accuracy for the product to be a success in the market.

Conventionally for action recognition, the skeletal joints of a human body are tracked with the help of necessary libraries and Kinect depth sensor. Kinect sensor demands a lot of computational power for transferring the data to a processor. Once the data is transferred by the Kinect sensor, the processor is kept idle and is not used to its maximum potential. The algorithm proposed in this paper makes use of the processor's computational power to its full potential to enhance the accuracy of the system. There are 4 key factors for the development of an embedded system, i.e., memory, speed, computational power, and efficiency. These key factors are interdependent and manipulating one of the key factors affects the other. Since we are not worried of the computational power, we increase the throughput of the system by parallel processing of the data available, improving the accuracy and also decreasing the memory requirement.

Initial research in the field of action recognition tracked 20 moving skeletal joints of the body, and features were obtained for all of these 20 joints. Later, it was found that some of the joints add to the redundancy in the features and affect the overall efficiency of the system. In order to reduce the redundancy in data, few of the joints were eliminated from the final features obtained to model the action. It was found that around 13 joints of the body were required for accurate action recognition [3, 4]. Although the redundancy was reduced, the feature set was huge and required the use of compression algorithms such as PCA, which again reduced the accuracy.

Our research indicated that only a few joints out of those 13 joints were required for modeling a particular action. Hence, all the other joints apart from the joints performing the action added to redundancy, increasing the memory required and also the computational power. The algorithm proposed tracks separate parts of the body in separate nodes running in parallel and acquires features only if the movement crosses a particular threshold. Each node has its own network and is responsible for analyzing the features tracked by that node. This decreases the features tracked by a neural network, eliminating the need for compression algorithms and also increases the accuracy of the system.

Fig. 1 ROS block diagram

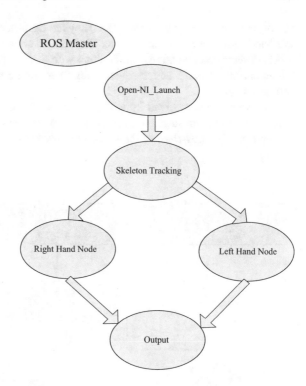

2   ROS Block Diagram

Robot Operating System generally known as ROS is a set of software libraries and tools which can be used to build embedded applications. It is a master–slave communication, and each process running under a ROS master is known as a ROS node. Inter-task communication is performed by ROS topics. A topic can either be published to a ROS master or can be subscribed from a ROS master. In the proposed algorithm, each task or node subscribes to a particular set of joints from the skeleton tracking libraries. As a prototype to show the accuracy of the system, only 2 nodes have been created which track 3 joints in the left and the right hands, respectively. The block diagram is given in Fig. 1.

The process in both the nodes is similar to each other, only difference being the joints tracked by the 2 nodes.

3   Process Flowchart

Each of the two nodes subscribes to 3 joints of the hand they are tracking, i.e., shoulder, elbow, and wrist. They receive the (x, y, z) coordinates from the ROS master and process the data to model the action being performed. For the purpose of prototyping, 4 actions are considered. They are right-hand wave, right-hand rotate, left-hand wave, and left-hand rotate. Since these actions are independent of each other, different combinations can be done with these 4 actions and the system as a whole can recognize 8 actions even though the feature set is not present for those actions as such.

The coordinates received by each node undergo the same process to create the feature set. The main stages of the process are preprocessing, feature extraction, and recognition. The flowchart of the process followed in both the nodes is represented by Fig. 2.

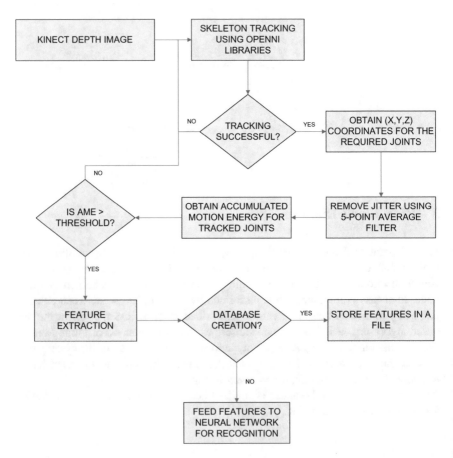

Fig. 2 Process flowchart

4 Preprocessing

Raw coordinates obtained from the Kinect sensor cannot be used in its original form to obtain the features. It has to be refined before the features are extracted for better accuracy. There are 2 important steps in preprocessing. They are jitter removal using 5-point average filter and calculation of accumulated motion energy (AME).

4.1 Jitter Removal Using 5-Point Average Filter

The raw coordinates obtained from the Kinect sensor have a lot of small variations commonly known as jitter or noise. These variations are erratic in nature and affect the motion patterns, making it difficult for the neural network to learn the pattern. Eliminating jitter is very essential for obtaining good results. To eliminate jitter, 5-point average filter can be used. The filter takes the coordinates over 5 frames, finds the average and fits the value to the middle frame. The mathematical equation for the filter is given by Eq. (1).

$$x_i^c = \frac{x_i^{c-2} + x_i^{c-1} + x_i^c + x_i^{c+1} + x_i^{c+2}}{5} \mid x_i^c \in X^c \tag{1}$$

where x_i^c is the (x, y, z) coordinates corresponding to a joint-i in the frame-c represented by the symbol X^c. For each joint in the frame-c, the values from frames $(c - 2)$, $(c - 1)$, $(c + 1)$, and $(c + 2)$ are taken along with the value from frame-c, summed up and averaged over the 5 frames to obtain the new value for that particular joint-i in the frame-c.

4.2 Accumulated Motion Energy (AME)

The main focus of the algorithm is to reduce the number of features being tracked so that the redundant data is not present while extracting features. One way to remove the redundant data is to neglect all unnecessary movements. This is done with the help of a concept known as accumulated motion energy which gives the total movement of all the joints in a particular frame from the initial frame [5]. The mathematical formula to calculate the accumulated motion energy or AME is given by Eq. (2).

$$\text{AME}(i) = \sum_{v=1}^{3} \sum_{j=1}^{i} (|f_v^j - f_v^{j-1}|) > \varepsilon \tag{2}$$

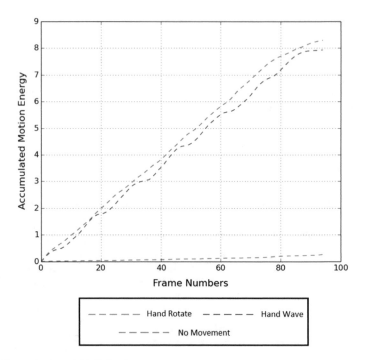

Fig. 3 Example plot of AME for 3 cases

For a frame-i, its 3D depth map is projected onto the three orthogonal planes which generate 3 projected frames $f_v \in \{1, 2, 3\}$. AME(i) is then calculated as the summation of motion energy maps. The motion energy maps for each frame are obtained by thresholding and accumulating between two consecutive projected frames. Once the AME is obtained for all frames that are being tracked, a threshold value is assigned such that if the motion energy value exceeds the set threshold value, the features are obtained for the action performed. If the AME value is less than the threshold set, then the values are neglected completely. This process helps us out in determining and pinpointing the part of the body which is involved in completing the action to be recognized. The features are obtained either from node tracking left hand or the node tracking right hand or sometimes both depending on the AME measured in both nodes. The threshold set for feature acquisition is based on trial and error. An example plot of AME is shown in Fig. 3.

5 Feature Extraction

Any signal such as image, speech, or text can be represented in terms of numbers, which is used by the neural network to analyze the pattern. Any set of numbers that can distinguish between 2 objects clearly such that there is no problem for the neural network to learn the pattern is known as the feature set for that object. In case of skeleton tracking and gesture recognition, the main features that researchers have worked with include position, velocity, displacement, angle, and orientation of a joint. In this project, position and displacement are considered to form the feature set. The static posture, the motion property, and the overall dynamics are considered to make up the feature set which is known as Eigen joints. The paper uses the difference in skeleton joints to characterize action information, which includes static posture feature f_{cc}, consecutive motion feature f_{cp}, and overall dynamics feature f_{ci} in each frame-c. The obtained features are then concatenated to form the feature vector $f_c = [f_{cc}, f_{cp}, f_{ci}]$ [5].

Generally for better results, normalization and PCA are applied once the feature set is obtained. This is done because of the huge feature set obtained due to the tracking of redundant joints. Since the algorithm used talks about tracking only relevant joints, the feature set is drastically reduced and there is no need for using compression methods such as PCA. The method to compute the features is represented in Fig. 4.

Assume N joints are tracked by each node. Hence, there are N joints in a frame X represented by $X = \{x_1, x_2, x_3, \ldots x_N\}$. To characterize the static posture feature of the current frame-c, pair-wise joint differences are computed within the frame-c. The mathematical formula for static posture feature is given by Eq. (3).

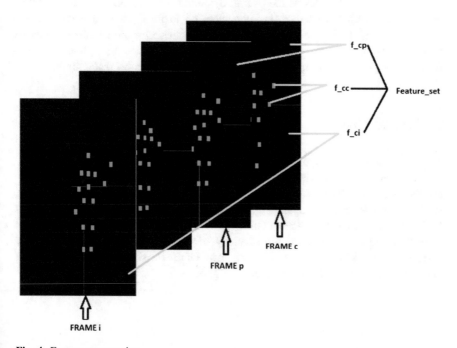

Fig. 4 Feature computation

$$f_{cc} = \left\{ x_i - x_j \big| i,j = 1,2,\ldots N; i \neq j \right\} \tag{3}$$

The consecutive motion feature is obtained by calculating the difference of joints in frame-c with frame-p. Each joint in frame-c and frame-p is considered pair-wise, and the difference is calculated. The mathematical formula for consecutive motion feature is given by Eq. (4).

$$f_{cp} = \left\{ x_i^c - x_j^p \big| x_i^c \in X_C; x_j^P \in X_P \right\} \tag{4}$$

where X_c is the current frame-c and X_p is the previous frame-p, each containing N joints. The next step is to calculate the overall dynamics feature. The overall dynamics feature basically represents the displacement of joint in a frame-c from the initial frame-i with respect to each other. The mathematical formula to calculate the overall dynamics feature is given by Eq. (5).

$$f_{ci} = \left\{ x_i^c - x_j^i \big| x_i^c \in X_C; x_j^i \in X_i \right\} \tag{5}$$

where X_c is the current frame-c and X_i is the initial frame-i. A joint-i is considered in frame-c, and a joint j is considered in frame-i. The difference between joints i and j is calculated which gives us the overall dynamics feature. Once the three feature vectors are calculated, all three are concatenated to form the final feature vector $f_c = [f_{cc}, f_{cp}, f_{ci}]$. If the user wishes to create a database for the actions in order to train the neural network, then the features are stored in a file. Otherwise, the features are given as an input to the neural network for action recognition.

6 Neural Network

The neural network is the most important part for recognizing the action performed. The concept of the classifier has to be used for our advantage to obtain better results, and this paper explains how it is done. To show how efficient the algorithm is, the paper presents a comparative study between two neural networks, SVM and GNB. In most research papers, GNB gives a better accuracy than SVM. The algorithm helps in increasing the accuracy of SVM classifier, and the accuracy obtained using SVM classifier is more than GNB classifier. The general block diagram of a neural network is given in Fig. 5.

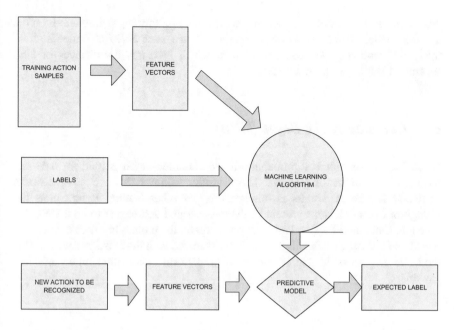

Fig. 5 Supervised learning model

6.1 Support Vector Machine (SVM)

Support vector machine, commonly known as SVM, is a discriminative classifier which separates two classes or set of data by introducing a hyperplane between them. The algorithm to introduce the hyperplane is such that the closest samples in two classes are separated by the maximum distance [6]. All the samples in a data set are plotted on a plane, and each class is differentiated by a hyperplane. Suppose a new sample has to be identified, it is plotted on the same plane and depending on which side of the hyperplane the sample lies, the class is decided. More the distance between two classes is, less is the probability of error. Obviously, as the number of classes increase, the distance of separation between these classes decrease and the chance of making a wrong classification is more [7].

The kernel function of the SVM is chosen based on the user's need. Sometimes, linear kernel gives more accuracy than a polynomial kernel or vice versa. It has to be made sure that the classes are a perfect fit for the kernel function. Over-fit or under-fit of classes has to be avoided.

In the paper, a SVM with linear kernel function is used. Although the paper talks about recognizing 8 actions, the SVM classifier has to differentiate only between 2 classes. This is achieved by using multiple networks working in coherence to detect the action. Each node has its own SVM network which is trained to recognize 2 actions. Overall system can recognize 4 actions (2 per node) for which the networks have been trained. Since the actions are independent of each other, 4 new actions whose database is not present can be recognized by performing permutations on the 4 actions whose database is already present. Hence, each SVM classifier is effectively differentiating between two classes, which increases the chances of high separation leading to high accuracy.

6.2   Gaussian Naïve Bayes (GNB)

Naïve Bayes classifier is a probability-based classifier which is used for data sets having high number of features. The classifier assumes that the presence of a particular feature in a class is unrelated to any other feature. It first plots the histogram of each feature present in the data set and assumes it to be a Gaussian function. Later based on the Naïve Bayes theorem, the probability is calculated and it is classified based on the highest probability match with the existing data set. This model is easy to build and is known to outperform many other highly efficient classifiers.

7   Results

The main aim of the paper was to improve the accuracy of recognition. One of the tools that can be used to calculate accuracy is the confusion matrix. Even though the system has two neural networks, trained to recognize just 2 actions, the confusion matrix is created considering both the networks as one main network. The network was trained with 18 samples for each action and tested for 18 new samples.

The overall accuracy of the system can be found out by the ratio of total number of correct decisions to the total number of samples.

Therefore, accuracy of the system using

$$GNB = (18 + 17 + 18 + 16)/(18 * 4) = 0.9583 \text{ or } 95.83\%$$

The accuracy of the 2 networks can be found separately in the same way. Calculations show that the accuracy of right-hand network is 97.22% and that of left-hand network is 94.44%. The overall accuracy of the system is the average of the accuracies of the 2 networks, which is something desirable. Generally as the number of classes increase, the accuracy decreases for the same number of samples. Since the networks work independently, by permutations and combinations of these 4 actions, the system can detect 8 actions in total without affecting the accuracy of the system. Although the number of actions that can be recognized is increasing, the accuracy of the system remains 95.83%. The confusion matrix using GNB classifier for 4 actions and 8 actions are shown in Figs. 6 and 7, respectively.

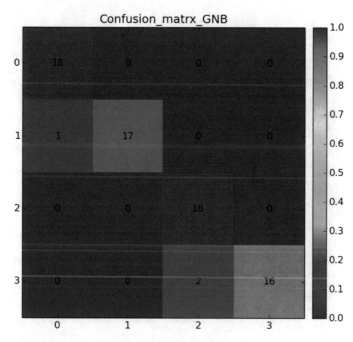

Fig. 6 Confusion matrix using GNB classifier for 4 actions

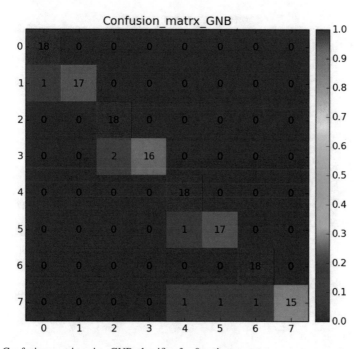

Fig. 7 Confusion matrix using GNB classifier for 8 actions

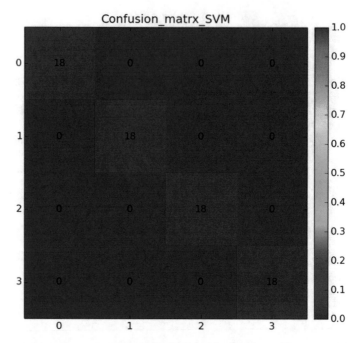

Fig. 8 Confusion matrix using SVM classifier for 4 actions

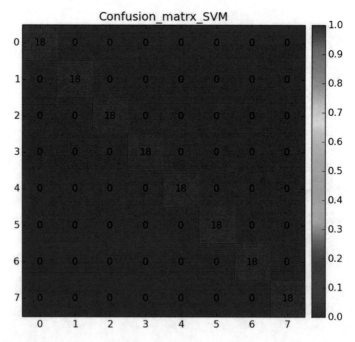

Fig. 9 Confusion matrix using SVM classifier for 8 actions

Where 0, 1, 2, 3, 4, 5, 6, 7 indicate right-hand wave (RHW), right-hand rotate (RHR), left-hand wave (LHW), left-hand rotate (LHR), RHW-LHW, RHW-LHR, RHR-LHW, and RHR-LHR, respectively.

Now, a similar process is followed with SVM classifier, and it is 100% for SVM classifier in all cases. The confusion matrix using SVM classifier for 4 actions and 8 actions are shown in Figs. 8 and 9, respectively.

8 Conclusion

Human–computer interaction is a field in which a lot of research is going on. One topic in this vast field of HCI is action recognition. One major problem research in action recognition faced was the accuracy of the recognition [8]. The accuracy falls down drastically as the number of actions to be recognized increases. In order to avoid the fall in accuracy, researchers made databases with more samples per action. But this increased the memory requirement of the system. The problem of low accuracy and high memory requirement always hindered the implementation of action recognition on an embedded platform.

This paper mainly aims at tackling the accuracy and the memory problems while compromising a little on the computational power required. Most of the embedded boards like Raspberry-pi or BeagleBone generally support Ubuntu operating system. The paper only uses the resources and libraries supported by Ubuntu OS for implementation of skeleton tracking and action recognition. With the help of ROS, parallel computing was introduced which helped in reducing the redundancy in features. For the first time, two neural networks working in coherence for one system was implemented which improved the accuracy drastically. The system is capable of detecting actions which are not present in the database although the networks used follow supervised learning algorithm. The paper incorporates the advantages of both supervised and unsupervised learning.

The raise in accuracy is clearly shown in the results section. By the implementation of the algorithm mentioned in this paper, the accuracy has been improved to almost 100%. The paper also shows that as the number of actions increases, the accuracy remains the same even for same number of samples. The paper effectively tackles the problems of low accuracy and high memory requirement and can be easily implemented on an embedded platform.

References

1. Cheng H, Hwang J (2008) Integrated video object tracking with applications in trajectory based event detection. J Vis Commun Image Represent 22(7):673–685
2. Pirsiavash H, Ramanan D (2012) Detecting activities of daily living in first-person camera view. In: Proceedings of the 2012 IEEE conference on computer vision and pattern recognition. IEEE

3. Papadopoulos GT, Axenopoulos A, Daras P (2014) Real time skeletal tracking based human action recognition using kinect data. In: Lecture notes in computer science, vol 8325, pp 473–483
4. Wang P, Li W, Ogunbona P, Gao Z, Zhang H (2014) Mining mid-level features for action recognition based on effective skeleton representation. eprint arXiv:1409.4014
5. Yang X, Tian Y (2014) Effective 3D action recognition using eigen joints. J Vis Commun Image Represent 25(1):2–11
6. Bobic A, Davis J (2001) The recognition of human movement using temporal templates. IEEE Trans Pattern Anal Mach Intell 23(3):257–267
7. Wang C, Shi XH (2015) A marker-less two-hand gesture recognition system using kinect depth camera. In: IEEE international conference on signal processing, communications and computing (ICSPCC)
8. Bian W, Tao D, Rui Y (2012) Cross-domain human recognition. IEEE Trans Syst Man Cybern (Part B) 42(2)

A Segmentation Approach Using Level Set Coding for Region Detection in MRI Images

Virupakshappa and Basavaraj Amarapur

Abstract Computer-aided diagnosis (CAD) systems for identifying brain tumor region in medical study have been investigated by various methods. This paper introduces an approach in computer-aided diagnosis for identification of brain tumor in early stages using level set segmentation method. The skull stripping and histogram equalization techniques are used as the processing techniques for the acquired image. The preprocessed image is used to segment region of interest using level set approach. The segmented image is fine-tuned by applying morphological operators. The proposed method gives better Mean Opinion Score (MOS) as compared to conventional level set method.

Keywords Image segmentation · MRI sample · Level set (LS) coding
Mean Opinion Score (MOS)

1 Introduction

MRI imaging modality produces extraordinary detailed images compared to any other modalities. For the diagnosis of various types of lesions and medical conditions, MRI is the method of choice. MRI has remarkable ability to answer the examination to any medical-related question (Fig. 1).

Medical image consists of minute variation and there exists co-resemblance between healthy biological part and affected part which makes human analysis a difficult task. Any human analysis requires a larger dataset which makes biological

Virupakshappa (✉)
Department of Computer Science and Engineering, Appa IET, Gulbarga,
Karnataka, India
e-mail: virupakshi.108@gmail.com

B. Amarapur
Department of Electrical and Electronics Engineering,
Poojya Doddappa Appa College of Engineering, Gulbarga, Karnataka, India
e-mail: bamarapur@yahoo.com

© Springer Nature Singapore Pte Ltd. 2018
A. K. Nandi et al. (eds.), *Computational Signal Processing
and Analysis*, Lecture Notes in Electrical Engineering 490,
https://doi.org/10.1007/978-981-10-8354-9_21

235

Fig. 1 MRI of brain showing
tumor

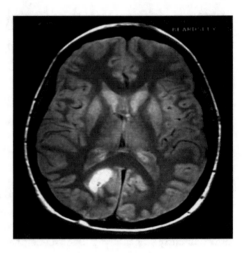

Fig. 1 MRI of brain showing
tumor

analysis a tough job. The tumor prediction in brain MRI image is a complicated problem, so it is very difficult job to design an automatic recognition system which could work on large datasets and predict the accurate results. Brain tumor is kind of death-dealing and incurable diseases in the world. Tumors are covered in the region of the brain which is crucial to maintain human body's critical activities, while they grow unconditionally to attack other portions of the brain, creating more and more tumors which are very minute in size to be unable to detect using any conventional imaging modalities. Brain tumor treatment using surgery or any radiation is very complex because of tumor location and its ability to spread quickly. It is analogous to finding the enemy hiding out in minefields.

According to recent survey, the occurrence of tumor has been on rise. Unfortunately, many of these are diagnosed too late, after symptoms start appearing. Compared to large tumor, it is easy to remove small tumor. It is very difficult to treat the tumor once symptoms appear. Using learning logic, it is very problematic to detect tumor in its early stages. This makes a barrier system, since brain is the most complex organ in the human body. If it is left untreated, tumor will spread to other parts of the body. Computer-aided surgical methods and sophisticated image-guided technologies become widely used in edge surgery [1–5]. A brain tumor is characterized by uncontrollable growth of cells. Patient's survival chances can be increased by finding and removing these tumors. Best way to treat the tumor is to detect and remove it completely before symptoms start appearing.

Detection of brain tumors at an early stage by human interaction may result in less interpretation as different types of tumors and their locations may not be predictive. For the realization of such automated system for early prediction of tumor, the major problem is with the processing of MRI images for estimations of brain tumors. It includes removal of noise, training of network, feature extraction of the brain images, and developing a suitable model to detect and classify the different types of tumors using various types of features. Usually brain tumors are not capable of spreading outside the brain. These kinds of tumors in brain need not to be treated, because of its self-limited growth. These sometimes cause problem because of their location, and radiation or surgery is helpful.

Effected (malignant) tumors are usually referred as brain tumor. This kind of tumor spread beyond the brain. Tumors of the brain that are affected always create a problem if untreated, so an aggressive approach is required. Brain tumors are again divided into two types: primary brain tumor originated inside the brain, and secondary or metastatic brain tumor comes into existence from other regions of the body and spread to the brain; usually, tumors occur when the cells in the body are divided without any control or order. If cells keep on contrasting, then it leads to aggregation of tissues and these aggregate of tissues is formed called as tumor. The word tumor usually referred to the effected tumor, which can attack adjoining tissue, and it can also extend to the other parts of the human body. A benign tumor will not spread beyond the brain.

A substantial methodological framework including new data analysis method was developed in [6] to meet the challenge of working with big data. Effected imaging phenotypes determined by MRI providing a mean of panoramic and noninvasive surveillance of oncogenic pathway activation for patient's treatment were presented. In [7], brain tumors are automatically detected with the help of MRI, and different steps of CAD system were presented. In [8], the combination of wavelet transform and the EEG signal used for primary brain detection is an approach to segmentation which was conferred. Ambiguity in the EEG signal is calculated by using proximate entropy. This observation leads to the process of tumor detection based on EEG analysis. In [9, 10] Describes a Matlab implementation for brain tumor detection and extraction from scanned MRI images of the brain, which include noise removal functions, segmentation and morphological operations used for coding in MRI images. The approach of automated segmentation of the proposed approach was outlined in this work. In [11], to classify the tumor type in MRI images of various patients with astrocytoma type of brain tumor, a method is developed by ANN approach called backpropagation network (BPN) and probabilistic neural network (PNN).

Many methods were developed for segmentation of regions from a given MRI sample, less focus is made on its localization. Localization of the mass elements in the MRI sample could minimize the overhead coming for processing the whole sample. With this objective, a simpler, however, robust approach of mass localization in MRI sample using recursive morphological approach with its fusion is suggested. In the approach of segmenting such region, a walker-based approach was suggested in [12]. This logic presents a simpler approach of region segmentation via boundary region tracking.

This paper proposes a new region detection approach for brain MRI images based on level set coding and a region localization algorithm. The proposed approach obtains the exact regions with perfect boundaries because of LS coding and also achieves improved segmentation accuracy due to the preprocessing and region localization logic. Simulations are carried out in various MRI images to show the performance of proposed approach.

The rest of this paper outlines the MRI processing in Sect. 2. Section 3 outlines the proposed approach of recurrent morphological coding. Section 4 outlines the experimental result obtained. Conclusion is given in Sect. 5.

2 Segmentation Approach

Level Set Segmentation: Initially, Osher and Sethian presented the method called as level set method for front propagation, applied to burning waves and ocean waves [11]. Recently, Malladi used this method for medical imaging applications [10]. The main intention of using level set approach is to embed a curve surrounded by that surface. Initially, for the formulation of LS method, it was necessary to define the boundary of the segmentation as part of a surface where image contour level is zero. Let implicit surface of the image be φ so that

$$\varphi(X, t) = \pm d \tag{1}$$

where X points a position in the space (domain), 't' is the time, and 'd' represents the distance among zero level set and the position X. If X is outside zero level set function, then sign ahead of d is positive. Otherwise, sign is negative. One important thing is the marking of positions where $\varphi = 0$.

The chain rule is used to evolve φ over time:

$$\varphi_t + \varphi_x x_t + \varphi_y y_t = 0 \tag{2}$$

$$\varphi_t + (x_t, y_t).\nabla_\varphi = 0 \tag{3}$$

Now, let us consider $(x_t, y_t) = s + n$, where s is some arbitrary vector and n is the vector normal to the front at point x. Then, the above equation can be modified as

$$\begin{aligned}
\varphi_t + (n + s).\nabla_\varphi &= 0 \\
\varphi_t + n.\nabla_\varphi + s.\nabla_\varphi &= 0 \\
\varphi_t + V_n.|\nabla_\varphi| + s.\nabla_\varphi &= 0
\end{aligned} \tag{4}$$

where v depicts the scalar, and V_n and s are the two values viewed as two independent forces that emerge the surface. When there is no feature present in the image using V_n, then formulation allows the default expansion or contraction of the level set; else, V_n can fall off toward 0 and levels on the actual edges are locked over vector stake.

3 Proposed Approach

This work implements a proficient system for extraction of tumor from input MRI image and recognizes the extracted data for many applications. The proposed method has four steps, namely acquisition of data, preprocessing, level set segmentation, and region localization as shown in Fig. 2.

3.1 Image Acquisition

Datasets are collected from various sources for different class of MRI which are considered for implementation of automated recognition system. Figure 3 shows the snapshot of considered database. This database can be used for further processing.

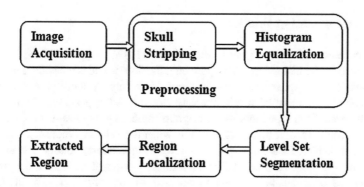

Fig. 2 Proposed approach

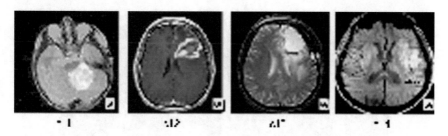

Fig. 3 An example of used MRI

3.2 Preprocessing

In the proposed method, preprocessing is carried out using the following two methods.

3.2.1 Skull Stripping

Skull stripping is used as preliminary step in many MRI sequences to remove extra-meningeal tissues from the MRI volume of head. This algorithm employs an edge detector to find the boundary between the brain and skull. In T1 scan, there exists a dark space between cerebral spinal fluid (CSF). This algorithm employs morphological operators to enhance the edge detection result and gives better separation of tissues. Finally, the result is improved by the application of aniso-tropic diffusion filter (Fig. 4).

3.2.2 Histogram Equalization

The occurrences of various gray levels and the relative frequency of the image are represented in the histogram. Histogram equalization provides a standard technique for altering the contrast and dynamic range of picture by modifying that picture such that histogram of picture's intensity has desired appearance. Compared to contrast stretching, operators of histogram modeling may use nonlinear and non-monotonic transfer functions to use between the values of input pixel intensity and output picture. This technique makes use of monotonic, nonlinear mapping in which the intensity of pixel in the picture is reassigned such that the uniform distributions of intensities are accommodated in the output. The result of this method on MRI image is shown in Fig. 5.

(a)   (b)

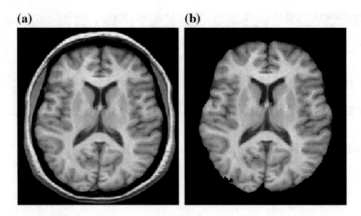

Fig. 4 **a** Original image, **b** skull-stripped image

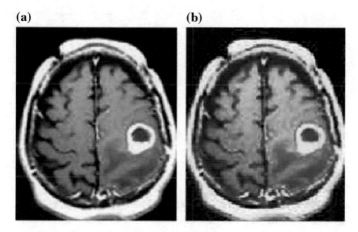

Fig. 5 **a** Input MRI image, **b** histogram-equalized image

3.3 Image Segmentation

Segmentation method is a process of subdividing an image into constituent parts of objects. The level of division of image depends upon the application domain, and the segmentation process should stop when the boundaries of the brain tumor are identified. The region of interest in our work is tumor, i.e., segmenting the tumor from its background. The tumor cells near the boundary are very fatty and look very dark, which creates problems in the edge recognition process. In order to overcome this problem, preprocessing to the acquired image was performed. First, skull stripping was done for removal of extra-meningeal tissues from the MRI volume. Second, histogram equalization approach is used to improve the gray level near the boundary.

3.4 Region Localization

The basic improvement required in the MRI image is an increase in contrast of the image. Human eye cannot percept the contrast present between the brain and the tumor region. Thus, in order to increase the contrast between the tumor region and normal brain, a segmentation filter is used to the MRI image, which produces perceptible improvement in the contrast of the image. Segmentation filters work to emphasize the changes, i.e., by expanding image intensity at the boundaries to emphasizing fine sharp details and enhancing the blurred detail. Most of the segmentation filters use a mask of 3 * 3 size pixel neighborhood. For every resulting pixel, mask estimates the weighted addition of respective input pixel in the image and its surrounding eight pixels. The weight of the middle pixel is positive, and for

Fig. 6 Structuring element

1	1	1
1	1	1
1	1	1

the bordering pixels, it is negative. The entire intensity of the image is unspoilt by organizing their sum of the weights equal to one. In order to extract the tumor region, we use morphological operators and logical operators. The non-tumor regions are characterized by separate vertical edges, diagonal edges, and horizontal edges, whereas in tumor regions all the edges are intermixed together. The edges of the tumors are generally short and interconnected with each other in various orientations, to connect isolated tumor, edges in each block of the image morphological erosion and dilation operators are used. The morphological operators are mainly used on the binary images. Erosion operator is used to weaken the edges of the prominence pixels. This operator basically shrinks the size of foreground pixels and results in holes for the larger area. Erosion operator expects two tuple data as input, such as input image and structuring element (Kernel). The structuring element is made up of a pattern represented as the coordinates values of various discrete sets of points with respect to any origin.

Figure 6 shows an example of structuring element. In every computation, the origin of the image is indicated by a circular-shaped ring surrounding any point. The origin need not to be present in the middle of the structuring element, but in most of the cases it is. The most commonly used structuring element is 3 * 3 grids with origin present at the center. The morphological operation translates each pixel position in the image in turn to the origin of the structuring element and then compares with underneath image pixel values. The comparison details and outcome depend on which morphological operator used.

4 Experimental Results

This section illustrates the performance evaluation of proposed approach. To test the proposed approach, various MRI images were used. For each and every MRI image, the proposed approach tends to segment the region with exact tumor. The obtained results are shown in Figs. 7 and 8.

Figure 7 represents the original MRI image used for testing.

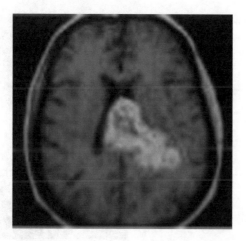

Fig. 7 Original test sample

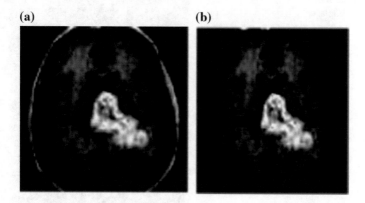

Fig. 8 **a** Intensity-mapped image, **b** skull-stripped image

Figure 8a shows the intensity-mapped MRI image. Intensity mapping is done here to obtain the uniform intensities at each and every pixel; thus, the region with abnormal intensities will be highlighted. This will reduce the segmentation complexity. After mapping intensity, the image is subjected to skull stripping as shown in Fig. 8b.

To enhance the illumination and to eliminate the noise effect, the histogram equalization is applied and the obtained histogram-equalized MRI image is shown in Fig. 9.

After obtaining the extracted region with outer skull region, the image is subjected to segmentation. By segmentation, the image will be binarized and the region with abnormalities will be in gray color, the remaining region is the black color, and the recurrent segmentation image is shown in Fig. 10.

Fig. 9 Histogram-equalized image

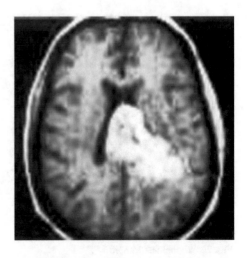

Fig. 10 Recurrent segmentation image

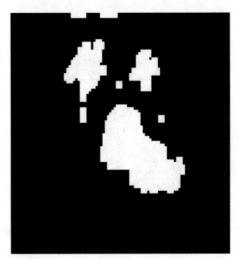

Region filling is done to avoid the edge cuttings. By region filling, the edges will be clearly highlighted and the regions will be obtained with exact boundaries. Figure 11 shows the region-filled image having regions with clear boundaries.

Figure 12 shows the final extracted region in which the tumor is existing. The figure also shows the region with clear edges.

Along with this experimental evaluation, the proposed and conventional approaches are subjected to subjective quality test experiment by evaluating the Mean Opinion Score (MOS). Subjective opinions were collected by a group of 20 experts; each expert gave scores for each test image: 1—bad, 2—poor, 3—fair,

Fig. 11 Region-filled image

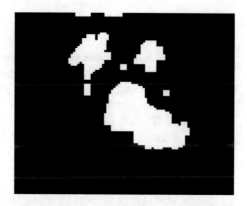

Fig. 12 Segmented mass regions

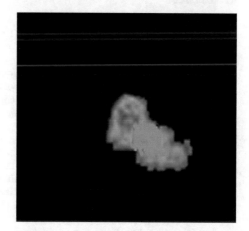

4—good, and 5—excellent. Estimation of Mean Opinion Score (MOS) is then done by calculating average of all experts' opinions [12]. Table 1 shows the summarized test results of MOS, in which preference will be given to higher values. Table 1 demonstrates that under the given conditions the proposed approach of segmentation is improved over the conventional method.

Table 1 MOS (Mean Opinion Score)

Noise	MOS	
	Conventional	Proposed
S1	2.35	3.54
S2	2.56	3.48
S3	2.85	3.65
S4	2.10	3.11
S5	2.47	3.24

5   Conclusion

This paper outlines an automatic detection method for the brain MRI image that is presented using the level set coding and region localization logic. It is found that during the recognition process the system results in better segmentation accuracy. The iteration count and the precision level of the segmentation are found to be improved by 50–60% in detection compared to the conventional edge segmentation. The outlined method works with 2D MRI images; in future, we will try to extend it for 3D MRI images.

References

1. Cline HE, Lorensen E, Kikinis R, Jolesz F (1990) Three-dimensional segmentation of MR images of the head using probability and connectivity. J Comput Assist Tomogr 14: 1037–1045
2. Vannier MW, Butterfield RL, Rickman DL, Jordan DM, Murphy WA, Biondetti PR (1985) Multispectral magnetic resonance image analysis. Radiology 154:221–224
3. Just M, Thelen M (1988) Tissue characterization with T1, T2, and proton-density values: results in 160 patients with brain tumors. Radiology 169:779–785
4. Just M, Higer HP, Schwarz M et al (1988) Tissue characterization of benign tumors: use of NMR-tissue parameters. Magn Reson Imaging 6:463–472
5. Gibbs P, Buckley DL, Blackband SJ, Horsman A (1996) Tumor volume determination from MR images by morphological segmentation. Phys Med Biol 41:2437–2446
6. Warfield SK, Dengler J, Zaers J et al (1995) Automatic identification of gray matter structures from MRI to improve the segmentation of white matter lesions. J Image Guid Surg 1:326–338
7. Warfield SK, Kaus MR, Jolesz FA, Kikinis R (1998) Adaptive template moderated spatially varying statistical segmentation. In: Wells WH, Colchester A, Delp S (eds) Proceedings of the first international conference on medical image computing and computer-assisted intervention. Springer, Boston, MA, 431–438
8. Bonnie NJ, Fukui MB, Meltzer CC (1999) Brain tumor volume measurement: comparison of manual and semi-automated methods. Radiology 212:811–816
9. Zhu H, Francis HY, Lam FK, Poon PWF (1995) Deformable region model for locating the boundary of brain tumors. In: Proceedings of the IEEE 17th annual conference on engineering in medicine and biology, vol 411. IEEE, Montreal, Quebec
10. Malladi R, Sethian JA, Vemuri B (1995) Shape modeling with front propagation: a level set approach. IEEE Trans Pattern Anal Mach Intell 17(2):158–175
11. Sethian JA (1999) Level set methods and fast marching methods: evolving interfaces in computational geometry, fluid mechanics, computer vision, and materials science. Cambridge University Press
12. Kim NS, Chang J-H (2000) Spectral enhancement based on global soft decision. IEEE Sig Process Lett 7(5):108–110

Off-line Odia Handwritten Character Recognition: A Hybrid Approach

Abhisek Sethy, Prashanta Kumar Patra and Deepak Ranjan Nayak

Abstract Optical character recognition (OCR) is one of the most popular and challenging topic of pattern recognition with a wide range of applications in various fields. This paper proposes an OCR system for Odia scripts which comprises of three stages, namely preprocessing, feature extraction, and classification. In the preprocessing stage, we have employed median filtering on the input image and subsequently we have applied normalization and skeletonization methods over images for extraction of boundary edge pixel points. In the feature extraction stage, initially the image is divided into 3×3 grids and the corresponding centroids for all the nine zones are evaluated. Thereafter, we have drawn the horizontal and vertical symmetric projection to the nearest pixel of the image which is dubbed as binary external symmetry axis constellation for unconstrained handwritten character. From which we have calculated the horizontal and vertical Euclidean distance for the same nearest pixel from centroid of each zone. Then we have calculated the mean Euclidean distance as well as the mean angular values of the zones. This is considered as the key feature values of our proposed system. Lastly, both kernel support vector machine (KSVM) and quadratic discriminant classifier (QDA) have been separately used as the classifier. To validate the proposed system, a series of experiments have been carried out on a standard database as NIT Rourkela Odia Database. From the database, we select 200 samples from each of the 47 categories. Simulation results based on a tenfold cross-validation approach indicate that the proposed system offers better recognition accuracy then other competent schemes.

A. Sethy (✉) · P. K. Patra
Department of Computer Science and Engineering, College of Engineering
and Technology Bhubaneswar, Bhubaneswar, Odisha, India
e-mail: abhisek052@gmail.com

P. K. Patra
e-mail: principalcet@cet.edu.in

D. R. Nayak
Department of Computer Science and Engineering, National Institute
of Technology Rourkela, Rourkela, Odisha, India
e-mail: depakranjannayak@gmail.com

© Springer Nature Singapore Pte Ltd. 2018   247
A. K. Nandi et al. (eds.), *Computational Signal Processing
and Analysis*, Lecture Notes in Electrical Engineering 490,
https://doi.org/10.1007/978-981-10-8354-9_22

Moreover, the recognition accuracy obtained by KSVM and QDA classifier is 96.5 and 97.4%, respectively.

Keywords Optical character recognition · Feature vector · BESAC KSVM · QDA

1 Introduction

Optical character recognition is one of the most notable areas of research in pattern recognition. In past decades, it has received considerable attention from more and more researchers for its importance in day-to-day life. OCR systems have not only been developed in recognition of printed, handwritten characters, but also for degraded characters. In OCR, all the stages are crucial starting from image acquisition to classification stage. However, a good recognition system needs a well-defined feature extraction procedure along with a proper classifier, in order to yield high performance rate [1]. Recognition of handwritten characters has been shown to be very challenging than printed one because of variations in writing skills, shapes, and orientations. A number of works have been proposed by researchers in recognizing handwritten scripts like Arabic, Chinese, and English. [2]. However, there are some languages in India for which automatic handwriting recognition systems have not been explored. Odia script language is one of these languages which have been derived from Devanagari scripts. Though this language is originated from Odisha, an eastern state of India, however, it is also spoken in some southern and northern parts of India. Because the Odia characters are complex in shape and structure, the recognition becomes more demanding. The most interesting property of this language is that it has no lower and uppercase letters. A less amount of works has been offered for handwritten Odia scripts, so the development of an ideal OCR system for handwritten Odia characters remains an open problem can be investigated. In past recent years, different authors made many attempts for analysis of Odia scripts which are reported in [3, 4].

In this paper, we have performed statistical analysis over the character images in order to obtain the feature vector values. Here, we have generated 3×3 grids for each character and evaluated the corresponding centroid for all the nine zones. Subsequently, we draw the horizontal and vertical projection angel to nearest pixel of the image. Moreover, we calculate horizontal and vertical Euclidian distance for same nearest pixel from centroid of each zone. Thereafter, these resultant values are considered as the key feature vector for the proposed recognition system. For the classification purpose, we have harnessed KSVM and RBF classifiers separately. The rest of this paper is summarized as follows: Sect. 2 highlights the related work done for Odia character; similarly adopted methodologies are described in Sect. 3. Section 4 represents the implementation stages, where classifier is introduced. Section 5 discussed about the analysis of simulation outcomes, and last Sect. 6 is used for conclusion and future work.

2 Related Work

In the last decades, several attempts have been made by different researchers in order to achieve a high recognition rate [4, 5]. Patra et al. [6] have applied Zernike moment and Mellin transform over printed characters and obtained 99% recognition rate using neural network. Pal et al. [7] introduced an approach where PCA was used in reducing the dimension of feature vectors generated using curvature. They have achieved 94.6% of accuracy on handwritten characters using quadratic classifier. Chaudhuri et al. [8] proposed a recognition model for printed document, in which water reservoir feature values are utilized and obtained 96.3% as the recognition rate. Padhi and Senapati in [9] performed a two-way approach for the recognition of printed character scripts. Their feature matrix contains the empirical values such as standard deviation and zone-based average centroid distance of images. They have listed two scenarios of classification, one for similar characters and other for distinct characters. Pujari et al. [10] have proposed a novel parallel thinning method for skeletonization of Odia characters which on the other hand preserves essential features. In addition, they have studied several other thinning algorithms. Later on, Mitra and Pujari in [11] have introduced a model based on directional features. They used fixed zoning to extract directional and SVM for classification. Nayak et al. [12] have proposed an edge detector based on nonlinear cellular automata for Odia character images. In [13], Sethy et al. employed wavelet transform for classification of Odia characters. PCA was used to reduce the feature dimensionality, and BPNN was used as classifier in this case and they accomplished 94.8% accuracy on a standard database. A new dimension of Boolean matching technique is proposed by Dash et al. in [14]. They had adopted a two-way strategic classification by considering histogram orientation of binary-coded external symmetry of the character images. For classification, they have applied random forest tree, SVM, and K-NN separately and achieved 89.92, 93.77, and 95.01% recognition rates, respectively. Table 1 lists most of the notable works on Odia handwritten character recognition.

Table 1 A summary of some related works

Authors & References	Feature extraction	Classifier	Recognition rate (%)
Dash et al. [14]	Binary-coded external symmetry	Random forest tree	89.92
		SVM	93.77
		K-NN	95.01
Sethy et al. [13]	2D-DWT and PCA	BPNN	94.8
Mitra and Pujari [11]	Directional decomposition	Multiclass SVM	95
Padhi et al. [9]	Standard deviation and zone centroid average distance	BPNN	92
Chaudhuri et al. [8]	Stroke	Quadratic classifier	96.3
Pal et al. [7]	Curvature feature and PCA	Feature-based tree classifier	94.6

From the aforementioned literature, it has been observed that most of the works were validated on smaller datasets. Further, dimension of features are more in some cases. Accuracy in some cases is very poor which can be increased.

3 Proposed Model for Recognition System

This section presents the proposed methodology adopted for handwritten Odia character recognition. Figure 1 depicts the generic steps of the proposed recognition system. As shown, the system comprises of various steps like image acquisition, preprocessing, feature extraction, and classification.

3.1 Image Acquisition

For validating the proposed model, we consider a standard database of Odia character, NIT Rourkela Odia database, which was developed at NIT, Rourkela [15]. This database contains 15,040 numbers of images for both Odia characters and numerals. However, for our experiment, we have chosen 200 samples for each of the 47 characters (i.e., 47 * 200 = 9400) from the database, and some of the sample images of the database are shown in Fig. 2.

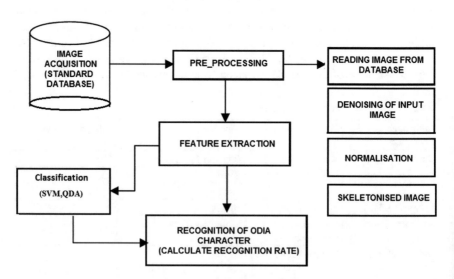

Fig. 1 Schematic block diagram of proposed recognition system

Fig. 2 A sample set of Odia handwritten characters

3.2 Image Preprocessing

Preprocessing of the character image is one of the most vital stages of OCR as it helps in extracting significant features from the character images. In this phase, we have first employed median filter over the images, and subsequently we harnessed normalization and skeletonization procedures over images and then extracted boundary edge pixel points.

3.3 Feature Extraction

Extracting precise features from the characters is the most important task of an OCR system. In this section, the proposed algorithm for evaluation of feature vector is presented. Initially, the preprocessed image is divided into 3 × 3 non-overlapping blocks. Thereafter, we have drawn the horizontal and vertical symmetric projection angel to the nearest pixels of the image which is dubbed as binary external symmetry axis constellation (BESAC) [14] for unconstrained handwritten character, from which we have calculated the horizontal and vertical Euclidean distance for the same nearest pixels from centroid of each block. Subsequently, the mean value of the Euclidean distance and the angular values of each block with respect to the midpoint of symmetry axes are determined and are considered as the key feature values of our proposed model. Following two algorithms, Algorithm I and Algorithm II, are used for evaluation of feature vector.

Algorithm I: Row and Column Symmetry axis
Input: Extract the boundary points (N) of character images
Output: Extract row symmetry axes $(RE_i: i = 1, 2,..., j)$
     Extract row symmetry axes $(CE_i: i = 1, 2,..., k)$
1. **For** $i = 1$ to N **do**
   While row chords creation is satisfied.
   Map the row chords accordingly so that sets of parallel row chords are grouped.
   End While.
2. **While** column chords creation is satisfied.
   Draw chords from each boundary point to other boundary point in same column.
   Map the column chords accordingly so that sets of vertical column chords are grouped.
   End While.
3. **End For**.
4. Find midpoint parallel row chords.
   Resulting axes consist of midpoints of row external chords.
   Provide potent row symmetry axes (RE_i)
   Resulting axes consist of midpoints of column external chords.
   Provide potent column symmetry axes (CE_i)
5. Return RE_i and CE_i

After performing the essential step for row and column symmetry axis which is given in **Algorithm I**, we have developed the constellation model according to their relative symmetric axis pixel position and midpoint pixel angle from center of the image. This constellation model generates two set of parameters for each row symmetry axis and column symmetry axis, where one parameter indicates mean value of the Euclidean distance of every symmetry axis pixel position to the centroid of each zone and other parameter indicates the angle between the respective images. Thereafter, we obtained four parameters of each image.

Algorithm II: Feature vector generation using BESAC
Input: Extract row symmetry axis $(RE_i: i = 1, 2, 3,..., j)$
     Extract row symmetry axis $(CE_i: i = 1, 2, 3,..., k)$
Output: Binary-coded rows and column symmetry axes
     (BESAC) feature (F).
1. Begin **For** $i = 1$ to j **do**
2. Find the midpoints of row symmetry axis.
   Compute angle from center of image to midpoints of row symmetry axis.
   Compute distance from centroid of image to boundaries of row symmetry axis.

3. **End For**
4. **For** $i = 1$ to k **do**
5. Find the midpoints of column symmetry axis.
   Compute angle from centroid to midpoints of column symmetry axis.
   Compute distance from centroid to column symmetry axis.
6. **End For**
7. Compute mean angle from centroid to midpoints of row symmetry axis.
8. Compute mean distance from centroid to boundaries of row symmetry axis.
9. Return F.

From the above algorithm, it is obvious that only four key features can be obtained for all the blocks of an image. For instance, a total of $4 * 9 = 36$ features can be extracted from a character image as the image is of size 81×81. Similar features can be extracted for all the 9400 images in the dataset, and a feature matrix of size 9400×36 is formed.

3.4 Classification

In order to accomplish classification task, the proposed model adopted two classifiers, namely kernel support vector machine (KSVM) and quadratic discriminant analysis (QDA) classifier for recognition of handwritten Odia characters. SVM or KSVM was originally developed for binary classification; however, to solve a multiclass problem using SVM or KSVM, many methods have been proposed. This paper uses "Winner-Takes-All (WTA)" strategy for multiclass classification. Further, out of many kernels such as linear, polynomial, radial basis function (RBF), we choose RBF kernel as it is the most widely used and has several advantages over others which include an extra parameter called γ. Next, a more generalized version of a linear classifier called QDA classifier is employed which discriminates classes with the help of quadric surfaces. QDA has a close resemblance to linear discriminant analysis.

4 Implementation

The overall implementation of the proposed model is listed as a pseudo code in Algorithm III. It has two main phases: off-line learning and online prediction.

Algorithm III: Proposed Pseudo Code for OCR

Off-line Learning:

1. Load input the images from the database and perform the preprocessing stage.
2. Divide the preprocessed images into 3×3 non-overlapping blocks.

3. Locate the centroid of each block.
4. Perform Algorithm I and Algorithm II and find the four key feature values which include mean Euclidian distance and mean angle for both symmetric axes of both horizontal and vertical w.r.t each of the nine blocks.
5. Apply KSVM/QDA on the feature matrix and evaluate the performance of the model on the testing data based on tenfold cross-validation procedure.

Online Prediction:

1. Provide query images from the user end.
2. Perform preprocessing on the query images.
3. Calculate statistical features using BESAC. Find out four key feature values.
4. Apply trained KSVM/QDA on the feature vectors generated.
5. Predict the class label.

5 Simulation Result and Discussion

All the experiments were performed over a system having specification as Windows 8, 64-bit operating system, and Intel (R) i7—4770 CPU @ 3.40 GHz, and all the simulations were conducted through MatLab 2014a on a standard database named as NIT Rourkela Odia Database. After preprocessing, four key feature vector values as mean Euclidian Distance and mean angles are evaluated from each of the nine non-overlapping blocks. Hence, there are in total 36 features generated for a single character image. After feature extraction from all the images, the feature matrix size becomes 9400×36. Now, this matrix is fed to KSM and QDA classifier separately for segregating Odia characters. A tenfold cross-validation procedure has been

Table 2 Classification rate achieved by KSVM

No.	CC	MC	CR (%)	No.	CC	MC	CR (%)
1	195	5	97.5	25	195	5	97.5
2	195	5	97.5	26	193	7	96.5
3	195	5	97.5	27	193	7	96.5
4	193	7	96.5	28	193	7	96.5
5	193	7	96.5	29	195	5	97.5
6	191	9	95.5	30	195	5	97.5
7	191	9	95.5	31	195	5	97.5
8	191	9	95.5	32	193	7	96.5
9	193	7	96.5	33	193	7	96.5
10	193	7	96.5	34	193	7	96.5
11	193	7	96.5	35	193	7	96.5
12	193	7	96.5	36	193	7	96.5

(continued)

Table 2 (continued)

No.	CC	MC	CR (%)	No.	CC	MC	CR (%)
13	193	7	96.5	37	193	7	96.5
14	191	9	95.5	38	193	7	96.5
15	191	9	95.5	39	195	5	97.5
16	191	9	95.5	40	195	5	97.5
17	193	7	96.5	41	193	7	96.5
18	193	7	96.5	42	193	7	96.5
19	193	7	96.5	43	194	6	97.0
20	193	7	96.5	44	194	6	97.0
21	193	7	96.5	45	194	6	97.0
22	191	9	95.5	46	194	6	97.0
23	191	9	95.5	47	194	6	97.0
24	191	9	95.5	Average classification rate = 96.5%			

Table 3 Classification rate achieved by QDA

No.	CC	MC	CR (%)	No.	CC	MC	CR (%)
1	195	5	97.5	25	194	6	97
2	195	5	97.5	26	194	6	97
3	195	5	97.5	27	195	5	97.5
4	195	5	97.5	28	195	5	97.5
5	195	5	97.5	29	195	5	97.5
6	195	5	97.5	30	195	5	97.5
7	195	5	97.5	31	195	5	97.5
8	195	5	97.5	32	194	6	97
9	195	5	97.5	33	195	5	97.5
10	195	5	97.5	34	195	5	97.5
11	195	5	97.5	35	195	5	97.5
12	195	5	97.5	36	195	5	97.5
13	195	5	97.5	37	194	6	97.0
14	195	5	97.5	38	194	6	97.0
15	195	5	97.5	39	194	6	97.0
16	195	5	97.5	40	194	6	97.0
17	194	6	97.0	41	195	5	97.5
18	194	6	97.0	42	195	5	97.5
19	194	6	97.0	43	195	5	97.5
20	195	5	97.5	44	195	5	97.5
21	195	5	97.5	45	194	6	97.0
22	195	5	97.5	46	194	6	97.0
23	195	5	97.5	47	194	6	97.0
24	195	5	97.5	Average classification rate = 97.4%			

employed to avoid over-fitting and to evaluate the performance of the system. The results of KSVM and QDA classifier are given in Tables 2 and 3, respectively. In tables, CC represents the number of correctly classified samples, MC represents the number of miss-classified samples, and CR represents the classification rate. The recognition rate obtained by KSVM and QDA is found to be 96.5 and 97.4%, respectively. It should be worth mentioning that the recognition rate achieved by QDA is better than KSVM. In case of KSVM, the values of C and γ are chosen to be 10^3 and 2^{-1}. The value of both these parameters is found experimentally.

6 Conclusion and Future Work

In this paper, we have presented a hybrid OCR system for recognition of off-line Odia characters. This system uses binary external symmetry axis constellation for feature extraction. For classification, KSVM and QDA are used. Experimental results on a standard dataset indicate the efficacy of the proposed system with least number of features. It is observed that the result obtained by QDA is better than KSVM. Though we found satisfactory results over a standard dataset, still the development is in its infancy. Many other feature extraction techniques and subsequently other modern classifiers like extreme learning, deep learning can be investigated in future for better recognition.

Acknowledgements The authors are sincerely thankful to the Department of Computer Science and Engineering, College of Engineering and Technology, Bhubaneswar. And we are also thankful to all the authors of references.

References

1. Govindan VK, Shivaprasad AP (1990) Character recognition—a review. Pattern Recogn 23
2. Mantas J (1986) An overview of character recognition methodologies. Pattern Recogn 19:425–430
3. Plamondon R, Srihari SN (2000) On-line and off-line handwritten recognition: a comprehensive survey. IEEE Trans PAMI 22
4. Pal U, Chaudhuri BB (2004) Indian script character recognition: a survey. Pattern Recogn 37
5. Dash KS, Puhan NB, Panda G (2016) Odia character recognition: a direction review. Artif Intell Rev
6. Patra PK, Nayak M, Nayak SK, Gabbak NK (2002) Probabilistic neural network for pattern classification. In: Proceeding of the international joint conference on neural networks, May 12–17, Hawaii
7. Pal U, Wakabayashi T, Kimura F (2005) A system for off-line Oriya handwritten character recognition using curvature feature. In: IEEE proceeding of the 10th international conference on information technology, September 21–23, China
8. Chaudhuri BB, Pal U, Mitra M (2001) Automatic recognition of printed Oriya script. In: IEEE proceeding of the sixth international conference on document analysis and recognition, September 10–13, Washington, USA

9. Padhi D, Senapati D (2015) Zone centroid distance and standard deviation based feature matrix for Odia handwritten character recognition. In: Proceeding of the international conference on frontiers of intelligent computing theory and applications, November 16–18, Kolkata, India

10. Pujari AK, Mitra C, Mishra S (2014) A new parallel thinning algorithm with stroke correction for Odia characters. Advanced computing, networking and informatics. Springer International Publishing Switzerland

11. Mitra C, Pujari AK (2013) Directional decomposition for Odia character recognition, mining intelligence and knowledge exploration. In: Proceeding in lecture notes in computer science, vol 8284. Springer, Berlin

12. Nayak DR, Dash R, Majhi B, Mohammed J (2016) Non-linear cellular automata based edge detector for optical character images. Simul Trans Soc Model Simul Int 1–11

13. Sethy A, Patra PK, Nayak DR (2016) Off-line Odia handwritten character recognition using DWT and PCA. In: Proceeding in international conference on advanced computing and intelligent engineering, December 21–23, Bhubaneswar, India

14. Dash KS, Puhan NB, Panda G (2016) BESAC: binary external symmetry axis constellation for unconstrained handwritten character recognition. Pattern Recogn Lett

15. Mishra TK, Majhi B, Sa PK, Panda S (2014) Model based Odia numeral recognition using fuzzy aggregated features. Frontiers of Computer Science. Springer, Berlin

A Method for Detection and Classification of Diabetes Noninvasively

S. Lekha and M. Suchetha

Abstract Diabetes a common ailment affecting the vast population of people requires continues monitoring of blood glucose levels so as to control this disorder. Presently, the common technique used to monitor these levels is through an invasive process of drawing blood. Although this technique achieves high accuracy, it encompasses all disadvantages associated with an invasive method. This inconvenience is felt more accurately in patients who frequently examine these levels through the day. Hence, there is a need for a noninvasive technique for predicting the glucose levels. This paper aims at analyzing the breath as a noninvasive technique to predict diabetes.

1 Introduction

High level of glucose in the body for a prolonged period of time leads to major health complication known as diabetes. Statistics from the World Health Organization (WHO) show that nearly 422 million people are affected with diabetes. The alarming increasing rate at which diabetes is affecting millions is a cause of concern, and there is a need to control this disorder. Diabetes is mainly detected and monitored by continuously measuring ones blood glucose level (BGL). Diabetes can be classified mainly into two classes, namely type-1 diabetes and type-2 diabetes. These conditions arise due to the fact that the body is unable to secrete enough insulin or is caused as a result of the body's inability to use the produced insulin [1]. Present technique of monitoring the BGL involves an invasive process of drawing blood from the forearm or the figure tip. Disposable strips containing sensing elements are used to quantify the glucose levels in blood. This

S. Lekha · M. Suchetha (✉)
School of Electronics Engineering, VIT University,
Chennai Campus, Chennai, India
e-mail: suchetha.m@vit.ac.in

S. Lekha
e-mail: s.lekha2014@vit.ac.in

© Springer Nature Singapore Pte Ltd. 2018
A. K. Nandi et al. (eds.), *Computational Signal Processing and Analysis*, Lecture Notes in Electrical Engineering 490,
https://doi.org/10.1007/978-981-10-8354-9_23

259

traditional technique is highly accurate; however, this causes a lot of discomforts, especially when numerous readings are collected throughout the day. This leads to a necessity to develop a noninvasive detection technique [2].

Recent studies have shown that urine, tears, saliva, sweat and breath samples show trace content of glucose in them. Hence by measuring these concentrations, one can predict diabetes. In this paper, breath analysis as a noninvasive diabetes monitoring technique is being examined. Exhaled human breath samples contain various volatile organic compounds (VOC) that present in them. Among these VOCs acetone, a ketone body, shows a good correlation to the BGL. It is monitored that as the glucose levels raise the acetone levels in the breath samples also raise. On considering the breath sample of the healthy samples, acetone levels are present in concentrations less than 0.76 ppmv, and for the diabetic samples, these levels are found higher than 1.71 ppmv [3]. Based on these ranges of acetone gas, a classifier is used to distinguish and predict diabetic patients from the healthy. Techniques such as gas chromatography, selected ion flow spectroscopy, cavity ring-down spectroscopy, and electronic nose are used in breath analysis to detect gas concentrations. Gas chromatography–mass spectroscopy [4] functions on the concept that the gas molecules present in breath have a difference in their chemical properties and this promotes their separation. Another technique called the selected ion flow spectroscopy works on the theory that when a neutral analyte molecule of a sample vapor meets the precursor ions, they experience a chemical ionization which relays on their chemical properties [5]. Cavity ring-down spectroscopy [6], an optical spectroscopic technique, determines the absolute extinction by samples that scatter or absorb light. Though these techniques achieved high precision, they still remain unsuitable for clinical application due to their complicated principle, low user-friendliness, and high cost. These drawbacks have been overcome with the electronic nose model [7]. This paper aims to use a commercially available electronic nose to detect the presence of acetone gas samples. In this study, we examine acetone levels (biomarker for diabetes) in breath samples and thus predict diabetes. This paper employs the SVM algorithm in order to categorize the collected data samples as healthy and diabetic breath samples. A classification algorithm with different kernels is programmed in MATLAB environment.

The paper is structured as Sect. 2 explains the concept of detecting diabetes through breath. In Sect. 3, the methodologies used in this analysis are explained while Sect. 4 reviews the results obtained from this analysis. Lastly, the conclusion is explained in Sect. 5.

2 Biomarker for Diabetes

Recent researches have shown that the volatile organic compounds present in breath samples act as good biomarkers to predict diseases. Acetone, one of the exhaled gases, shows a good correlation to the BGL. Hence, acetone is considered to be a dependable biomarker for detecting diabetes. When a person is diabetic, it is seen

Table 1 Levels of acetone concentration in breath

Subjects	Concentration of acetone (ppm)
Healthy	0.22–0.80
Type-2 diabetic	1.76–3.73
Type-1 diabetic	As high as 21

that the body cells cannot make use of the glucose in blood. Due to this inability, the body cells utilize fat as the required energy. During this conversion, acetone is one of the by-products produced. These concentrations are then excreted from the body through breath and urine. It is studied that these concentration levels are higher for a diabetic patient. Table 1 shows the variations in the concentration levels [6].

3 Methodology and System Description

In this paper, breath analysis as an alternative technique to measure BGL is examined. Figure 1 depicts the entire framework carried out in this study. The system consists of two main sections. Initially, the acetone concentrations are to be detected by a sensor system. Next, these collected data samples are to be analyzed and classified.

Fig. 1 Flow diagram of the system

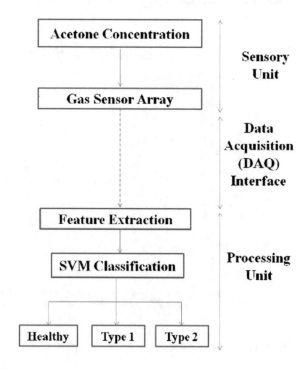

3.1 Structure of the Device

The approach begins with sensing the acetone concentrations. In order to measure these concentrations commercially available gas sensors are used in this analysis. MQ-3 and MQ-5 tin dioxide (SnO_2)-based gas sensors show a high sensitivity to VOC. Hence, an array of these gas sensors is used in our analysis to detect the concentration of acetone in breath samples. According to the concentration levels presented in Table 1, acetone samples for healthy and diabetic patients are considered and the readings from these gas sensors are collected.

Figure 2 shows the example of one of the gas sensor readings. In order to collect the data signals, an Arduino Uno board was been used to interface these sensor reading with the computer. It is observed that as the concentration of acetone gas changes, there is a change observed in the sensor's resistance. The resistance of the sensor (R_s) is calculated as

$$R_s = \frac{V_c - V_{RL}}{V_{RL}} \times R_L \tag{1}$$

where R_L corresponds to the load resistance, V_c is the circuit voltage, and V_{RL} is the output load voltage. The sensor's response is recorded for 1000 s, and this data is further analyzed for diabetes prediction [8]. In this analysis, a total of 54 data samples were collected and the sensor's response was calculated. Around 30 samples of type 2 and type 1 samples are collected.

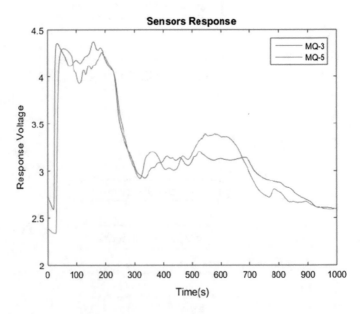

Fig. 2 Plot of the sensor response

3.2 *Data Analysis and Classification*

After obtaining the data signals from the gas sensors, feature extraction and classification algorithms are programmed in MATLAB environment to distinguish the collected data into the previously mentioned three classes. In order to classify, the data support vector machine (SVM) with different kernels is programmed.

(1) Feature extraction: After collecting the data signals, features are to be extracted before classification can be done. In this analysis, mainly two features are calculated. The first includes the maximum analog voltage response of sensor, and the second includes the sensor's response calculated by

$$S = \frac{R_s}{R_a} \tag{2}$$

where R_s is the resistance of the sensor, when it is in the presence of acetone samples and R_a represents the resistance of the sensor in the presence of air.

(2) Classification: A SVM algorithm is programmed to classify the data samples acquired through the sensory unit. SVM [9], a supervised learning algorithm, constructs hyperplanes which are used to segregating the data [10]. In this analysis, the sensor data is classified into a multi-class SVM classifier. SVM, which is a kernel-based classifier, is suitable for both linear and nonlinear problems. Initially, a few of the collected acetone samples are used to train the classifier. Once the SVM algorithm is trained, acetone data samples are then tested and finally, the performance of the classifier is calculated to cross-validate the classifier.

4 Results and Discussions

As mentioned in the precious section, two features are extracted from the raw data signals and with the aid of the SVM classifier, the diabetic samples are segregated from the normal samples. Figure 3 shows the scatter plot of all the feature points that is analyzed in this study for data classification. Nearly, 54 data samples were collected from the sensors. Initially, half of the data samples were used to train the classifier and the rest of the samples were utilized to test the classifier. The scatter plot clearly distinguishes the acetone concentrations into the three respective classes.

In the analysis carried out, a multi-level SVM is programmed in MATLAB environment. The classifier mainly function works on the equation given by

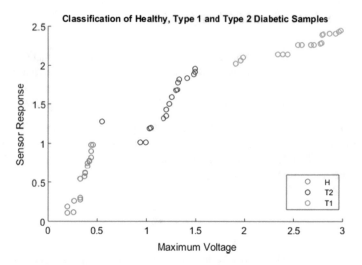

Fig. 3 Scatter plot of the classified data

$$F(x) = \sum_{l=1}^{n} w_i * x_i + b \tag{3}$$

where w_i is the weight vector that signifies the hyperplane which separates the classes and b is the bias. A one versus all SVM approach is used in the multi-level classifier. The approach utilizes the binary classifier to separate each class from the rest. Classification based on different kernels was programmed, and the performance has been evaluated. The different kernels such as the linear, quadratic, polynomial, multilayer perceptron (MLP), and Gaussian radial basis function (RBF) kernels are evaluated. In order to evaluate the performance, three parameters were considered in our analysis which includes the accuracy, sensitivity, and the specificity of the classifier. These parameters were calculated by

$$Accuracy = \frac{TP + TN}{TP + FP + TN + FN} \tag{4}$$

$$Sensitivity = \frac{TN}{TP + FN} \tag{5}$$

$$Specificity = \frac{TN}{FP + TN} \tag{6}$$

where TP indicates the true positive value (that is, the positive class data values which are predicted as positive), TN is the true negative value (negative class data values which are predicted as negative), FP represents the false positive value (positive class

Table 2 Performance evaluation

S. No.	Kernal	Performance in %		
		Accuracy	Sensitivity	Specificity
1	Linear	96.25	97.5	95
2	Quadratic	97.5	100	95
3	Polynomial	97.5	97.5	97.5
4	RBF	98.333	95	100
5	MLP	98.75	100	97.4359

data values which are predicted as negative) while FN indicates the false negative value (negative class data values which are predicted as positive) [11].

Accuracy expresses the percentage of correctly classified features by the total number of classified features. Sensitivity is the fraction of correctly classified positive features that is the healthy samples by the true positive features. Specificity is the fraction of correctly classified negative features that is the diabetic samples by the true negative features. Table 2 compares the performance of the classifier with different Kernels.

5 Conclusion

Initially, the paper investigates the levels of acetone concentrations present in human breath samples for monitoring diabetes. In this analysis, various acetone levels were collected and with the aid of the SVM classifier, the data signals were classified to predict diabetes. The acetone concentrations were initially quantifier with the help of commercially available gas sensor, and these sensor responses were recorded with the help of Arduino board. Next, the samples were classified into three classes labeled as healthy, type-1 diabetes, and type-2 diabetes. The multi-level SVM classifier algorithm was programmed in MATLAB platform, and various kernel functions were programmed to classify the collected acetone samples. Finally, the performance of the classifier is evaluated and a maximum accuracy of 98.75%.

References

1. Alberti KGMM, Zimmet PZ (1998) Denition, diagnosis and classification of diabetes mellitus and its complications. Part 1: diagnosis and classification of diabetes mellitus. Provisional report of a WHO consultation. Diabet Med 15(7):539–553
2. Makaram P, Owens D, Aceros J (2014) Trends in nanomaterial-based non-invasive diabetes sensing technologies. Diagnostics 4(2):27–46
3. Wang P, Tan Y, Xie H, Shen F (1997) A novel method for diabetes diagnosis based on electronic nose. Biosens Bioelectron 12(9):10311036

4. Deng C, Zhang J, Yu X, Zhang X, Zhang X (2004) Determination of acetone in human breath by gas chromatography mass spectrometry and solid-phase microextraction with on-fiber derivatization. J Chromatogr 810:269–275
5. Moorhead K, Lee D, Chase JG, Moot A, Ledingham K, Scotter J, Allardyce R, Senthilmohan S, Endre Z (2007) Classification algorithms for SIFT-MS medical diagnosis. In: Proceedings of the 29th annual international conference of the IEEE EMBS, Cit Internationale, Lyon, France, 23–26 August
6. Wang C, Mbi A, Shepherd M (2010) A study on breath acetone in diabetic patients using a cavity ringdown breath analyzer: exploring correlations of breath acetone with blood glucose and glycohemoglobin A1C. IEEE Sens J 10(1):54–63
7. Guo D, Zhang D, Li N, Zhang L, Yang J (2010) A novel breath analysis system based on electronic olfaction. IEEE Trans Biomed Eng 57(11):2753–2763
8. Lee DS et al (2003) GaN thin films as gas sensors. Sens Actuators B: Chem 89(3):305–310
9. Chang CC, Lin CJ (2011) LIBSVM: a library for support vector machines. ACM Trans Intell Syst Technol 2(3):127
10. Burges CJ (1998) A tutorial on support vector machines for pattern recognition. Data Min Knowl Discov 2(2):121–167
11. Muthuvel K, Suresh LP, Veni SK, Kannan KB (2014) ECG signal feature extraction and classification using harr wavelet transform and neural network. In: 2014 International Conference on circuit, power and computing technologies (ICCPCT), 20 Mar 2014, pp 1396–1399, IEEE

Sound-Based Control System Used in Home Automation

K. Mohanaprasad

Abstract Technological advancements have led to humans using their senses of sight, touch, and speech for various applications, control, and security. Voice control is the most powerful tool as it renders man the power to limitless applications. However, not only do the characteristics of the sound produced change variably across individuals, the characteristics of the same sound produced by one individual on repetition may also vary. Thus, there is the need for a system that can work upon any kind of sound and give the desired output for the user. The purpose of this project is to develop an audio-based application control system which uses speech, claps, snaps or any combination of the three, as inputs. This form of voice control enables users with any amputation or physical disability like blindness or deafness to perform tasks, like switching a fan on/off or sending an emergency text for help, with ease. For this project, we have developed two algorithms on MATLAB that help in detecting peaks in the input audio signal. These peaks help us to arrive at a decision as to which appliance or device is to be controlled. The decision is then sent to an AT89S52, 8051 microcontroller that is connected to a relay module. Thus, multiple appliances like air conditioner, television, telephone, light bulbs can be controlled using voice.

1 Introduction

Majority of the voice-based applications works on a two-step process. The first involves translating the command input into text. As the pitch and intensity of the words produced by each person's vocal tracts are different, the more successful applications are dependent on a massive library of speech recordings. The next step is fairly simpler; an application is executed based on the text deciphered.

K. Mohanaprasad (✉)
SENSE, VIT University, Chennai, India
e-mail: kmohanaprasad@vit.ac.in

© Springer Nature Singapore Pte Ltd. 2018
A. K. Nandi et al. (eds.), *Computational Signal Processing and Analysis*, Lecture Notes in Electrical Engineering 490, https://doi.org/10.1007/978-981-10-8354-9_24

267

But instead of a voice-to-text-based system, a voice control system would be more direct, simple, and elegant. This will provide an advantage to people who are dumb, or are unable to speak properly. For example, a person who lisps will not be able to use a voice-to-text device properly, but he will be able to control the speed of a fan using a voice control system.

This inspired us to create a system that can also be used by people with other types of disabilities like blindness, deafness, or any physical amputations like loss of limbs. These issues cultivated the motivation to develop this project.

The main aim of this project is to design a prototype by which people can do everyday operations (like turning on/off a light/fan) by just giving a voice command. To do this, we must determine the peak impulses in each signal and therefore determine how many input impulses have been given.

2 Literature Survey

The idea behind this project is to remove the need to physically operate device. The user can just give a voice signal in the form of a clap/finger snap to switch on/off a device.

The reason behind this emerged from the fact that people with disabilities in their vision are dependent upon people around them or need to struggle to do simple everyday tasks. In making a voice-controlled device, we remove this dependency. Also, in old age homes, it becomes very difficult for the elderly to walk all the way to the switchboard in order to operate appliances. In such situations, it will make their life very comfortable if they could operate it with just a clap or the snap of a finger.

Some of the applications and publications through which the evolution of the proposed system could be sourced are as follows:

"Design & Implementation of Smart House Control Using LabVIEW," which was published in 2012 in the International Journal of Soft Computing and Engineering [1], describes the hardware implantation of a smart home system using the LABview software. It clearly presents the true essence and aim of home automation, i.e., the consolidation of control or monitoring signals from appliances, fittings, or basic services.

In the paper titled "Voice Recognition Wireless Home Automation System Based On Zigbee" published in "IOSR Journal of Electronics and Communication Engineering" in 2013 [2], the author designs a wireless system based on voice recognition which uses Zigbee, an IEEE 802.15.4 standard for data communications with business and consumer devices. The system requires low data rate and power consumption, but works best only for the old and disabled who live alone and necessitate a secure network. Also, it works on the concept of a Zigbee network

[3] which requires us to train the words for recognition, test the recognition software, detect errors in the seven-segment display used (which happens when voice commands are too short or too long), clearing memory from time to time so as to avoid overload of data, finding results, uploading the results from host side, downloading them on the user side, resetting data, etc. As it can be seen, this is a very long and tedious process which is neither time effective nor efficient in reaching our end objective.

In the same year, another paper was published titled "Low-Cost Self-assistive Voice-Controlled Technology for Disabled People" in the International Journal of Modern Engineering Research [4]. In this paper, the author describes the design of a voice-controlled wheelchair [5] for the elderly and the physically challenged. However, the voice recognition system needs a library so that the input voice command is matched with every keyword present in the archive [6, 7].

The paper "A Voice-Input Voice-Output Communication Aid for People With Severe Speech Impairment," published in the IEEE Transactions On Neural Systems And Rehabilitation Engineering, VOL. 21, NO. 1, JANUARY 2013 [8], describes a voice-input voice-output communication aid for people who have moderate-to-severe dysarthria, the most common speech disorder. The device designed helps produce intelligible speech output from highly disordered speech. It also requires considerably low training data. However, the code is of high computational complexity and the device is not cost efficient as it requires a laptop or a personal data assistant.

In the paper titled "Voice Command-Based Computer Application Control Using MFCC" published in International Journal of Innovative Research in Science, Engineering and Technology in 2015, the voice control system requires testing of the acquired voice with the trained voice using MFCCs for speaker recognition. This system shall however not work for the deaf-mute population.

Till date, the existing voice control systems are constrained in many ways [9–11]. So far no system exists that supports universal voice commands on a device able to control any application. The most technological advanced voice assistants like Siri and Google only support built-in functions like messaging, calling, Web searching.

The major hurdle we encountered in this project was to develop an algorithm that would take in any form of voice or sound created by the individual and performs the required action. So, we worked backward from what we required and felt that calculating the peaks of a signal would simplify our process. We determined ways in which we could simplify the process of finding the peaks. So, we down-sampled our initial signal, considered only the absolute values, and set a threshold value so as to eliminate all other samples.

The algorithm is designed in a way to avoid the need for a library as that would not only lead to computational complexities, but also require a lot of time and funding. Thus, the resulting project should be efficient and cost-effective at the same time.

3 Methodology

3.1 Form of Audio Input

Human voice can be simply defined as the sound produced using the vocal folds for talking, screaming, singing, crying, etc. The characteristics like amplitude, frequency, and phase differ across voices of different human beings. Because of this uniqueness, voice can be used in security measures, like user identification in an automated teller machine (ATM). However, if a person says "Hello!" five times in a row, there will be differences in the pitch, loudness, and frequency. This establishes the fact that setting parameters dependent on the human voice and deriving accurate and precise results from them is a very straining task. The reason Google is able to search through voice is because that powerful search engine has 7000 h of recording with which it can compare the input audio signal and accordingly deliver the search results. At present, very few voice-based applications have their game going.

Secondly, the contribution of noise by the environment plays a crucial role in making voice-based applications really tough. Noise varies from place to place and adds unpredictable fluctuations.

Moreover, during speaker recognition or speech reconstruction that uses the mel frequency cepstral coefficients (MFCCs), a lot of training is done on the vector pools like speaker adaptation for instance. Even though working with the help of feature vectors like MFCCS, linear predictive coding coefficients and filter banks produce more desirable results, the complexity for coding increases drastically.

Even if these problems were tackled, secondary issues like human pronunciation, presence of a huge number of languages, dialects pose a problem for system designing. And even if a universal accepted system based on voice control was developed, what about the people who are unable to speak at all?

Therefore, speech should not be the only way to produce an audio input for the system. Any form of sound, in the form of non-speech, claps, or snaps, should be used to operate the appliance or perform the required application.

3.2 Parameter Extraction

It is easier to work upon signals made of the same sound than on processing signals of different sounds. These sounds can be either speech: word or letter, non-speech like "la," or a clap. The challenge here lies in extracting the number of peaks from the audio input signal. The simplest approach would be to use the command "findpeaks" in MATLAB to find out the spikes in the signal. However, that would also include the peaks contributed by noise. Another approach would be to pre-define a threshold value, and whenever the signal has amplitude greater than the threshold, a counter would accordingly increment. However, the value of the threshold will be dependent on the system and surroundings. Its value will be higher in a noisy environment, and vice versa in a quiet neighborhood.

These complications have been tackled using two different approaches. The two algorithms corresponding to these approaches shall be discussed in the next chapter of design and implementation.

4 Design

4.1 Software

Two separate algorithms have been developed to achieve the same objective, i.e., finding the number of peaks in the audio input signal. Once the number is derived, the corresponding command is delivered to the microcontroller. After receiving the character, the microcontroller switches the corresponding appliance on/off using relays. A hex file is created using Keil that helps in sending a character to the AT89S52.

Algorithm One:

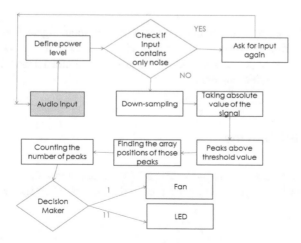

Algorithm Two:

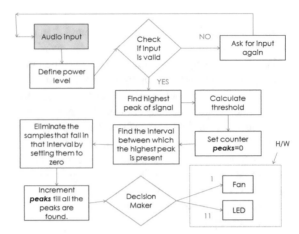

The sound is first processed in the software used on the user's PC and a decision is made according to one of the two algorithms used. The decision is then converted into a language that is understandable by the microcontroller using Keil.

This signal is then transmitted in a bit-wise manner through the USB-to-TTL converting device which is attached to the PC's USB port. The signal is received by the microcontroller using ports P3.0 and P3.1. Once the signal is received, it passes on the signal to the corresponding relay pin depending on the way it has been programmed. (P3.0 → Relay 1 → Coded as "Q" using Keil.)

Depending on what decision has been made, either relay 1 or relay 2 will be activated and the corresponding device is switched on/off.

5   Results and Analysis

The project design in theory is based on receiving input signal from user end and the output being received in the form of either the LED or fan switching on/off. Now, let us take a look at each step of the process and understand why each step is essential to achieving our final target.

5.1   Algorithm One

We will first take a look at the working of the first algorithm.

The input is first provided by the user in the form of either finger snaps or claps as shown.

Fig. 1 Input audio signal of clap

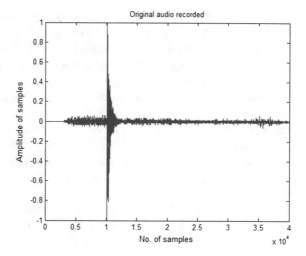

From Fig. 1, it can clearly be made out that the input given is a single snap or clap. So, the samples in that particular sample space show high-amplitude values.

But there are too many samples to evaluate in order to find the number of peaks in the signal. Hence, to make the processing of the signal a little easier, we then down-sample the sample space by providing a step size of 32.

As can be inferred from Fig. 2, the number of samples has been reduced to 1250 samples (40,000/32).

It is interesting to note at this point that the highest peak from the set of impulses initially obtained has been lost due to down-sampling of the data. So, there might be questions raised regarding the integrity of the result displayed.

Fig. 2 Down-sampling on audio input

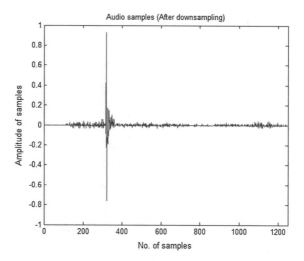

To find out why this algorithm works despite the loss of critical data, we need to understand our final objective behind this algorithm. The idea is to define a peak such that it is above a particular threshold value of amplitude. Also, it is important to understand that when a clap or snap is made, it might seem that there should only be one impulse shown in our graph.

While it is true in an ideal scenario where our clap lasts for just a time period large enough to be a part of a single simple, practically speaking, when a clap is made, a bunch of impulses are generated very close to one another.

This goes to show that despite having lost out on the highest peak, the data obtained manages to capture a peak which is above the pre-set threshold value therefore leading to no loss in information and maintaining the integrity of the result.

We then find the absolute value of our down-sampled signal as shown below. This is done so as to ensure that our processing is more convenient (Fig. 3).

Next, we set a threshold value above which any sample recorded is considered a peak.

In our design, we have set the threshold to be half of the highest peak recorded.

If we notice the previous graph, it is fairly evident that there is no peak that is greater than half of the highest peak. This is shown in the final step of the program (Fig. 4).

Following this, the software takes a decision to be "1" meaning that the command to be sent to the microcontroller would be to switch on the fan (Fig. 5).

Now, in our second algorithm, we follow a slightly different approach.

We first find the highest peak in the spectrum. Now, we set half the maximum amplitude to be the threshold value above which a sample can be categorized as a peak.

Now, we first take the input from the user.

As we see from (Fig. 6), the number of peaks that were given by the user is 3.

Fig. 3 Absolute value of input

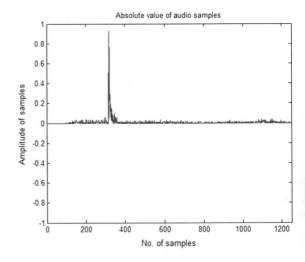

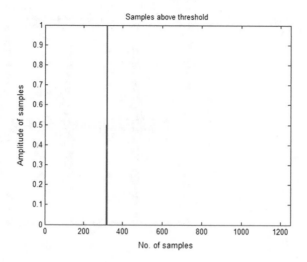

Fig. 4 Sample positions above threshold

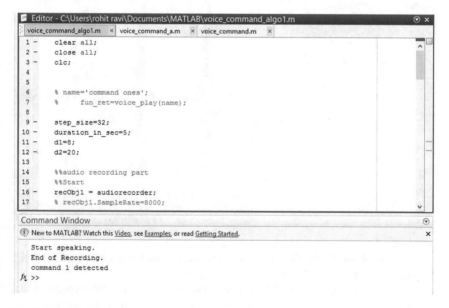

Fig. 5 Output window

Now, to find this out using the software, we find each peak and then remove it while raising a counter to calculate peaks.

The conditions for the peak are set as greater than half the highest sample amplitude. By doing so, we can eliminate all external noises from the program.

Fig. 6 Audio input from
user: three ones ("1...1...1")

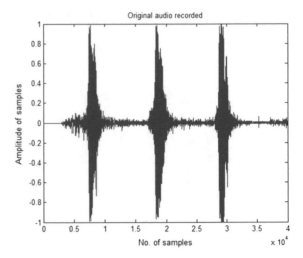

Fig. 7 Signal after
elimination of one interval

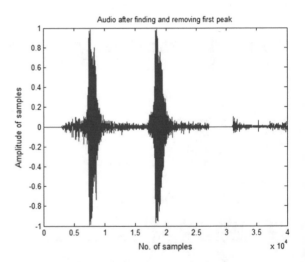

Now, as we see in Fig. 7, the peak is calculated and is then removed so that it does not cause errors in calculation of total peaks.

If we notice carefully, we will see that not only the peak has been removed, but also a bunch of samples to the left and right of the peak has been removed.

This is done under the assumption that we would not be able to give in voice inputs in such quick duration. (4000 samples would mean one-tenth of a second; this shows that unless consecutive inputs are given in under one-tenth of a second, the final output will continue to be in integrity.)

This procedure is followed until we have only one peak left as shown (Fig. 8).

Once we have only one peak left, the software includes the last peak and reaches at a final peak count.

Fig. 8 Removal of interval in second iteration

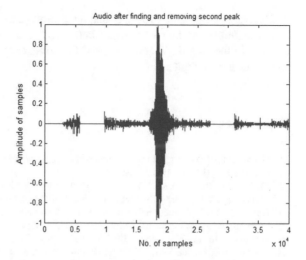

Now, once the decision has been reached, it needs to be transmitted onto the microcontroller.

So, in order to interface the program to the microcontroller, we use a software called "Keil" to convert the data into a form which the microcontroller can understand.

The hex files are then generated and burnt onto the microcontroller. The microcontroller is connected to the PC using a USBB-to-TTL converter which is a device which provides serial, asynchronous communication between the PC and the microcontroller.

Once the decision is passed onto the microcontroller, it further passes the information onto the relay module which is connected to it.

The relay module will then switch on/off the corresponding appliance it is connected to, based on the decision it receives from the microcontroller and the current state of the appliance.

6   Conclusion

In this project, we have created a direct voice to control system which a person having any form of physical disability or amputation can use. Whether the person has a speech impairment, is a quadriplegic, or suffers from ALS disorder, he or she will be able to control any appliance like opening the fridge or perform applications like sending a text with the use of any form of voice: speech, clap, snap, etc.

Not only does the code have low computational complexity, but also can be adapted on to various platforms like Android for mobile functions. The system's simple design makes it cost-effective, thereby making it feasible to be extrapolated to larger markets. This concept can be extended to network security purposes as well.

Lastly, we can conclusively say that the prototype design, although being simple and straightforward in nature when compared to the existing models, is highly efficient in meeting the requirements of the common people by being effective as well as cheap to implement.

References

1. Hamed B (2012) Design & implementation of smart house control using LabVIEW. Int J Soft Comput Eng (IJSCE) 1(6):98–105
2. Thakur DS, Sharma A (2013) Voice recognition wireless home automation system based on Zigbee. IOSR J Electron Commun Eng (IOSR-JECE) 6(1):65–74
3. Devi YU (2012) Wireless home automation system using Zigbee. Int J Sci Eng Res 3:1–5
4. Puviarasi R, Ramalingam M, Chinnavan E (2013) Low cost self-assistive voice controlled technology for disabled people. Int J Mod Eng Res (IJMER) 3:2133–2138
5. Simpson RC, Levine SP (2002) Voice control of a powered wheelchair. IEEE Trans Neural Syst Rehabil Eng 10(2):122–125
6. Laisa CP, Almeida NS, Correa AGD, Lopes RD, Zuffo MK (2013) Accessible display design to control home area networks. IEEE Trans Consum Electron 59(2):422–427
7. Gnanasekar AK, Jayavelu P, Nagarajan V (2012) Speech recognition based wireless automation of home loads with fault identification. In: IEEE international conference on communications and signal processing (ICCSP), vol 3. pp 128–132
8. Hawley MS, Cunningham SP, Green PD, Enderby P, Palmer R, Sehgal S, O'Neill P (2013) A voice-input voice-output communication aid for people with severe speech impairment. IEEE Trans Neural Syst Rehabil Eng 21(1):23–31
9. Baig F, Beg S, Khan MF (2013) Zigbee based home appliances controlling through spoken commands using handheld devices. Int J Smart Home 7(1):19–26
10. Jeyasree T, Gayathri I, Kamin Uttamambigai SP (2013) Implementation of wireless automation of home loads based on microsoft SDK. Int J Sci Technol 2(5):402–404
11. Anamul Haque SM, Kamruzzaman SM, Ashraful Islam M (2006) A system for smart-home control of appliances based on timer and speech interaction. In: Proceedings of the 4th international conference on electrical engineering & 2nd annual paper meet, vol 2. Jan 2006, pp 128–131

An Empirical Mode Decomposition-Based Method for Feature Extraction and Classification of Sleep Apnea

A. Smruthy and M. Suchetha

Abstract Background: Sleep apnea is a breathing disorder found among thirty percentage of the total population. Polysomnography (PSG) analysis is the standard method used for the identification of sleep apnea. Sleep laboratories are conducting this sleep test. Unavailability of sleep laboratories in rural areas makes the detection difficult for ordinary people. There are different methods for detecting sleep apnea. Past researches show that electrocardiogram-based detection is more accurate among other signals. This paper investigates the idea of electrocardiogram (ECG) signals for the recognition of sleep apnea. Methods: In this paper, the classification of healthy and apnea subjects is performed using electrocardiogram signals. The proper feature extraction from these signal segments is executed with the help of empirical mode decomposition (EMD). EMD algorithm decomposes the incoming signals into different intrinsic mode functions (IMFs). Four morphological features are extracted from these IMF levels. These features include the morphological characteristics of QRS complex, T and P waves. The classification of healthy and apnea subjects is done using the machine learning technique called support vector machine. Result: All the experiments are carried out by using St. Vincents University Hospital/University College Dublin Sleep Apnea Database (UCD database). This database is available online in physionet. It is observed from the results that by using empirical mode decomposition; it could be possible to extract the proper morphological features from this ECG segments. This technique also enhances the accuracy of the classifier. The overall sensitivity, specificity, and accuracy achieved for this proposed work are 90, 85, and 93.33%, respectively.

Keywords Sleep apnea · Polysomnography · Electrocardiogram Empirical mode decomposition · Empirical mode Functions

A. Smruthy (✉) · M. Suchetha
School of Electronics Engineering, VIT University, Chennai Campus,
Chennai, India
e-mail: smruthy.a2014@vit.ac.in

M. Suchetha
e-mail: suchetha.m@vit.ac.in

© Springer Nature Singapore Pte Ltd. 2018
A. K. Nandi et al. (eds.), *Computational Signal Processing and Analysis*, Lecture Notes in Electrical Engineering 490,
https://doi.org/10.1007/978-981-10-8354-9_25

1 Introduction

Sleep-disordered breathing is common nowadays. Obstructive sleep apnea is a kind of sleep disorder in which the partial or complete collapsing of upper airway occurs. As a result, a reduction in blood oxygen saturation develops and which leads to the sudden awakening from the sleep. The common signs of sleep apnea are the excessive daytime sleepiness and snoring. This condition is hereditary for some people. Most of the time the person is not able to recognize the condition as it is occurring at night times. The unidentified sleep apnea condition can lead to the memory loss, heart attack, and stroke or even accidents. Polysomnography (PSG) analysis is the standard method used for the identification of sleep apnea. In this method, the extraction of physiological signals such as ECG, EMG, EOG, SpO2, and EEG is done with the help of a trained technician. This method is time consuming and expensive. Obstructive sleep apnea can be treated by using continuous positive airway pressure. In this method, a predefined quantity of pressure is continuously applying through the nose. This method avoids the collapsing of the upper airway. Features from oronasal signals [1–4], photoplethysmography and actigraphy [2, 5–9] are also utilized for investigating sleep apnea. Bsoul et al. developed a technique for the detection of sleep apnea-based on electrocardiogram signals [10]. A new feature called ECG-derived respiratory information is used for the classification of healthy and apnea subjects. Support vector-based machine learning technique is used for classification. Since this method is more accurate our proposed method also using ECG signal features for the classification of healthy and apnea subjects.

2 Materials and Methods

The proposed method consists of four stages. In the first stage, the incoming ECG signal segments are decomposed into different intrinsic mode functions by using empirical mode decomposition. Next step is the selection of appropriate features from these decomposed levels. Selections of the morphological features are done by taking the spectrum of the decomposed levels. ECG signal frequency is limited to 0–300 Hz. The decomposed levels, which belong to the frequency of ECG signals, are selected for further analysis. These selected levels are added together to form the reconstructed signal. Next phase includes the feature extraction from this reconstructed signal. Four features are extracted from this reconstructed signal. These four features are the R peak value, RR interval, peak values of P and T waves. The final stage is the classification of healthy and apnea subjects using support vector classifier. Figure 1 represents the block diagram of the proposed method.

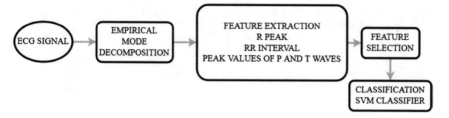

Fig. 1 Overview of the proposed work

2.1 Empirical Mode Decomposition

Empirical mode decomposition (EMD) [11, 12] is mainly used for the decomposition of an input signal adaptively in a group of AM–FM components. The Fourier or wavelet-based data representation methods usually require prior information about the basis functions for the representation of a signal. In EMD-based data representations, there is no need of a prior knowledge about the basis function. EMD is specially used for the signals which have a nonlinear and a non-stationary nature. In EMD, the signal is decomposed into various intrinsic mode functions. Each IMF level obtained after the EMD decomposition has the same length as the original signal. Each of these signals is represented in the time domain. EMD is useful for the events which may happen in a specific interval of time. This kind of information is obvious in an EMD investigation, however very covered up in the Fourier space or in wavelet coefficients. The steps involved in the decomposition are [11]

- Consider input signal as $x(t)$
- Identify the local maxima, local minima of the signal and connect them through a cubic spline [13].
- Compute the mean of these two envelops $m(t)$ [13]. IMF calculation $h(t) = x(t) - m(t)$ [13].
- Repeating the iteration on the residual $m(t)$ such that no IMF can be extracted further and represents a monotonic function [13].
- Finally, the input signal can be written as a sum of IMFs and residue.

-
$$x(t) = \sum_{k=1}^{n} C_k(t) + R(t) \qquad (1)$$

where $C_k(t)$ represents the sum of the intrinsic mode functions and $R(t)$ represents the residual part.

Table 1 Normal values of the ECG signal features [12]

S. No.	Features	Amplitude (mV)	Duration (ms)
1	R peak	1	80–120
2	RR interval	–	0.4–0.2 s
3	P wave	01–0.2	60–80
4	T wave	0.1–0.3	120–160

2.2 Feature Extraction and Feature Selection

The proposed work includes four features [14]. These four features are the R peak value, RR interval, peak values of P, and T waves. The normal values of these features are represented in Table 1. The mean value of all these features is used for the classification of healthy and apnea subjects. Features are selected based on the performance obtained in the classification phase.

2.3 Classification

Classification of healthy and apnea subjects is performed using support vector machine (SVM) classifier [15]. This classification is based on the supervised learning method. SVM classifier is used for the classification, regression, and outlier detection. The main aim of this proposed work is to classify the healthy and apnea subjects using support vector machine classifier.

3 Result and Discussion

Twenty-five subject's overnight PSG recordings are collected from St. Vincents University Hospital/University College Dublin Sleep Apnea Database (UCD database). This database is available online in physionet. MATLAB software is used for simulation works. Each subject's sampled ECG signal is decomposed into various intrinsic mode functions with the help of the empirical mode decomposition algorithm. Spectrum of these decomposed levels is calculated with the help of Fast Fourier Transform (FFT). The selections of appropriate IMFs are done by comparing the ECG signal frequency 0–300 Hz with the decomposed spectrum. The levels which fall into the range of ECG signal frequency are selected for the reconstruction of the input signal. The different decomposed levels are shown in Fig. 2. The selected levels added together to form the reconstructed signal. After reconstruction, four features are selected from this reconstructed signal. Peak detection from this reconstructed signal is performed by comparing with the adjacent sample values of the reconstructed signal. R wave peak value, RR interval,

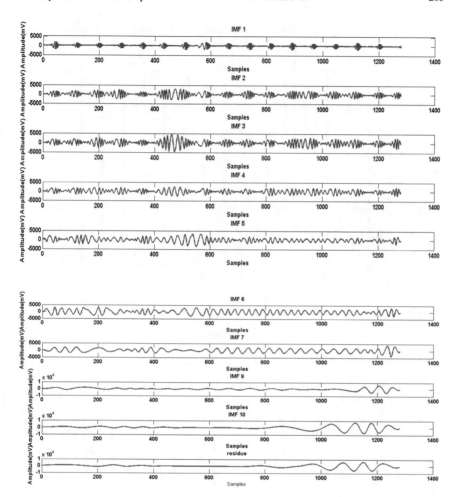

Fig. 2 Different intrinsic mode functions obtained after empirical mode decomposition

Peak values of P and T waves are calculated. The extracted features are represented in Fig. 3.

The mean value of these features is estimated and these values are given as the input of SVM classifier. Figure 4 shows the classified result of healthy and apnea subjects.

The performance of the proposed work is evaluated based on the sensitivity, specificity, and accuracy of the classifier. All these performance parameters are calculated as

$$\text{Sensitivity} = \frac{\text{TP}}{\text{TP} + \text{FN}} \tag{2}$$

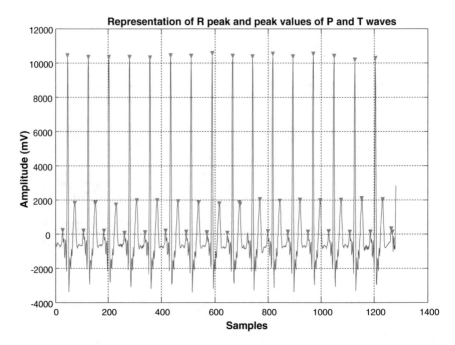

Fig. 3 Representation of R peak and peak values of P and T waves

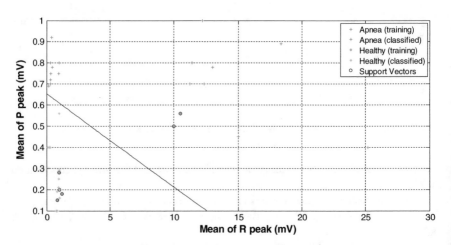

Fig. 4 Classified output of healthy and apnea subjects using support vector machine classifier

$$\text{Specificity} = \frac{\text{TN}}{\text{TN} + \text{FP}} \tag{3}$$

$$\text{Accuracy} = \frac{TN + TP}{TN + TP + FN + FP} \qquad (4)$$

where TP, TN, FP, FN are true positive, true negative, false positive, false negative estimations, respectively. The overall sensitivity, specificity, and accuracy obtained after the classification are 90, 85, and 93.33%, respectively. The results show that by using empirical mode decomposition it could be possible to extract the proper features from the ECG signal segments. This method enhances the accuracy of the proposed method comparing with the other methods.

4 Conclusion

The proposed work represents the classification of healthy and apnea subjects by using ECG signal features with empirical mode decomposition. In this work, the EMD algorithm helps to extract the appropriate features from the ECG signal segments. Out of the four features two features are selected for the classification since these features perform well comparing to the other features. The sensitivity, specificity, and accuracy obtained for this work are 90, 85, and 93.33%, respectively.

References

1. Koley BL, Dey D (2013) Real-time adaptive apnea and hypopnea event detection methodology for portable sleep apnea monitoring devices. IEEE Trans Biomed Eng 60 (12):3354–3363
2. Ciołek M et al (2015) Automated detection of sleep apnea and hypopnea events based on robust airflow envelope tracking in the presence of breathing artifacts. IEEE J Biomed Health Inform 19(2):418–429
3. Jin J, Sánchez-Sinencio E (2015) A home sleep apnea screening device with time-domain signal processing and autonomous scoring capability. IEEE Trans Biomed Circ Syst 9(1):96–104
4. Gutierrez´-Tobal GC, Alvarez D, del Campo F, Hornero R (2016) Utility of adaboost to detect sleep apnea-hypopnea syndrome from single-channel airflow. IEEE Trans Biomed Eng 63(3):636–646
5. Karmakar C, Khandoker A, Penzel T, Schobel C, Palaniswami M (2014) Detection of respiratory arousals using photoplethysmography (ppg) signal in sleep apnea patients. IEEE J Biomed Health Inform 18(3):1065–1073
6. Domingues A, Paiva T, Sanches JM (2014) Sleep and wakefulness state detection in nocturnal actigraphy based on movement information. IEEE Trans Biomed Eng 61(2):426–434
7. Mora GG, Kortelainen JM, Hernandez EPR, Tenhunen M, Bianchi AM, Méndez MO, Guillermina Guerrero (2015) Evaluation of pressure bed sensor for automatic SAHS screening. IEEE Trans Instrum Meas 64(7):1935–1943
8. Ahmad S, Batkin I, Kelly O, Dajani HR, Bolic M, Groza V (2013) Multiparameter physiological analysis in obstructive sleep apnea simulated with Mueller maneuver. IEEE Trans Instrum Meas 62(10):2751–2762

9. Hwang SH, Lee HJ, Yoon HN, Lee YJ, Lee YJ, Jeong DU, Park KS et al (2014) Unconstrained sleep apnea monitoring using polyvinylidene fluoride film-based sensor. IEEE Trans Biomed Eng 61(7):2125–2134

10. Bsoul M, Minn H, Tamil L (2011) Apnea medassist: real-time sleep apnea monitor using single-lead ecg. IEEE Trans Inf Technol Biomed 15(3):416–427

11. Huang NE, Shen Z, Long SR, Wu MC, Shih HH, Zheng Q, Yen NC, Tung CC, Liu HH (1998) The empirical mode decomposition and the hilbert spectrum for nonlinear and non-stationary time series analysis. In: Proceedings of the royal society of London A: mathematical, physical and engineering sciences, vol 454. The Royal Society, pp 903–995

12. Xie G, Zhang B, Wang J, Mao C, Zheng X, Cui Y, Gao Y (2010) Research on the propagation characteristic of non-stationary wind power in microgrid network based on empirical mode decomposition. In: 2010 international conference on power system technology (POW-ERCON), pp 1–6, IEEE

13. Li H, Feng X, Cao L, Liang H, Chen X (2016) A new ecg signal classification based on WPD and ApEn feature extraction. Circ Syst Sign Process 35(1):339–352

14. Sivaranjni V, Rammohan T (2016) Detection of sleep apnea through ecg signal features. In: 2016 2nd international conference on advances in electrical, electronics, information, communication and bio-informatics (AEEICB), pp 322–326, IEEE

15. Khandoker AH, Palaniswami M, Karmakar CK (2009) Support vector machines for automated recognition of obstructive sleep apnea syndrome from ecg recordings. IEEE Trans Inf Technol Biomed 13(1):37–48

Kapur's Entropy and Active Contour-Based Segmentation and Analysis of Retinal Optic Disc

D. Shriranjani, Shiffani G. Tebby, Suresh Chandra Satapathy, Nilanjan Dey and V. Rajinikanth

Abstract Retinal image scrutiny is essential to detect and supervise a wide variety of retinal infections. Segmentation of region of interest (ROI) from the retinal image is widely preferred to have a clear idea about the infected section. In the proposed work, a new two-stage approach is presented for automatic segmentation of the optic disc (OD) in retinal images. This approach includes the chaotic bat algorithm (CBA) assisted Kapur's multi-thresholding as the preprocessing stage and active contour (AC) segmentation as the post-processing stage. This method initially identifies the suitable value of threshold to enhance the OD in the chosen retinal image. The enhanced OD is then processed using the gray scale morphological operation, and finally, the OD is extracted using AC segmentation process. To test the proposed approach, optic disc images of different category are acquired from the RIM-ONE database. Experimental results demonstrate that the average Jaccard index, Dice coefficient, precision, sensitivity, specificity, and accuracy are greater than 83.74, 93.66, 98.18, 92.85, 98.43, and 97.28%, respectively. Hence, the proposed work is extremely significant for the segmentation of OD and can be used as the automated screening tool for the OD related retinal diseases.

Keywords Retinal images · Optic disc · Bat algorithm · Active contour
Performance evaluation

D. Shriranjani (✉) · S. G. Tebby · V. Rajinikanth
Department of Electronics and Instrumentation, St. Joseph's College of Engineering,
Chennai 600119, Tamil Nadu, India
e-mail: ranjani_shri@rediffmail.com

S. C. Satapathy
Department of Computer Science and Engineering, P.V.P. Siddhartha Institute
of Technology, Vijayawada 520007, Andhra Pradesh, India

N. Dey
Department of Information Technology, Techno India College of Technology,
Kolkata 700156, India

© Springer Nature Singapore Pte Ltd. 2018 287
A. K. Nandi et al. (eds.), *Computational Signal Processing
and Analysis*, Lecture Notes in Electrical Engineering 490,
https://doi.org/10.1007/978-981-10-8354-9_26

1 Introduction

The deepest coating of the eye is retina, and computerized inspection of essential retinal regions is extremely significant to determine, examine and to offer efficient healing for eye illness. Normally, retinal pictures are widely considered in identification and handling of retinal vasculature [1–3], diabetic retinopathy [4], ocular hypertension (OHT) [5], optic nerve disorder [5], macular edema [6], glaucoma [7], and cardiovascular diseases [8].

Due to its importance, retinal image examination is widely discussed by the most of the researchers [9–12]. Lesion, structural and bodily deviation in optic disc, and retinal vessels are largely scrutinized to perceive a range of infection, and this difference comprises the modification in silhouette, dimension, and exterior of the retinal sections. Hence, optic disc segmentation is an important task in retinal disease examination.

The objective of this paper is to develop an automated segmentation tool for the optic disc examination. This paper proposes a novel approach with chaotic bat algorithm (CBA) assisted Kapur's multi-thresholding and active contour segmentation (ACS) for the optic disc extraction from digitized retinal images.

Proposed approach is tested on the following category of the optic disc images, such as normal (114), deep (14), moderate (12), early stage (12), and OHT (11) found in RIM-ONE dataset [13, 14]. The proposed work is executed using above said dataset, and the mined disc is validated against the ground truth. Results of this work verify that implemented method is competent in mining the disc of retina.

2 Methodology

The literature offers a variety of image thresholding schemes [15–17]. In the proposed work, heuristic algorithm supported scheme is considered to mine the optic disc from the retinal image. Initially, a preprocessing work is executed on the image to improve its visibility, and a post-processing action is carried out to extract the optic disc. Finally, the mined optic disc is compared with expert's ground truth and the vital parameters such as the image similarity index and the statistical parameters are computed.

2.1 Bat Algorithm

The bat algorithm (BA) is the one of the most triumphant nature inspired algorithm by Yang [18, 19]. In the literature, traditional and improved forms of BA are successfully applied to solve image processing tasks [20, 21]. In this paper, modern version of the BA, known as the Chaotic BA (CBA) discussed in [22], is adopted.

In CBA, the new location of the BA is reorganized based the following expression:

$$X_{new} = X_{old} + (A^t \oplus IM) \tag{1}$$

where X_{new} is the reorganized location of bats, X_{old} is the preliminary location, A is the sound intensity, and IM is chaotic operator known as Ikeda Map [23].

In this work, the CBA parameters are assigned as follows:

Total number of agents = 30, number of runs = 1000, search dimension = 3, sound intensity varies from 1 to 20 insteps of 0.05, and the IM parameter is assigned based on [22, 23].

2.2 Kapur's Entropy

Image thresholding based on Kapur's entropy was firstly discussed in 1985 to examine the gray scale picture based on the entropy of histogram [24]. This procedure finds the optimal threshold (T) based on its maximized overall entropy.

Let $T = [t_1, t_2, ..., t_{k-1}]$ is a vector of the image thresholds.

Then, Kapur's entropy can be expressed as;

$$J_{max} = f_{kapur}(T) = \sum_{j=1}^{k} H_j^C \quad \text{for } C\{1, 2, 3\} \tag{2}$$

Normally, entropy is calculated separately based on its individual threshold (t) value. Other details regarding Kapur's function are comprehensively discussed in [24, 26, 27].

2.3 Morphological Function

Image morphology functions are normally adopted to improve the visual appearance. The basic morphology operations, such as line structuring element (*strel*) and image fill (*imfill*), are considered to improve the edges and appearance of multi-threshold retinal image.

2.4 Active Contour Segmentation

Active contour segmentation (ACS) is used to extract essential region from pre-processed image [28, 29]. In ACS, an adjustable snake is chosen to follow identical pixel groups available in the image with respect to minimal energy scheme.

Snake's energy value is expressed as:

$$\frac{\min}{C}\left\{E_{\mathrm{GAC}}(C) = \int_0^{L(C)} g(|\nabla I_0 C(s))|\,\mathrm{d}s\right\} \tag{3}$$

where ds—the Euclidean element of length and $L(C)$—the span of curve C which assures $L(C) = \int_0^{L(C)} \mathrm{d}s$. The constraint g is a border pointer, and it may vanish based on the object frontier shown below:

$$g(|\nabla I_0|) = \frac{1}{1 + \beta|\nabla I_0|^2} \tag{4}$$

where I_0 denotes base picture and β is a random numeral. The energy function quickly diminishes based on boundary.

Scientifically, it is represented as:

$$\partial_t C = \left(kg - \langle \nabla_g, N \rangle\right)N \tag{5}$$

where $\partial_t C = \partial C/\partial t$ denotes the twist of snake, t—time of iteration, and k, N are the bend and normal for the snake. Now, the snake's structure is constantly adjusted till reaching E_{GAC}.

2.5 Performance Evaluation

Performance of medical image segmentation procedures is regularly established using a relative investigation with the expert's ground truth (GT). In this paper, performance of the proposed work is evaluated by computing the picture quality measures like Jaccard, Dice, false positive (FP) rate, and false negative (FN) rate [30] and also the statistical measures, such as precision, sensitivity, specificity, and accuracy existing in the literature [31].

3 Result and Discussion

The experimental results of the implemented procedure are presented in this section. Figure 1 depicts the overview of the heuristic algorithm-based retinal image segmentation procedure, and their outcomes are presented in Fig. 2.

The proposed approach is demonstrated using the deep category of retinal optic disc existing in RIM-ONE. Initially, the RGB scaled test data are preprocessed using a tri-level thresholding procedure using the CBA and Kapur's entropy, and the result is shown in Fig. 2b. Thresholding will enhance the optic disc by

Fig. 1 Block diagram of the proposed segmentation scheme

Fig. 2 Test data and its
corresponding outcome

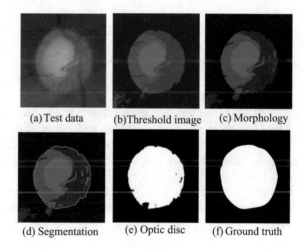

(a) Test data   (b) Threshold image   (c) Morphology

(d) Segmentation   (e) Optic disc   (f) Ground truth

eliminating the unwanted blood vessel from the test image. Later, the traditional morphological operation is implemented to obtain a smooth retinal optic as presented in Fig. 2c. Finally, the ACS scheme is implemented to mine the optic disc illustrated in Fig. 2d. The extracted optic disc depicted in Fig. 2e is then validated against the ground truth shown in Fig. 2f. For RIM-ONE dataset, each image has five ground truths offered by the experts. and the extracted optic disc should be individually evaluated with each ground truth.

Tables 1 and 2 present the image similarity measure values and statistical parameters for the considered test image depicted in Fig. 2a. Table 1 presents the details, such as Jaccard, Dice, FP, FN, optic disc (OD) pixels, and GT pixels. Similarly, Table 2 shows the values of precision, sensitivity, specificity, and accuracy for all the five ground truths, and the average values confirm that the

Table 1 Image similarity values of the deep test image

GT	Jaccard	Dice	FP	FN	OD pixels	GT pixels
GT1	0.8424	0.9144	0.1581	0.0244	20,921	18,454
GT2	0.8797	0.9360	0.0348	0.0897		22,135
GT3	0.8996	0.9471	0.0491	0.0562		21,071
GT4	0.8856	0.9393	0.0902	0.0345		19,817
GT5	0.8207	0.9015	0.2058	0.0105		17,502
Avg	0.8656	0.9277	0.1076	0.0431	20,921	19,796

Table 2 Image statistical measures of the deep test image

GT	Precision	Sensitivity	Specificity	Accuracy
GT1	0.9920	0.9295	0.9808	0.9548
GT2	0.9620	0.9779	0.9243	0.9507
GT3	0.9791	0.9714	0.9561	0.9637
GT4	0.9875	0.9529	0.9721	0.9624
GT5	0.9971	0.9158	0.9926	0.9534
Avg	0.9835	0.9495	0.9652	0.9570

proposed approach is efficient in offering better segmented optic disc with improved quality parameters.

Similar procedure is implemented for all the test images existing in the RIM-ONE database, such as normal, early, moderate, and OHT, and its results are presented in Table 3. This table shows the sample of the test image picked from the normal, early, moderate, and OHT dataset and the corresponding preprocessed and post-processed output.

Table 4 depicts the average value of the optic disc similarity measures for the entire test database with the ground truths. From this table, it can be noted that average Jaccard is greater than 83.74%, and the Dice coefficient is higher than

Table 3 Sample results from the test image dataset

	Test image	Segmented image	Morphology	Optic disc
Normal				
Early				
Moderate				
OHT				

Table 4 Average similarity measures for the test dataset

Image	Jaccard	Dice	FP	FN
Deep	0.8814	0.9506	0.0277	0.0629
Normal	0.8374	0.9383	0.0149	0.0433
Early	0.8614	0.9528	0.0164	0.0385
Moderate	0.8572	0.9473	0.0182	0.0398
OHT	0.8395	0.9366	0.0178	0.0305

Table 5 Average statistical measures for the test dataset

Image	Precision	Sensitivity	Specificity	Accuracy
Deep	0.9893	0.9604	0.9875	0.9733
Normal	0.9816	0.9632	0.9843	0.9728
Early	0.9874	0.9544	0.9856	0.9761
Moderate	0.9959	0.9285	0.9972	0.9829
OHT	0.9927	0.9317	0.9868	0.9808

93.66%. Similarly, FP and FN ratios are completely negligible. This table also confirms that the OD and GT pixels are nearly identical. Other major measures known as the statistical parameters are given in Table 5. This table confirms that the precision, sensitivity, specificity, and accuracy are greater than 98.18, 92.85, 98.43, and 97.28%, respectively. From these results, it is confirmed that proposed heuristic algorithm-based approach is efficient in extracting the optic disc, which is similar to the ground truths offered by the experts.

Normally, the medical field requires a substantial amount of computer assisted signal and image processing techniques in the clinical level. The usage of modern procedures is always desirable to have a clear analysis of the disease in its premature stage. From this article, it can be confirmed that the proposed heuristic algorithm-based tool is capable of mining the disc in RGB retinal image. In future, this methodology can be implemented to analyze the clinical RGB retinal images.

4 Conclusion

In this work, a computer supported scheme is discussed to extract and analyze the optic disc of the RGB retinal image database. This work implements a novel technique by integrating the heuristic multi-thresholding and active contour segmentation. During the experimental work, entire RIM-ONE database is considered. The optic disc region is extracted from the entire dataset categorized as normal, early, moderate, deep, and OHT, later confirmed with ground truths supplied by expert members. Result also verifies that picture resemblance measures and statistical parameters are superior.

References

1. Raja NSM, Kavitha G, Ramakrishnan S (2011) Assessment of retinal vasculature abnormalities using slantlet transform based digital image processing, biomed 2011. Biomed Sci Instrum 47:88–93
2. Keerthana K, Jayasuriya TJ, Raja NSM, Rajinikanth V (2017) Retinal vessel extraction based on firefly algorithm guided multi-scale matched filter. Int J Mod Sci Technol 2(2):74–80
3. Raja NSM, Kavitha G, Ramakrishnan S (2012) Analysis of vasculature in human retinal images using particle swarm optimization based Tsallis multi-level thresholding and similarity measures. Lect Notes Comput Sci 7677:380–387
4. Hajeb SH, Rabbani H, Akhlaghi MR (2012) Diabetic retinopathy grading by digital curvelet transform. Comput Mathem Methods Med 2012: 11 p. (Article ID 761901)
5. Ross JE, Bron AJ, Reeves BC, Emmerson PG (1985) Detection of optic nerve damage in ocular hypertension. Br J Ophthalmol 69(12):897–903
6. Sreejini KS, Govindan VK (2013) Automatic grading of severity of diabetic macular edema using color fundus images. In: IEEE third international conference on advances in computing and communications (ICACC), pp 177–180
7. Samanta S, Ahmed SS, Salem MAM, Nath SS, Dey N, Chowdhury SS (2014) Haralick features based automated glaucoma classification using back propagation neural network. Adv Intell Syst Comput 327:351–358
8. McClintic BR, McClintic JI, Bisognano JD, Block RC (2010) The relationship between retinal microvascular abnormalities and coronary heart disease: a review. Am J Med 123(4): 374.e1–374.e7
9. Dey N, Roy AB, Das A, Chaudhuri SS (2012) Optical cup to disc ratio measurement for glaucoma diagnosis using Harris corner. In: Third international conference on computing communication and networking technologies (ICCCNT), IEEE, 2012. https://doi.org/10.1109/icccnt.2012.6395971
10. Chakraborty S, Mukherjee A, Chatterjee D, Maji P, Acharjee S, Dey N (2014) A semi-automated system for optic nerve head segmentation in digital retinal images. In: 2014 international conference on information technology (ICIT), pp 112–117
11. Nath SS, Mishra G, Kar J, Chakraborty S, Dey N (2014) A survey of image classification methods and techniques. In: International conference on control, instrumentation, communication and computational technologies (ICCICCT), pp 554–557
12. Varsha Shree TD, Revanth K, Sri Madhava Raja N, Rajinikanth V (2018) A hybrid image processing approach to examine abnormality in retinal optic disc. Procedia Comput Sci 125:157–164
13. Fumero F, Alayon S, Sanchez JL, Sigut J, Gonzalez-Hernandez M (2011) RIM-ONE: an open retinal image database for optic nerve evaluation. In: 2011 24th international symposium on computer-based medical systems (CBMS), pp 1–6
14. http://medimrg.webs.ull.es/research/retinal-imaging/rim-one/
15. Bose S, Mukherjee A, Chakraborty S, Samanta S, Dey N (2013) Parallel image segmentation using multi-threading and k-means algorithm. In: 2013 IEEE international conference on computational intelligence and computing research (ICCIC), Madurai, pp 26–28
16. Pal G, Acharjee S, Rudrapaul D, Ashour AS, Dey N (2015) Video segmentation using minimum ratio similarity measurement. Int J Image Min 1(1):87–110
17. Tuba M (2014) Multilevel image thresholding by nature-inspired algorithms: a short review. Comput Sci J Moldova 22(3):318–338
18. Yang XS (2011) Nature-inspired metaheuristic algorithms, 2nd edn, Luniver Press, Frome, UK
19. Yang XS (2013) Bat algorithm: literature review and applications. Int J Bio-Inspired Comput 5(3):141–149
20. Rajinikanth V, Aashiha JP, Atchaya A (2014) Gray-level histogram based multilevel threshold selection with bat algorithm. Int J Comput Appl 93(16):1–8

21. Rajinikanth V, Couceiro MS (2015) Optimal multilevel image threshold selection using a novel objective function. Adv Intell Syst Comput 340:177–186
22. Satapathy SC, Raja NSM, Rajinikanth V, Ashour AS, Dey N (2016) Multi-level image thresholding using Otsu and chaotic bat algorithm. Neural Comput Appl. https://doi.org/10.1007/s00521-016-2645-5
23. Abhinaya B, Raja NSM (2015) Solving multi-level image thresholding problem—an analysis with cuckoo search algorithm. Adv Intell Syst Comput 339:177–186
24. Kapur JN, Sahoo PK, Wong AKC (1985) A new method for gray-level picture thresholding using the entropy of the histogram. Comput Vision Graph Image Process 29:273–285
25. Bhandari AK, Kumar A, Singh GK (2015) Modified artificial bee colony based computationally efficient multilevel thresholding for satellite image segmentation using Kapur's, Otsu and Tsallis functions. Expert Syst Appl 42:1573–1601
26. Lakshmi VS, Tebby SG, Shriranjani D, Rajinikanth V (2016) Chaotic cuckoo search and Kapur/Tsallis approach in segmentation of T.cruzi from blood smear images. Int J Comput Sci Inform Secur (IJCSIS) 14:51–56
27. Manic KS, Priya RK, Rajinikanth V (2016) Image multithresholding based on Kapur/Tsallis entropy and firefly algorithm. Indian J Sci Technol 9(12):89949
28. Houhou N, Thiran J-P, Bresson X (2008) Fast texture segmentation model based on the shape operator and active contour. In: IEEE conference on computer vision and pattern recognition, CVPR 2008, Anchorage, AK, pp 1–8 (2008)
29. Bresson X, Esedoglu S, Vandergheynst P, Thiran J-P, Osher S (2007) Fast global minimization of the active contour/snake model. J Math Imaging Vis 28(2):151–167
30. Chaddad A, Tanougast C (2016) Quantitative evaluation of robust skull stripping and tumor detection applied to axial MR images. Brain Inform 3(1):53–61
31. Moghaddam RF, Cheriet M (2010) A multi-scale framework for adaptive binarization of degraded document images. Pattern Recogn 43(6):2186–2198

Segmentation of Tumor from Brain MRI Using Fuzzy Entropy and Distance Regularised Level Set

I. Thivya Roopini, M. Vasanthi, V. Rajinikanth, M. Rekha
and M. Sangeetha

Abstract Image processing is needed in medical discipline for variety of disease assessments. In this work, Firefly Algorithm (FA)-assisted approach is implemented to extract tumor from brain magnetic resonance image (MRI). MRI is a clinically proven procedure to record and analyze the suspicious regions of vital body parts. The proposed approach is implemented by integrating the fuzzy entropy and Distance Regularized Level Set (DRLS) to mine tumor region from axial, sagittal, and coronal views' brain MRI dataset. The proposed approach is a three-step procedure, such as skull stripping, FA-assisted fuzzy entropy-based multi-thresholding, and DRLS-based segmentation. After extracting the tumor region, the size of the tumor mass is examined using the 2D Minkowski distance measures, such as area, area density, perimeter, and perimeter density. Further, the vital features from the segmented tumor are extracted using GLCM and Haar wavelet transform. Proposed approach shows an agreeable result in extraction and analysis of brain tumor of chosen MRI dataset.

Keywords Brain MRI · Firefly · Fuzzy entropy · DRLS · Feature extraction

1 Introduction

In recent years, image processing procedures are widely adopted in medical field to identify, analyze, and treatment planning for a variety of diseases. Among them, cancer is one of threatening disease in human community and requires the image/signal processing procedure at the initial stage to recognize, to classify, and to plan

I. Thivya Roopini (✉) · M. Vasanthi · V. Rajinikanth · M. Sangeetha
Department of Electronics and Instrumentation, St. Joseph's College of Engineering,
Chennai 600119, Tamil Nadu, India
e-mail: thivyaroopini7@gmail.com

M. Rekha
Department of Science, Velammal Vidhyashram, Sholinganallur, OMR, Chennai 600119,
Tamil Nadu, India

© Springer Nature Singapore Pte Ltd. 2018   297
A. K. Nandi et al. (eds.), *Computational Signal Processing
and Analysis*, Lecture Notes in Electrical Engineering 490,
https://doi.org/10.1007/978-981-10-8354-9_27

the appropriate treatment process to control and cure the disease [1, 2]. From the literature, it can be observed that a number of image processing techniques are proposed by the researchers to investigate the cancer with the help of clinical images [3].

Even though classical image processing methods are available, heuristic algorithm-based approaches are widely implemented due to its simplicity and adaptability [4, 5]. The literature also offers a significant application of heuristic algorithm-based image processing schemes on the grayscale and RGB images [6–12]. From these studies, one can survey that heuristic and traditional approach-based image processing scheme will provide better result.

In this paper, an automated tool is proposed and implemented to examine the tumor section from the brain MRI dataset. In order to improve the accuracy, integration of the heuristic multi-thresholding and traditional segmentation procedure is considered.

In this work, firefly algorithm (FA)-assisted fuzzy entropy (FE)-based multi-level thresholding and the distance regularized level set (DRLS)-based segmentation are implemented to mine the tumor core from the brain MRI dataset.

To justify the superiority of the proposed approach, the extracted tumor core is examined using the 2D Minkowski distance measures [13], and the vital image features are also extracted using GLCM and Haar wavelet transform [14, 15]. The proposed approach is tested on various shaped brain MRI, such as axial, sagittal, and coronal views, and the results are recorded. This study confirms that proposed approach is efficient in extracting the tumor core from the brain MRI dataset irrespective of its view.

2 Methodology

This section presents the methodologies and its implementation to complete the assigned task.

2.1 Skull Elimination

Skull stripping is the chief process in brain image segmentation. It is essential to eliminate the skull and the related area from brain MRI for quantitative analysis. In this paper, it is performed using an image filter which separated the skull and the rest of the image sections by masking the pixels based on its intensity levels. In brain MRI, the skull/bone section will have the maximum threshold value (threshold > 200) with respect to other brain sections. The image filter will separate the brain section based on a chosen threshold value. Then by employing the solidity property, the skull is stripped from the test image [16].

2.2 Firefly Algorithm

FA was firstly discussed by Yang [17, 18]. In FA, blinking lighting patterns spawned by fireflies are modeled by means of an appropriate arithmetical equation. Here, FA proposed in [8, 10] is adopted.

Modernized location in the search universe is expressed as;

$$X_i^{t+1} = X_i^t + \beta_0 e^{-\gamma d_{ij}^2}(X_j^t - X_i^t) + \alpha_1 \cdot \text{sign}(\text{rand} - 1/2) \oplus A \cdot |s|^{\alpha/2} \quad (1)$$

where X_i^{t+1} is modernized location of ith firefly, X_i^t is early location of ith firefly, $\beta_0 e^{-\gamma d_{ij}^2}(X_j^t - X_i^t)$ is appeal among fireflies, A is a arbitrary term, β is the spatial supporter, α is sequential supporter, and $\Gamma(\beta)$ is Gamma utility. In this paper, FA values are allocated as follows: search dimension, $D = T$; strength of firefly is preferred as 30, iteration number is fixed as 1200, and terminating measure is the maximal $f(T)$.

The FA execution is as follows:

START;
    Assign necessary values, such as D, $f(T)$;
    Create preliminary locality for agents based on x_i ($i = 1, 2, \ldots n$)
    Decide strength of ith agent related to ith $f(T)$ cost
If Current iteration < Max iteration;
For $i = 1,2, \ldots, n$;
For $j = 1,2, \ldots, n$;
    If illumination is $j > i$,
    Estimate Cartesian space & shift i on the way to j;
    End if;
    Replicate previous procedure till Current iteration = Max iteration;
    Compute illumination and modernize agent locations;
End for j;
End for i;
    Arrange agents in downhill form with respect to its rank and provide solution;
End if;
Register $f(T)$ and T.
STOP;

2.3 Fuzzy–Tsallis Entropy

In recent years, entropy-assisted methods are largely considered by the researchers for image processing applications [9, 19]. Tsallis function is one of the widely considered entropy for image multi-thresholding. In this paper, recently discovered hybrid procedure known as the fuzzy–Tsallis entropy is adopted [20–22]. A well-known trapezoidal membership function is chosen to guess the membership of n segmented sections μ_1, μ_2, ..., μ_n, with the help of $(n - 1)$ thresholds. The partisanship factor has "$2 \times (n - 1)$" unidentified parameters, specifically a_1, c_1... a_{n-1}, c_{n-1} where $0 \le a_1 \le c_1 \le \cdots \le a_{n-1} \le c_{n-1} \le L - 1$. The optimal threshold for the test image is then discovered with the help of the FA.

Finally, $(n - 1)$ levels of thresholds are discovered using the fuzzy parameters as follows:

$$t_1 = \frac{(a_1 + c_1)}{2}, \ t_2 = \frac{(a_2 + c_2)}{2}, \ldots, \ t_{n-1} = \frac{(a_{n-1} + c_{n-1})}{2} \tag{2}$$

2.4   Distance Regularized Level Set

In image segmentation applications, level set approaches have been largely considered to mine vital supporting data from test picture. This paper implements the current edition of level set approach known as the Distance Regularized Level Set (DRLS) technique to extract tumor region from the brain MRI dataset [23].

This is scientifically represented as:

$$\text{Considering the curve growth is } = \frac{\partial C(s, t)}{\partial t} = FN \tag{3}$$

where C—the curve vector consisting the spatial (s) and temporal (t) parameters, F—the haste variable, and N—the deepest standard vector to the curve C. Additional information regarding the DRLS can be found in [24].

2.5   Performance Assessment

In brain-related disease examination, to plan the preface treatment procedure, it is important to evaluate the dimension and orientation of the tumor mass. The volume of the mass is then checked using the two-dimensional (2D) Minkowski distance measures available in the literature [13]. Additionally, the tumor area and statistical properties are also computed using the gray-level co-occurrence matrix (GLCM) and Haar wavelet transform existing in the literature [14, 15]. These statistical features can be considered to train and check the computerized classifier systems in future.

3 Result and Discussion

This division shows the experimental results of implemented work. In this study considers 256 × 256-sized brain MR images, such as axial (A), sagittal (S), and coronal (C) views recorded with the flair and T2 modalities and are obtained from Cerebrix and Brainix datasets [25].

Initially, the skull region from the test image is removed using the image filter discussed in Sect. 2. Then, FA-assisted fuzzy–Tsallis entropy-based multi-thresholding process is implemented on the test data in order to group similar pixels. After the thresholding, the DRLS procedure is then applied to mine the tumor section from threshold picture. Finally, the extracted tumor core is analyzed using the 2D Minkowski measures and GLCM features. All the experimental work is implemented using MATLAB 2012 software.

Figure 1 depicts chosen test image (axial) and its results. Figure 1a presents a flair modality test image, Fig. 1b, c depicts the outcome of the skull stripping operation, Fig. 1d illustrates the result of the FA and fuzzy–Tsallis tri-level thresholding process, and Fig. 1e, f shows the DRLS segmentation and extracted tumor region.

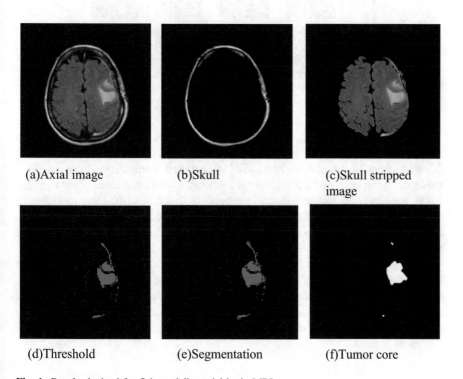

(a)Axial image   (b)Skull   (c)Skull stripped
                                            image

(d)Threshold   (e)Segmentation   (f)Tumor core

Fig. 1 Result obtained for flair modality axial brain MRI

Table 1 T2 modality-based test images

Similar image processing technique is applied for the other pictures depicted in Table 1. Here, T2 modality-based sagittal (slice numbers 8, 12, and 16) and coronal (slice numbers 10, 14, and 16) brain MRI is presented. Compared to the axial view, the skull region existing in these images is complex, and more care should be taken while implementing the skull removal procedure. From this table, one can observe that proposed approach efficiently removes the skull section without troubling the soft brain region. This table also presents the thresholding result and the extracted tumor region.

After extracting the possible tumor core, the dimension and its orientation in the brain MRI are computed using the Minkowski and GLCM function and its equivalent results are shown in Tables 2 and 3, respectively. In future, these results can be used to develop an automated classification tool, which classifies the tumor as malignant and benign, which will help medical expert to plan the brain tumor treatment process.

Table 2 2D Minkowski features

Image	Area	Area density	Perimeter	Perimeter density
A	2038	0.0277	205.26	0.0207
C	2046	0.0283	217.11	0.0219
	2814	0.0316	261.36	0.0322
	1174	0.0110	184.52	0.0189
S	2618	0.0297	236.19	0.0318
	1847	0.0191	175.39	0.0181
	1022	0.0153	142.18	0.0126

Table 3 GLCM features

Image	Area	Major axis	Minor axis	Eccentricity	Equiv. diameter
A	1017	42.2842	35.2246	0.3028	31.3927
C	994	27.1947	22.2975	0.2294	29.0045
	1302	42.8119	33.0486	0.3347	35.3065
	847	20.0663	14.7468	0.1955	23.0925
S	1084	42.3357	34.8254	0.3005	31.0528
	948	25.5822	17.5217	0.1865	22.0477
	879	19.0475	15.3947	0.1692	19.0036

4 Conclusion

This paper proposes an automated scheme to extract and analyze the abnormal section from the brain MRI dataset. This procedure implements a heuristic algorithm and fuzzy–Tsallis entropy-based procedure to enhance the tumor section and the level set-based approach to extract the tumor core. To exhibit the superiority of the proposed procedure, various brain MRI views, such as axial, sagittal, and coronal views recorded with the flair and T2 modalities, are considered. The experimental result demonstrates that proposed approach is very efficient in extracting the tumor core from the chosen brain MRI dataset. In order to support the automatic classification, well-known dimensional analyses, such as Minkowski and GLCM function are also implemented. In future, the efficiency of the proposed approach can be tested using the real-time clinical brain MRI dataset.

References

1. Abdel-Maksoud E, Elmogy M, Al-Awadi R (2015) Brain tumor segmentation based on a hybrid clustering technique. Egypt Inf J 16(1):71–81
2. Abdullah AA, Chize BS, Zakaria Z (2012) Design of cellular neural network (CNN) simulator based on matlab for brain tumor detection. J Med Imaging Health Inf 2(3):296–306

3. Bauer S, Wiest R, Nolte LP, Reyes M (2013) A survey of MRI-based medical image analysis for brain tumor studies. Phys Med Biol 58(13):97–129

4. Sezgin M, Sankar B (2004) Survey over image thresholding techniques and quantitative performance evaluation. J Electron Imaging 13:146–165

5. Tuba M (2014) Multilevel image thresholding by nature-inspired algorithms: a short review. Comput Sci J Moldova 22:318–338

6. Raja NSM, Rajinikanth V (2014) Brownian distribution guided bacterial foraging algorithm for controller design problem. Advances in intelligent systems and computing, vol 248, pp 141–148 (2014)

7. Rajinikanth V, Raja NSM, Latha K (2014) Optimal multilevel image thresholding: an analysis with PSO and BFO algorithms. Aust J Basic Appl Sci 8:443–454

8. Raja NSM, Rajinikanth V, Latha K (2014) Otsu based optimal multilevel image thresholding using firefly algorithm. Model Simul Eng 2014, Article ID 794574:17

9. Lakshmi VS, Tebby SG, Shriranjani D, Rajinikanth V (2016) Chaotic cuckoo search and Kapur/Tsallis approach in segmentation of T.cruzi from blood smear images. Int J Comput Sci Inf Secur (IJCSIS) 14, CIC 2016, 51–56

10. Rajinikanth V, Couceiro MS (2015) RGB histogram based color image segmentation using firefly algorithm. Procedia Comput Sci 46:1449–1457

11. Sarkar S, Das S (2013) Multilevel image thresholding based on 2D histogram and maximum Tsallis entropy—a differential evolution approach. IEEE Trans Image Process 22(12):4788–4797

12. Rajinikanth V, Raja NSM, Satapathy SC (2016) Robust color image multi-thresholding using between-class variance and cuckoo search algorithm. In: Advances in intelligent systems and computing, vol 433, pp 379–386

13. Legland D, Kiêu K, Devaux M-F (2007) Computation of Minkowski measures on 2D and 3D binary images. Image Anal Stereol 26:83–92

14. Manickavasagam K, Sutha S, Kamalanand K (2014) An automated system based on 2 d empirical mode decomposition and K-means clustering for classification of plasmodium species in thin blood smear images. BMC Infect Dis 14(3):1

15. Manickavasagam K, Sutha S, Kamalanand K (2014) Development of systems for classification of different plasmodium species in thin blood smear microscopic images. J Adv Microsc Res 9(2):86–92

16. Chaddad A, Tanougast C (2016) Quantitative evaluation of robust skull stripping and tumor detection applied to axial MR images. Brain Inform 3(1):53–61

17. Yang XS (2009) Firefly algorithms for multimodal optimization. In: Lecture notes in computer science, vol 5792, pp 169–178 (2009)

18. Yang XS (2009) Firefly algorithm, Lévy flights and global optimization. In: Proceedings of the 29th SGAI international conference on innovative techniques and applications of artificial intelligence (AI'09), pp 209–218. Springer, Berlin (2009)

19. Manic KS, Priya RK, Rajinikanth V (2016) Image multithresholding based on Kapur/Tsallis entropy and firefly algorithm. Indian J Sci Technol 9(12):89949

20. Tang Y, Di Q, Guan X, Liu F (2008) Threshold selection based on Fuzzy Tsallis entropy and particle swarm optimization. Neuro Quantol 6(4):412–419

21. Sarkar S, Das S, Paul S, Polley S, Burman R, Chaudhuri SS (2013) Multi-level image segmentation based on fuzzy-Tsallis entropy and differential evolution. In: IEEE international conference on fuzzy systems (FUZZ), pp 1–8. https://doi.org/10.1109/fuzz-ieee.2013.6622406

22. Rajinikanth V, Satapathy SC Segmentation of ischemic stroke lesion in brain MRI based on social group optimization and fuzzy-Tsallis entropy. Arabian J Sci Eng

23. Li C, Xu C, Gui C, Fox MD (2010) Distance regularized level set evolution and its application to image segmentation. IEEE Trans Image Process 19(12):3243–3254

24. Vaishnavi GK, Jeevananthan K, Begum SR, Kamalanand K (2014) Geometrical analysis of schistosome egg images using distance regularized level set method for automated species identification. J Bioinf Intell Control 3(2):147–152

25. Brain Tumor Database (CEREBRIX and BRAINIX). http://www.osirix-viewer.com/datasets/

Effect of Denoising on Vectorized Convolutional Neural Network for Hyperspectral Image Classification

K. Deepa Merlin Dixon, V. Sowmya and K. P. Soman

Abstract The remotely sensed high-dimensional hyperspectral imagery is a single capture of a scene at different spectral wavelengths. Since it contains an enormous amount of information, it has multiple areas of application in the field of remote sensing, forensic, biomedical, etc. Hyperspectral images are very prone to noise due to atmospheric effects and instrumental errors. In the past, the bands which were affected by noise were discarded before further processing such as classification. Therefore, along with the noise the relevant features present in the hyperspectral image are lost. To avoid this, researchers developed many denoising techniques. The goal of denoising technique is to remove the noise effectively while preserving the important features. Recently, the convolutional neural network (CNN) serves as a benchmark on vision-related task. Hence, hyperspectral images can be classified using CNN. The data is fed to the network as pixel vectors thus called Vectorized Convolutional Neural Network (VCNN). The objective of this work is to analyze the effect of denoising on VCNN. Here, VCNN functions as the classifier. For the purpose of comparison and to analyze the effect of denoising on VCNN, the network is trained with raw data (without denoising) and denoised data using techniques such as Total Variation (TV), Wavelet, and Least Square. The performance of the classifier is evaluated by analyzing its precision, recall, and F1-score. Also, comparison based on classwise accuracies and average accuracies for all the methods has been performed. From the comparative classification result, it is observed that Least Square denoising performs well on VCNN.

K. D. M. Dixon (✉) · V. Sowmya · K. P. Soman
Center for Computational Engineering and Networking (CEN),
Amrita School of Engineering, Amrita Vishwa Vidyapeetham,
Amrita University, Coimbatore 641112, Tamil Nadu, India
e-mail: deepamerlin123@gmail.com

V. Sowmya
e-mail: v_sowmya@cb.amrita.edu

K. P. Soman
e-mail: kp_soman@amrita.edu

© Springer Nature Singapore Pte Ltd. 2018
A. K. Nandi et al. (eds.), *Computational Signal Processing
and Analysis*, Lecture Notes in Electrical Engineering 490,
https://doi.org/10.1007/978-981-10-8354-9_28

Keywords Least square denoising · Total variation-based denoising
Wavelet denoising · Vectorized convolutional neural network · Hyperspectral
image classification

1 Introduction

Hyperspectral remote sensing acquires images of earth surface in narrow and continuous spectral bands in the electromagnetic spectrum. The high-dimensional hyperspectral images contain enormous amount of spectral, spatial, and radiometric information stored in hundreds of bands. The abundant information in hyperspectral images attracted researchers to use it in multiple areas of application such as, in the field of remote sensing, environmental monitors, military surveillance, soil type analysis. [1]. While capturing the hyperspectral image (HSI), it is inevitably corrupted with noise due to stochastic error, dark current, and thermal electronics [1]. The presence of noise in HSI will degrade the performance of previously listed applications. Hence, HSI denoising is an important preprocessing step before any further processing such as classification.

HSI classification is one of the major areas of research in the field of remote sensing [2]. The commonly used classifiers for HSI classification are probability-based multinominal logistic regression (MLR), support vector machine (SVM), minimum spanning forest (MSF) etc. [2]. Recently, deep learning-based methods achieve benchmark results in many fields [3, 4]. In deep learning, the convolutional neural network (CNN) gives better classification rate on vision-related tasks [3, 5]. CNNs are biologically inspired deep learning models that use a neural network trained from raw image pixel values to classifier outputs [3]. Hence, CNNs can be used for HSI classification [6, 7]. In this work, the 3D hyperspectral data is converted to 2D for the ease of handling the data. The data is fed to the network as pixel vector, thus called Vectorized Convolutional Neural Network (VCNN).

The aim of this work is to determine the effect of denoising on VCNN. For analyzing the effect of denoising on VCNN, the network is trained with raw data (without denoising) and denoised data. In this paper, three different denoising techniques are utilized such as Least Square denoising [8], Total Variation-based denoising [8], and Wavelet denoising [8].

This paper explains the classification of hyperspectral images using VCNN in Sect. 2, and a detail explanation of the experimental procedure is given in Sect. 3. Sections 4 and 5 describe the dataset and experimental results and analysis.

2 VCNN for Hyperspectral Image Classification

The 3D hyperspectral data ($m \times n \times b$) with hundreds of spectral bands can be illustrated as 2D curves (1D array) [3]. In this work, CNN is used for the purpose of classification. CNN is a feed-forward neural network which is very similar to ordinary neural network. A typical CNN contains varying number of convolutional layers followed by a maxpooling layer and finally connected to a fully or partially connected layer. The number of layers and number of filters decide the performance of these networks. The number of layers and number of filters will vary depending on the data. The computation required will increase by increasing the number of layers and filters. Therefore, while designing the network, compact designs are to be selected. In this work, the HSI data is fed to the network as pixel vectors, hence called Vectorized Convolutional Neural Network (VCNN).

2.1 Architecture of VCNN for HSI Classification

The architecture VCNN is shown in Fig. 1. The network consists of five layers: an input layer (L1), a convolutional layer (C1), a maxpooling layer (P1), a fully connected layer (F1), and an output layer (L2). The trainable parameters in this network are the weights and biases in each layer [1]. All the trainable parameters of the VCNN are experimentally fixed. In the hyperspectral image, there are 16 different pixel samples and each pixel sample can be represented as a 2D image whose height is equal to 1 [3]. The size of input layer is ($1 \times n1$), where $n1$ is the number of bands. Followed by the input layer, there is a convolutional layer (C1) which filters the input data with 20 kernels of size ($k1 \times 1$). The layer C1 contains ($20 \times 1 \times n2$) nodes where $n2 = ((n1 - k1/s) + 1)$. Here 's' represents the stride and it is equal to one. There are ($20 \times (k1 + 1)$) trainable parameters between the layer L1 and C1. The layer next to the layer C1 is a maxpooling layer (P1) with kernels of size ($1 \times k2$). This layer P1 contains ($20 \times 1 \times n3$) number of nodes where $n3 = n2/k2$. The outputs of the layer P1 are fed to the fully connected layer (F1). It has $n4$ number of nodes and ($20 \times n3 + 1$) $\times n4$ number of trainable parameters. The output layer (L2) contains $n5$ number of nodes with ($n4 + 1$) $\times n5$ number of trainable parameters. So the VCNN have a total of $20 \times (k1 + 1) +$ ($20 \times n3 + 1$) $\times n4 + (n4 + 1) \times n5$ number of trainable parameters.

3 Experimental Procedure

The proposed work is used to analyze the effect of denoising on VCNN for HSI classification. The network is trained with raw hyperspectral data and with denoised hyperspectral data. The denoising techniques used in this work are Total Variation

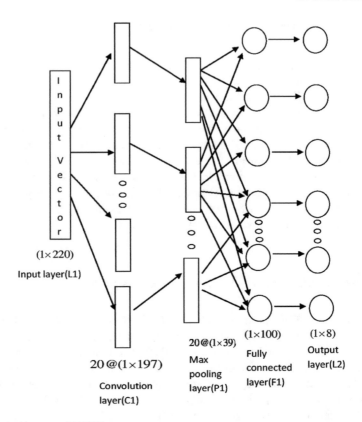

Fig. 1 Architecture of VCNN

(TV) [9], Wavelet [10, 11], and Least Square (LS) [8]. The neural network acts as a classifier. Similar networks were already evaluated for classification without performing denoising [3]. The experiment is performed on the subset of Indian pines dataset. The details of the selected classes are given in Table 1. The flow of the proposed method is shown in Fig. 2. In this method, the VCNN is trained for both raw hyperspectral data and denoised hyperspectral data. Each band of HSI image is vectorized to form a 2D image of size ($b \times mn$) where b is the number of bands, and m and n are the number of rows and columns present in each band. The pixelwise classification of HSI is done using the ground truth. Each pixel is normalized in the range of $[-1, +1]$ before giving to the VCNN. The input is given to the VCNN as pixel vectors. For training the VCNN, 80% of data from each class is selected randomly and the rest 20% is used for testing the network.

In detail, the first convolutional layer of VCNN was implemented with a 1D filter. The neural network is expected to learn these filter weights over the training process such that they extract essential features from the signal which are capable of distinguishing between the different hyperspectral data. The rectified linear activation function was chosen as the activation function for the convolutional layers.

Table 1 Subset of Indian pines

No	Class	Samples
CN	Corn-notill	1428
CM	Corn-mintill	830
GP	Grass pasture	483
HW	Hay-windrowed	478
SN	Soybean-notill	972
SM	Soybean-mintill	2455
SC	Soybean-clean	593
WD	Woods	1265

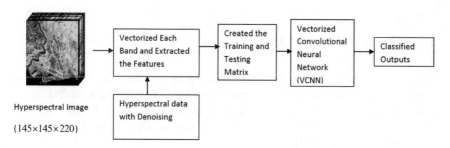

Fig. 2 Flow of work for analyzing the effect of denoising on VCNN

Further, the information extracted from these features are used to predict the label corresponding to the data point by adjusting the weights and biases across the next maxpooling layer and fully connected layers; the first is implemented with a rectified linear activation function and the second with a softmax function for activation. The error function for backpropagation was chosen as the difference between the predicted and original class labels. The optimization function was solved using the Gradient descent method. Further drop out was done during training to avoid overfitting. The whole network was implemented and tested on the tensorflow framework [12].

4 Dataset Description

The commonly used standard hyperspectral dataset Indian pines are used in this work [2]. The data used in the experiment consists of 220 spectral bands with 145 × 145 pixels. It has spectral and spatial resolution of 10 and 20 nm, respectively. The ground truth includes 16 classes. The dataset that is used for this project is a subset of Indian pines containing eight classes. The details of the selected classes are given in Table 1.

5 Experimental Results and Analysis

The hyperspectral data which is very prone to noise is denoised using techniques such as TV, Wavelet, and LS denoising. The signal-to-noise ratio (SNR) values for each denoising techniques is measured and recorded for analyzing the amount of removal of noise from the signal or image [8]. The LS denoising gives a high SNR when compared to other denoising techniques [8]. Then the effect of denoising on VCNN is analyzed using the methods given below.

For the VCNN network, $n1$ and $n5$ are the input spectral channel size and number of output classes, respectively. Here $k1 = 24$, $n2 = 197$, $n3 = 197$, $k2 = 5$. The nodes in the fully connected layer $n4$ are fixed as 100 [3].

5.1 Classification Accuracy

The classification accuracy obtained for Indian pines dataset on VCNN is measured using ground truth provided along with the data. The classification accuracy is measured and recorded before applying denoising and after performing different denoising techniques. The classwise accuracy and average accuracy can be interpreted as in Eqs. (1) and (2).

$$\text{classwise accuracy}, \text{CA} = \frac{\text{No. of pixels which are correctly classified in each class}}{\text{Total no of pixels in each class}} \times 100 \tag{1}$$

$$\text{Average Accuracy}, \text{AA} = \frac{\sum_{C=1}^{N} CA_C}{n} \tag{2}$$

where CA_c is the classwise accuracy of cth class and n is the total number of classes in the hyperspectral image. The classification accuracy obtained for different denoising techniques is given in Table 2. From Table 2, it is observed that Least Square denoising gives higher average accuracy than other denoising techniques. Classification accuracy obtained for Indian pines dataset before and after applying Least Square denoising is 85 and 90%, respectively. For TV and Wavelet denoising, the classification accuracy obtained is 87 and 88%, respectively. From the comparative classification result, it is observed that Least Square denoising performs better on VCNN when compared with the existing denoising techniques.

The classifier (VCNN) performance on various denoising techniques is evaluated by taking its precision, recall, and F1-score. The detailed classification report for all the cases is given in Tables 3, 4, and 5. On comparing Tables 3, 4, and 5, it is observed that LS denoising performs better on VCNN. The precision [13], recall

Table 2 Classwise accuracy (%) for raw data and denoised hyperspectral data

Class	Raw data	TV	Wavelet	LS
CN	89	95	91	93
CM	78	85	91	92
GP	84	89	87	88
HW	88	95	91	89
SN	79	91	86	92
SM	85	79	85	90
SC	83	82	85	87
WD	91	87	91	91
Average accuracy (%)	85	87	88	**90**

The highest average accuracy value is marked as bold

Table 3 Precision of the classifier obtained for raw data, TV denoised, Wavelet, and LS

Data	Class								Average
	CN	CM	GP	HW	SN	SM	SC	WD	
Raw data	0.68	0.89	0.93	0.97	0.89	0.83	0.88	0.97	0.88
TV	0.65	0.88	0.90	1.00	0.89	0.98	0.94	0.99	0.90
Wavelet	0.69	0.89	0.93	0.98	0.88	0.95	0.86	0.99	0.90
LS	0.74	0.93	0.96	1.00	0.91	0.94	0.94	0.98	**0.92**

The highest average accuracy value is marked as bold

Table 4 Recall of the classifier obtained for raw data, TV denoised, Wavelet, and LS data

	Class								Average
	CN	CM	GP	HW	SN	SM	SC	WD	
Raw data	0.89	0.78	0.84	0.88	0.79	0.85	0.83	0.91	0.85
TV	0.95	0.85	0.89	0.95	0.91	0.79	0.82	0.87	0.87
Wavelet	0.91	0.91	0.87	0.91	0.86	0.85	0.85	0.91	0.88
LS	0.93	0.92	0.88	0.89	0.92	0.90	0.87	0.91	**0.90**

The highest average accuracy value is marked as bold

Table 5 F1-score of the classifier obtained for raw data, TV denoised, Wavelet, and LS

Data	Class								Average
	CN	CM	GP	HW	SN	SM	SC	WD	
Raw data	0.77	0.83	0.88	0.92	0.83	0.84	0.85	0.94	0.85
TV	0.77	0.86	0.89	0.97	0.90	0.87	0.88	0.92	0.88
Wavelet	0.79	0.90	0.90	0.94	0.87	0.90	0.86	0.95	0.88
LS	0.82	0.92	0.91	0.94	0.92	0.92	0.90	0.95	**0.91**

The highest average accuracy value is marked as bold

[13], and F1-score [13] of LS denoised hyperspectral data on VCNN are slightly higher when compared to existing denoising methods.

6 Conclusion

In this paper, the effect of denoising on VCNN is analyzed. The experiment is carried out on the subset of Indian pines dataset, the effectiveness of the technique is proposed based on the accuracy measurement parameters like the classwise accuracy, average accuracy, and the performance of the classifier is evaluated on the basis of precision, recall, and F1-score. From the experimental results obtained, it is evident that preprocessing of the hyperspectral image with Least Square denoising performs well on VCNN by providing good classification rate when compared to other denoising techniques such as TV and Wavelet.

References

1. Zhao YQ, Yang J (2015) Hyperspectral image denoising via sparse representation and low-rank constraint. IEEE Trans Geosci Remote Sens 53(1)
2. Aswathy C, Sowmya V, Soman K (2015) Admm based hyperspectral image classification improved by denoising using legendrefenchel transformation. Indian J Sci Technol 8(24):1
3. Hu W, Huang Y, Wei L, Zhang F, Li H (2015) Deep convolutional neural networks for hyperspectral image classification. J Sens
4. Athira S, Mohan R, Poornachandran P, Soman KP (2016) Automatic modulation classification using convolutional neural network. International Science Press, IJCTA, pp 7733–7742
5. Ren JS, Xu L (2015) On vectorization of deep convolutional neural networks for vision tasks. arXiv preprint arXiv:1501.07338
6. Chen Y, Jiang H, Li C, Jia X, Ghamisi P (2016) Deep feature extraction and classification of hyperspectral images based on convolutional neural networks. IEEE Trans Geosci Remote Sens 54(10):6232–6251
7. Slavkovikj V, Verstockt S, De Neve W, Van Hoecke S, Van de Walle R (2015) Hyperspectral image classification with convolutional neural networks. In: Proceedings conference. 23rd ACM international conference on Multimedia, 2015, pp 1159–1162
8. Srivatsa S, Ajay A, Chandni C, Sowmya V, Soman K (2016) Application of least square denoising to improve admm based hyperspectral image classification. In: proceedings conference, 6th international conference on adavances in computing and communications, ICACC 2016, vol 93, pp 416–423
9. Rudin LI, Osher S, Fatemi E (1992) Nonlinear total var iation based noise removal algorithms. Physica D 60(1):259–268
10. Donoho DL, Johnstone JM (1994) Ideal spatial adaptat ion by wavelet shrinkage. Biometrika 81(3):425–455

11. Bhosale NP, Manza RR (2014) Image denoising based on wavelet for satellite imagery: a review. Int J Modern Eng Res 4(2014): 63–68
12. Abadi M et al (2015) TensorFlow: large-scale machine learning on heterogeneous systems. Software available from tensorflow.org. Available: http://tensorflow.org/
13. Fawcett T (2006) An introduction to ROC analysis. Pattern Recogn Lett 27(8):861–874

Classification of fNIRS Signals for Decoding Right- and Left-Arm Movement Execution Using SVM for BCI Applications

A. Janani and M. Sasikala

Abstract Brain–computer interface-based systems help people who are incapable of interacting with the external environment using their peripheral nervous system. BCIs allow users to communicate purely based on their mental processes alone. Signals such as fNIRS corresponding to the imagination of various limb movements can be acquired noninvasively from the brain and translated into commands that can control an effector without using the muscles. The present study aims at classifying Right-Arm and Left-Arm movement combination using SVM. The study also aims at analyzing the efficacy of two different features, namely average signal amplitude and the difference between the average signal amplitudes of ΔHbO and ΔHbR on the accuracies obtained. The combination of these two features is also explored. The results of the study indicate that chosen features yield average accuracies between 70 and 76.67% calculated for all the subjects. The difference of mean amplitudes of ΔHbO and ΔHbR is investigated as one of the features for fNIRS-BCI application, and it yields an average accuracy of 70%. It indicates the possibility of using this feature for evaluating the binary BCI system for practical communication use. Two-feature combination improved the average accuracy value from 70 to 76.67%. The results obtained from the study suggest that distinct patterns of hemodynamic response arising out of Right-Arm and Left-Arm movements can be exploited for the development of BCI which are best described by the features used in the present study.

Keywords Functional near-infrared spectroscopy · Brain–computer interface Support vector machine

A. Janani (✉) · M. Sasikala
Department of Electronics and Communication Engineering,
Centre for Medical Electronics, Anna University, Chennai, India
e-mail: jananinambi@gmail.com

M. Sasikala
e-mail: sasikala@annauniv.edu

© Springer Nature Singapore Pte Ltd. 2018
A. K. Nandi et al. (eds.), *Computational Signal Processing and Analysis*, Lecture Notes in Electrical Engineering 490,
https://doi.org/10.1007/978-981-10-8354-9_29

1 Introduction

A brain–computer interface (BCI) system offers motor-disabled people a new lease of life by enabling them to communicate with the outside world purely based on their thought processes. It bypasses the peripheral nervous system and the muscles, which are non-functional in people who are suffering from motor neuron diseases. BCI systems are also used as a means for restoration of motor functions in the form of a neuro-prosthesis to help motor-disabled people to perform simple tasks [1]. There have been several modalities such as EEG, MEG, fMRI which are attractive for BCI applications for their ability to measure the brain signal noninvasively. Functional near-infrared spectroscopy (fNIRS) method can be used for BCI applications for its many advantages. fNIRS is beneficial over the other modalities for BCI application because of its easiness to use, portability, safety, low noise, and no susceptibility to electrical noise. The feasibility of using fNIRS for BCI application was first reported in [2]. fNIRS measures hemodynamic changes in the cerebral cortical tissue of the brain. The principle of NIRS uses two chromophores, namely oxygenated hemoglobin (HbO) and deoxygenated hemoglobin (HbR) that are sensitive to two different wavelengths in the near-infrared range, 700–1000 nm, known as "optical window" in which the biological tissue is partially transparent to light. The light absorption by water molecules and hemoglobin is relatively less in the optical window compared to other wavelength ranges. Hence, it is possible to detect the light traversed through the brain tissue using pairs of optical source and detector placed on the scalp [3]. The relative changes in the concentration of oxygenated hemoglobin and deoxygenated hemoglobin signify the neuronal activation relevant to the action being performed.

The viability to develop a two-class NIRS-based BCI system has been studied in [4, 5]. Extraction of meaningful features from the signal is important to discriminate the brain activities corresponding to different movements to develop a BCI system. The commonly used features in fNIRS-based BCI studies so far have been the mean values of hemoglobin concentration [4, 6–8], signal slope [5, 8], variance [7]. Apart from extracting the mean values of hemoglobin concentration from the fNIRS signal, the difference of mean amplitudes of ΔHbO and ΔHbR is also investigated as one of the possible features to classify Right-Arm and Left-Arm movements in this study.

The objective of the present study is to classify Right-Arm and Left-Arm movement combination using SVM classifier. Two different features have been extracted, namely average signal amplitude and the difference between the average signal amplitudes of ΔHbO and ΔHbR from the fNIRS signal, and their individual effects on the classification accuracies obtained have been analyzed. The combination of both the features of fNIRS signals has also been analyzed across all the subjects.

2 Methodology

2.1 fNIRS Data Collection and Participants

Three NIRS datasets are chosen from a study of hybrid EEG-fNIRS in an asynchronous SMR-based BCI involving healthy subjects [6]. The dataset consists of trials lasting 12 s each. The experiment consisted of four movement executions (Right-Arm raising, Left-Arm raising, right-hand gripping, and left-hand gripping). Out of which, Right-Arm raising and Left-Arm raising movements are considered for the study. The subjects were instructed to rest for the initial 6 s and to perform the task for the following 6 s. The datasets were obtained using an NIRScout 8–16 system (NIRx Medizintechnik GmbH, Germany) equipped with 12 sources and 12 detectors combined in 34 channels of acquisition. The sources and detectors placed maximum at 3.4 cm from each other were distributed evenly on the motor cortex, the sampling frequency being 10.42 Hz, and the wavelengths used were 760 and 850 nm. The system and the experimental protocol are described in [6]. Six channels evenly placed around C3 and C4 according to 10–20 international system on the left and right motor cortex are chosen for the purpose of analysis.

2.2 Data Processing

fNIRS raw signals are offset corrected to have zero mean. This step is done to remove very low-frequency signal artifacts and drift [9]. The light attenuation representing optical density (OD) values of both the wavelengths is baseline compensated by subtracting the mean value of the initial rest period from the signal. For all the subjects, the OD values at 760 and 850 nm are converted into changes in concentration of deoxygenated hemoglobin (ΔHbR) and oxygenated hemoglobin (ΔHbO), respectively, by using modified Beer–Lambert law with the help of the following equation [10].

$$\begin{bmatrix} \Delta HbR \\ \Delta HbO \end{bmatrix} = (d)^{-1} \begin{bmatrix} \varepsilon_{HbR\lambda_1} & \varepsilon_{HbO\lambda_1} \\ \varepsilon_{HbR\lambda_2} & \varepsilon_{HbO\lambda_2} \end{bmatrix}^{-1} \begin{bmatrix} \Delta OD(\Delta t, \lambda_1)/DPF(\lambda_1) \\ \Delta OD(\Delta t, \lambda_2)/DPF(\lambda_2) \end{bmatrix} \tag{1}$$

Differential pathlength factor of 760 nm ($DPF(\lambda_1)$ = 7.15),
Differential pathlength factor of 850 nm ($DPF(\lambda_2)$ = 5.98).

Extinction coefficient (ε) describes how strongly a chromophore absorbs light at a particular wavelength and is separately given for HbR and HbO. For HbR, ε values are 0.7861/1.6745 (higher/lower wavelength) and similarly for HbO are 1.1596/0.6096 (higher/lower wavelength). The concentration changes of HbR and

HbO contain high frequency as well as low-frequency noise corresponding to cardiac interference and respiration. The signals are filtered using a fourth-order IIR Butterworth band-pass filter with cut-off frequencies of 0.01–0.2 Hz in order to remove noise associated with unwanted physiological oscillations and high-frequency instrument noise.

2.3   Feature Extraction

Two different features are extracted, namely average signal amplitude of ΔHbO and difference between the average signal amplitudes of ΔHbO and ΔHbR from the fNIRS signal. The difference between the average signal amplitudes of ΔHbO and ΔHbR is analyzed for its efficacy for fNIRS-BCI applications in this study. The combination of both of these features is also used for classification. These features are calculated across the six channels evenly placed around C3 and C4. Features are extracted over the time interval of 6–12 s after stimulation onset for all the subjects and for both the tasks. Scaling is performed before classification. This step is done to standardize the range of independent features of the data. Nonscaled data tend to over-fitting that negatively impacts the classifier model to generalize. Scaling is done using the following equation [10].

$$x' = \frac{x - \min(x)}{\max(x) - \min(x)} \qquad (2)$$

where x' represents the feature values rescaled between -1 and 1, $x \in R^n$ are the original values of the features, and $\max(x)$ and $\min(x)$ represent the largest and smallest values, respectively.

2.4   Classification

SVM is used to classify Right-Arm and Left-Arm movement combination. The classifier is trained on the extracted and scaled features. SVM is the most commonly used classifier for fNIRS-BCI studies apart from LDA [8, 10, 11]. It is a supervised learning algorithm that designs a decision function to optimally separate the data into two classes. SVM is advantageous for online BCI applications as well for its simplicity and lower computational requirements.

   In this study, the linear kernel of the SVM algorithm is chosen. Here, 80% of the total data are used for training and the remaining for testing. Classification accuracies are determined for the designed classifier model of Right-Arm and Left-Arm movement combination. Classification accuracies indicate the number of times that the classifier model correctly predicts the class of the trials in the testing data. Average accuracies are calculated across all the subjects for the feature types used in the study.

3 Results

SVM classified Right-Arm and Left-Arm movement combination with average accuracies ranging from 70 to 76.67% calculated for all the subjects. In the present study, the difference of mean amplitudes of ΔHbO and ΔHbR is investigated as one of the candidate features and it yields an average accuracy of 70% for the SVM classifier. The average signal amplitude of ΔHbO feature of subject 3 yielded the maximum accuracy of 90% for Right-Arm and Left-Arm binary classification. The results obtained from the investigated feature show that the accuracy values are consistent across the subjects. The requirement for a binary BCI system to be used for practical communication purpose is to have an accuracy value greater than 70% [12]. This target is achieved by the features used in the present study. The two-feature combination of the average signal amplitude of ΔHbO and difference between the average signal amplitudes of ΔHbO and ΔHbR is also analyzed, and it shows improvement in the classification accuracies compared to their individual accuracies. The accuracy results obtained from using different features of various subjects are presented in Table 1. Figure 1 shows the averaged signal for Right-Arm movement over C3 and C4 channels. It is observed that ΔHbO during Right-Arm movement is more over C3 channel which is located on the contralateral hemisphere. Similarly, ΔHbO during Right-Arm movement is less over C4 channel which is ipsilateral to the movement performed. Since there are differential activations due to two different tasks over the chosen channels, there is a possibility to discriminate the tasks with sensible accuracy that is shown in the obtained results. Figure 2 shows the 2-D feature space of Right-Arm and Left-Arm movement tasks for average signal amplitude and difference of mean amplitudes of ΔHbO and ΔHbR. The accuracy values of SVM obtained from using different features of all the subjects are plotted in Fig. 3. The improvement in the accuracy values when both the features are combined is also shown.

Table 1 Classification accuracies by SVM for different features of fNIRS signal across all subjects

Subjects	Features		
	Average signal amplitude of ΔHbO	Difference in mean amplitudes of ΔHbO and ΔHbR	Combination of both the features
1	60	70	80
2	70	70	60
3	90	70	90
Average	73.33%	70%	76.667%

Fig. 1 Averaged signal over the task period of 6 s for Right-Arm movement over C3 and C4 channels. The dashed line indicates the end of the task period

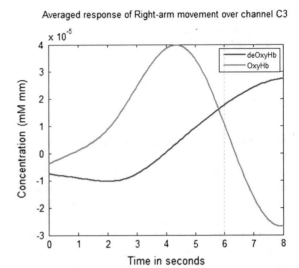

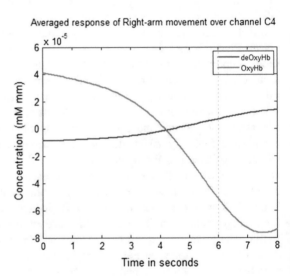

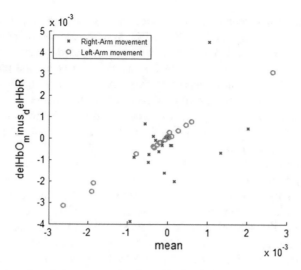

Fig. 2 2D scatter plot of the average signal amplitude of ΔHbO and the difference of mean amplitudes of ΔHbO and ΔHbR of subject 3

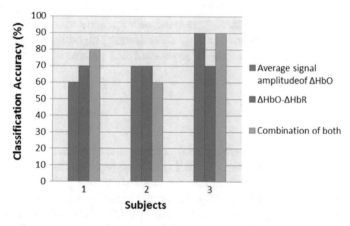

Fig. 3 Classification accuracies SVM from different feature types for Right-Arm and Left-Arm movement combination

4 Discussion

Different patterns of activation arise out of right and left limb movements that can be classified by algorithms for the design of a binary BCI system. By using this concept, the objective of the present study has been framed. The study has been carried out on three subjects to classify Right-Arm and Left-Arm movement execution using SVM.

This study also compares two features, namely average signal amplitude of ΔHbO that is commonly used in fNIRS-BCI systems and the difference of mean amplitudes of ΔHbO and ΔHbR that has been proposed. Right- and Left-Arm movements are analyzed over C3 and C4 channels located on the primary motor cortex. These channels are chosen as they are strongly associated with hand movement [13]. The spatial patterns observed over the chosen channels on the primary motor cortex are important for discrimination of the two different movements. Figure 1 depicts the pattern exhibited by Right-Arm movement over C3 and C4 channels. The proposed feature has shown betterment in the results when combined with average signal amplitude of ΔHbO with an average accuracy of 76.67%. Hence, the results show that this two-feature combination can be used for designing a binary fNIRS-BCI system. However, the classification accuracies differed from subject to subject. The trial-to-trial variability in the hemodynamic signal and also the subject-to-subject differences could have contributed to the variations in the accuracy values. Training could be provided for BCI users to generate distinct brain signals related to tasks using neurofeedback that eventually improves the accuracy values.

The present study involved few subjects that might also be the reason behind the low accuracy values. A large number of subjects might improve the results.

Various other features could be extracted from the fNIRS signal, and their combinations can be investigated to provide better classification accuracies. As a part of further work, more advanced machine learning algorithms would be pondered to provide solutions to three or more class BCI problems.

5 Conclusion

Distinct patterns of hemodynamic response arising out of different movements can be exploited for the development of BCI which are best described by the features used in the present study. Besides the common feature used in fNIRS-BCI studies such as the mean of ΔHbO, the difference of mean amplitudes of ΔHbO and ΔHbR is investigated for its potentiality to be used in fNIRS-BCI systems. It yields an average accuracy of 70% for SVM classifier. The two-feature combination has also been investigated, and it results in an improved accuracy value of 76.67%. The results indicate the possibility of using the studied features for evaluating the multi-class capability of the BCI system. The results obtained in this study demonstrate that the chosen features when combined together can discriminate fNIRS signals corresponding to Right- and Left-Arm movement tasks for the development of a binary BCI system.

References

1. Wolpaw JR, Birbaumer N, McFarland DJ, Pfurtscheller G, Vaughan TM (2002) Brain-computer interfaces for communication and control. Clin Neurophysiol 113:767–791

2. Coyle S, Ward T, Markham C, McDarby G (2004) On the suitability of near-infrared (NIR) systems for next-generation brain-computer interfaces. Physiol Meas 25:815–822. https://doi.org/10.1088/0967-3334/25/4/003

3. Jobsis FF (1977) Noninvasive, infrared monitoring of cerebral and myocardial oxygen sufficiency and circulatory parameters. Science 198:1264–1267

4. Sitaram R, Zhang HH, Guan CT, Thulasidas M, Hoshi Y, Ishiawa A et al (2007) Temporal classification of multichannel near-infrared spectroscopy signals of motor imagery for developing a brain-computer interface. Neuroimage 34:1416–1427. https://doi.org/10.1016/j.neuroimage.2006.11.005N

5. Naseer N, Hong K-S (2013) Classification of functional near-infrared spectroscopy signals corresponding to the right-and left-wrist motor imagery for development of a brain-computer interface. Neurosci Lett 553:84–89. https://doi.org/10.1016/j.neulet.2013.08.021

6. Buccino AP, Keles HO, Omurtag A (2016) Hybrid EEG-fNIRS asynchronous brain-computer interface for multiple motor tasks. PLoS ONE 11(1)

7. Holper L, Wolf M (2011) Single trial classification of motor imagery differing in task complexity: a functional near-infrared spectroscopy study. J Neuroeng Rehabil 8:34. https://doi.org/10.1186/1743-0003-8-34

8. Hong KS, Naseer N, Kim YH (2015) Classification of prefrontal and motor cortex signals for three-class fNIRS-BCI. Neurosci Lett 587:87–92

9. Strait M, Scheutz M (2014) What we can and cannot (yet) do with functional near infrared spectroscopy. Front Neurosci 8

10. Naseer N, Qureshi NK, Noori FM, Hong KS (2016) Analysis of different classification techniques for two-class functional near-infrared spectroscopy-based brain-computer interface. Comput Intell Neurosci, 5480760, https://doi.org/10.1155/2016/5480760

11. Robinson N, Zaidi AD, Rana M, Prasad VA, Guan C, Birbaumer N et al (2016) Real-time subject-independent pattern classification of overt and covert movements from fNIRS signals. PLoS ONE 11(7):e0159959. https://doi.org/10.1371/journal.pone.0159959

12. Hwang HJ, Lim JH, Kim DW, Im CH (2014) Evaluation of various mental task combinations for near-infrared spectroscopy-based brain-computer interfaces. J Biomed Opt 19(7):077005–077005

13. Coyle SM, Ward TE, Markham CM (2007) Brain-computer interface using a simplified functional near-infrared spectroscopy system. J Neural Eng 4:219–226

Estimation of Texture Variation in Malaria Diagnosis

A. Vijayalakshmi, B. Rajesh Kanna and Shanthi Banukumar

Abstract Malaria parasite has been visually inspected from the Giemsa-stained blood smear image using light microscope. The trained technicians are needed to screen the malaria from the microscope; this manual inspection requires more time. To reduce the problems in manual inspection, nowadays pathologist moves to the digital image visual inspection. The computer-aided microscopic image examination will improve the consistency in detection, and even a semiskilled laboratory technician can be employed for diagnosis. Most of the computer-aided malaria parasite detection consists of four stages namely preprocessing of blood smear images, segmentation of infected erythrocyte, extracting the features, detection of parasite and classification of the parasites. Feature extraction is one of the vital stages to detect and classify the parasite. To carry out feature extraction, geometric, color, and texture-based features are extracted for identifying the infected erythrocyte. Among these clause of features, texture might be considered as a very fine feature, and it provides the characteristics of smoothness over the region of interest using the spatial distribution of intensity. The proposed work demonstrates the merit of the texture feature in digital pathology which is prone to vary with respect to change in image brightness. In microscope, brightness of the image could be altered by iris aperture diameter and illumination intensity control knob. However, the existing literature failed to mention the details about these illumination controlling parameters. So the obtained texture feature may not be considered as distinct feature. In this paper, we conducted an experiment to bring out the deviation of texture feature values by changing the brightness of the acquired image by varying the intensity control knob.

A. Vijayalakshmi (✉) · B. Rajesh Kanna
School of Computing Science and Engineering, VIT University, Chennai, India
e-mail: vijayalakshmi.av@vit.ac.in

B. Rajesh Kanna
e-mail: rajeshkanna.b@vit.ac.in

S. Banukumar
Department of Microbiology, Tagore Medical College & Hospital, Chennai, India
e-mail: shanthibanukumar@gmail.com

© Springer Nature Singapore Pte Ltd. 2018
A. K. Nandi et al. (eds.), *Computational Signal Processing and Analysis*, Lecture Notes in Electrical Engineering 490,
https://doi.org/10.1007/978-981-10-8354-9_30

Keywords Malaria diagnosis · Image brightness · Gray-level co-occurrence
matrix · Digital pathology · Digital microscopy

1   Introduction and Background

Malaria is an infectious disease caused by Plasmodium parasite. There are four
types of parasites namely Plasmodium vivax, Plasmodium falciparum, Plasmodium
ovale, and Plasmodium malariae which cause malaria disease [1]. These parasites
invade human red blood cells, it passes through four stages of development life
cycle, and the stages are ring, trophozoite, schizont, and gametocytes. Symptoms of
the malaria disease can be identified from clinical examination followed by the
laboratory test to confirm the malaria affected victim. There are several laboratory
methods are available like microscopic examination of stained thin or thick blood
smear, quantitative buffy coat (QBC) test, rapid diagnosis test, and molecular
diagnosis methods [1]. WHO recommends microscopic examination of thin
Giemsa-stained blood smear as the gold standard for diagnosing malaria parasite
[1]. However, microscopic examination is a quite time-consuming task, laborious
process, and trained technicians are being employed to accurately detect the malaria
parasite. To supplement this cognitive task, research fraternity trusted digital image
processing approaches for efficient malaria diagnosis. Almost, all the image-based
malaria diagnoses follows four functional pipelines namely image acquisition,
image preprocessing, erythrocyte segmentation, feature extraction and classification
[2] as illustrated in Fig. 1.

1.1   *Image Acquisition*

It is the process of collecting malaria blood smear images from the digital camera
mounted on the microscope. The stained smears are observed from microscope with
the magnification of 100× objective lens. For computer-aided malaria parasite
detection, thick and thin blood smear images are used. Images obtained from thin
blood smear are used to identify the type of malaria and its severity stages [3],
whereas thick blood smear is used for quantifying the infected erythrocytes [4].
Though several dyes have been used to stain erythrocyte, popularly used dyes are
Giemsa and Leishman [5].

Fig. 1 Steps for malaria parasite detection and classification

1.2 Preprocessing of Blood Smear Images

It is the process of removing the unnecessary details present in the acquired malaria parasite image for better visualization and for further analysis. In the existing literature, image filters like median [5], geometric mean [6], Gaussian [2], Laplacian [7], wiener [8], low pass [9], SUSAN [10] were used to eliminate the noise. And to enhance the contrast adaptive or local histogram equalization [11], dark stretching [12], partial contrast stretching algorithm, and histogram matching [13] are used. Most of the literature suggested Gray World Assumption could be the ideal choice, because it is used to eliminate the variation in the brightness of the image [14].

1.3 Segmentation of Infected Erythrocyte

It is a process of isolating the red blood cell (erythrocyte) and eliminating the other details such as white blood cell, platelet, and other artifacts from the preprocessed image. The isolated segment may include infected and non-infected malaria parasites. Further, to discriminate the infected and non-infected erythrocytes, finer segmentation techniques have been used. From the previous studies, we listed the various segmentation techniques utilized for isolating erythrocytes. They are circle Hough transform [15], Otsu threshold [8], pulse-coupled neural network [16], rule-based algorithm [17], Chan-Vese algorithm [11], granulometry [10], edge detection algorithm [18], Marker-controlled watershed algorithm, and normalized cut algorithm [4, 5]. In most of the literature, watershed transformation algorithm was used to segment the erythrocytes. Similarly to discriminate the infected erythrocytes, few techniques are used such as histogram threshold, moving k-means clustering, fuzzy rule-based system, annular ring ratio method, Zack threshold, and N-cut algorithm.

1.4 Extracting the Features

It is the important step to gather the insights from the erythrocytes. And it is a set of process for deriving size, color, stippling, surface, and pigment features of the infected erythrocyte for subsequent human interpretation to detect the infected malaria parasite. From the segmented erythrocytes, we can extract the coarse-level features like size and color. The size of infected region is being inferred from the efficient image-based models and its area estimation technique [19, 20]. Area and color are first-level eliminating features to differentiate the species in the malaria parasite. It is a vital feature to differentiate the species in the malaria parasite. The existing literature refers some morphological features, which are retrieved using Hu moment [4], Chain code, Bending energy, Roundness ratio [21], Relative shape [22],

Shape features [5, 6, 10, 14] and Area granulometry [23]. From the previous literature, the features used to extract the intensity of the malaria images are histogram, color histogram, entropy, color channel intensity, and color autocorrelogram. Among all, texture is a finer feature used by most of research fraternity to discriminate infected and non-infected malaria. It provides the spatial distribution of intensity over the region of interest and gives the quantitative evaluation of the images. Some texture features like gray-level co-occurrence matrix (GLCM), gray-level run-length matrix (GLRLM), gray-level size-zone matrix (GLSZM), local binary pattern, flat texture, gradient texture, Laplacian texture, and wavelet feature [6] has been used in previous studies.

1.5 Detection and Classification of the Parasites

It is the process of extracting the information about infected parasite and categorizing its types. The various classification algorithms used in the literature are Bayesian classifier [24], feedforward backpropagation neural network [4, 10, 21], classification and regression tree [5], K-nearest neighbor classifier [9, 14, 16], logistic regression [5], AdaBoost algorithm [25], Naïve Bayes tree [26], support vector machine [3, 6, 7, 9, 17], and multilayer perception network [5, 9].

2 Motivation and Challenges

In microscope, image brightness variation occurs due to the change in the magnification, iris aperture diaphragm, and illumination intensity control knob. As per the WHO standard, the magnification of the malaria blood smear is fixed to 100× objective lens. Hence, the variation of the brightness can occur only because of the manual adjustments of the iris aperture diaphragm and illumination intensity control knob. Moreover, theory of the light microscope defines that brightness depends on the amount of light passing through the condenser controlled via the iris aperture. For low magnification factor (eg.10x), iris aperture is tuned to higher diameter, inversely for the higher magnification, iris aperture is small. It gathers optimum light so we will get bright images. But tuning the illumination intensity controller only makes the variation of brightness in the image. We tried to observe the impact of this brightness variation on the global texture feature.

Image texture provides information about the spatial arrangement of intensities of an image. These texture features are used in the digital pathology to detect or classify the malaria parasites. The characteristics of the texture features behave differently for infected and non-infected erythrocyte regions. So the texture descriptor plays an important role in identifying the malaria parasite. However, when extracting the infected region texture properties, the nearest region property also influences the texture feature [27]. These artifacts influence the definite variation in the texture feature metric. We observed that especially in image-based malaria diagnosis, the previous computer-aided microscopic researches ignored to incorporate this variation in their formulations. From this proposed work, we tried to bring out the existence of non-uniformity in texture descriptor.

3 Experiment

In this session, we describe the experiment which has been conducted by the authors. The purpose of this experiment is to identify the variation in texture values of malaria blood smear images related to its brightness. Texture is one of the important features to improve the sensitivity and specificity of image based malaria detection techniques. We considered the global features of the image rather than segmented erythrocyte. Here, we used Olympus CX21i bright field microscope to view the malaria blood smear images. And Canon EOS1200D digital camera was used to digitize the captured view of microscope. The blood smear is examined with the total magnification of $10\times$ in ocular lens and $100\times$ in objective lens. The procedure is divided into four steps namely smear collection, microscope setup, focusing, and image acquisition with varied brightness, which have been explained below.

3.1 Smear Collection

Giemsa staining primarily uses two solutions: eosin and methylene. Eosin helps to change the parasite nucleus into red, and methylene will change the cytoplasm into blue. WHO recommends Giemsa staining is the reliable method to detect the malaria parasite for early case detection of malaria disease. Therefore, we have collected few Giemsa-stained blood smears infected with *P. falciparum-* and *P. vivax* from Tagore Medical College & Hospital, Chennai, India. The pathologist works for Tagore Medical College & Hospital had labeled the infected regions in the acquired image and these labeled samples were used to evaluate the efficiency of the proposed experiment.

3.2 Microscope Setup

Before proceeding to the required microscopic adjustment, we need to properly clean the microscope. For cleaning the ocular lens, objective lens, and condenser lens, lens paper or clean cotton must be used. Raise the condenser knob to check the amount of light the condenser gathers into microscopic stage. Then, adjust the iris diaphragm to control the amount of light passing through condenser. In order to examine the slides with the magnification of 10× in ocular lens and 100× in objective lens, we need to rotate the nose piece and fix it.

3.3 Focusing

To focus the specimen, we have to place the smear on the microscopic stage. Then, add a drop of liquid paraffin on the stained region of the smear to increase the resolving power of microscope. Now, move the microscopic stage in the upward direction so as to make a contact on the liquid paraffin with the objective lens. Ensure that the appropriate light has to pass through the stained region of the smear.

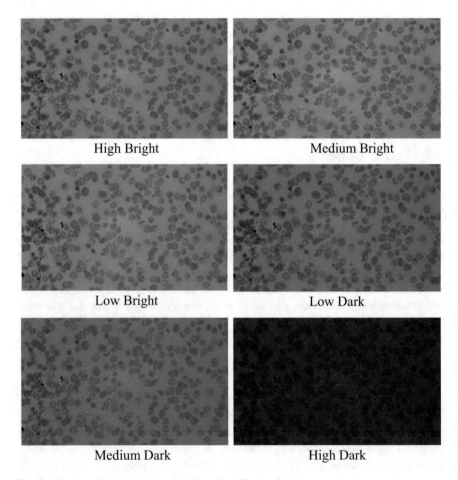

Fig. 2 Microscopic image captured with various illuminations

Adjust the coarse adjustment knob, and look through the ocular eyepiece until the isolated erythrocyte gets visibility. Thereafter, tune the fine adjustment knob to focus the clear details of the erythrocyte image.

3.4 Image Acquisition with Varied Brightness

After completing the above three steps, we need to alter the brightness for same specimen with varied brightness in image acquisition. There are two ways to change the brightness variations in microscope, either increase/decrease the diameter of the iris aperture or adjusting the illumination intensity control (condenser) knob. In existing practice of microscopy, the iris aperture diameter is always inversely proportional to the magnification factor, since, we kept the objective lens in 100× magnification oil immersion and fixed the diameter of the iris aperture as constant throughout the experiment. We made the brightness variation only by adjusting the illumination control knob to vary the light emitted from the light source. In our experiment, the amount of light emitted from the light source could be varied by allowing high illumination to capture high brightness image and low illumination to high dark image and other images are captured in between these illuminations with fixed intervals. Here, the high and low illuminated images are labeled as High Bright (HB), High Dark (HD) images; the intermediate illuminated images are labeled as Medium Bright (MB), Low Bright (LB), Low Dark (LD), Medium Dark (MD) and are shown in Fig. 2.

4 Performance Analysis

In this analysis, we try to figure out the variation of texture descriptor value for the same image with its six classes of brightness level as described in experiment section. We derived GLCM for every image and estimated the global texture feature like energy, contrast, and homogeneity from the constructed GLCM. The obtained texture feature value indicates the measure of closeness of the distribution of GLCM, the local intensity variation of GLCM, and uniformity of the image illumination distribution.

It is observed from the previous literature, Gray World Assumption (GWA) provides the efficient image preprocessing technique to nullify the brightness and contrast variation in image-based malaria diagnosis. Hence, in this proposed analysis, we once again extracted the aforementioned texture feature for the earlier images (six classes) after preprocessing with GWA. However, we found that there is no significant nullification of invariant texture feature after GWA preprocessing. Figures 3, 4, and 5 show the evidence in persistence of variation in GLCM global texture descriptor, even after employing GWA preprocessing.

Fig. 3 Variation in GLCM texture property—contrast

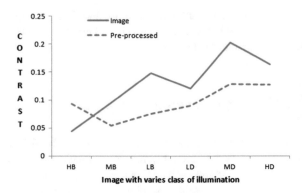

Fig. 4 Variation in GLCM texture property—homogeneity

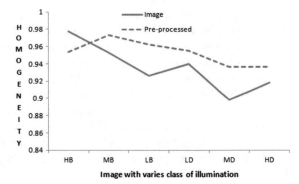

Fig. 5 Variation in GLCM texture property—energy

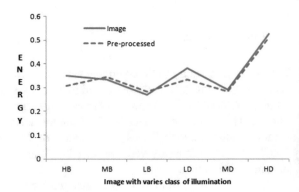

5 Conclusion

Though we have detected the variation of texture feature by varying the image brightness, we are in the process of categorizing the texture features which are sensitive and insensitive for malaria diagnosis. To normalize the variation of sensitive texture features, we are also in the process of formulating a model to nullify the texture variance.

References

1. Tangpukdee N, Duangdee C, Wilairatana P, Krudsood S (2009) Malaria diagnosis: a brief review. Korean J Parasitol 47(2):93–102. https://doi.org/10.3347/kjp.2009.47.2.93
2. Chayadevi M, Raju G (2014) Usage of art for automatic malaria parasite identification based on fractal features. Int J Video Image Process. Netw Secur 14:7–15
3. Linder N, Turkki R, Walliander M et al (2014) A malaria diagnostic tool based on computer vision screening and visualization of Plasmodium falciparum candidate areas in digitized blood smears. PLoS One 9:e104855
4. Khan NA, Pervaz H, Latif AK, Musharraf A (2014) Unsupervised identification of malaria parasites using computer vision. In: Proceedings of 11th international joint conference on computer science and software engineering (JCSSE), Pattaya, Thailand, pp 263–267
5. Das DK, Mukherjee R, Chakraborty C (2015) Computational microscopic imaging for malaria parasite detection: a systematic review. J Microsc 260(1):1–19
6. Das DK, Maiti AK, Chakraborty C (2015) Automated system for characterization and classification of malaria-infected stages using light microscopic images of thin blood smears. J Microsc 257:238–252
7. Ghosh M, Das D, Chakraborty C, Ray AK (2011) Plasmodium vivax segmentation using modified fuzzy divergence. In: Proceedings of international conference on image information processing, Shimla, India, pp 1–5
8. May Z, Aziz SSAM, Salamat R (2013) Automated quantification and classification of malaria parasites in thin blood smears. In: Proceedings of international conference on signal and image processing applications (ICSIPA), Melaka, Malaysia, pp 369–373
9. Diaz G, Gonzalez FA, Romero E (2009) Asemi-automatic method for quantification and classification of erythrocytes infected with malaria parasites in microscopic images. J Biomed Inform 42:296–307
10. Soni J (2011) Advanced image analysis based system for automatic detection of malarial parasite in blood images using susan approach. Int J Eng Sci Technol 3:5260–5274
11. Purwar Y, Shah SL, Clarke G, Almugairi A, Muehlenbachs A (2011) Automated and unsupervised detection of malarial parasites in microscopic images. Malar J 10:364
12. Hanif NSMM, Mashor MY, Mohamed Z (2011) Image enhancement and segmentation using dark stretching technique for plasmodium falciparum for thick blood smear. In: Proceedings of 7th international colloquium on signal processing and its applications (CSPA), Penang, Malaysia, pp 257–260
13. Abbas N, Mohamad D (2013) Microscopic RGB color images enhancement for blood cells segmentation in YCBCR color space for k-means clustering. J Theor Appl Inf Technol 55:117–125
14. Tek FB, Dempster AG, Kale I (2010) Parasite detection and identification for automated thin blood film malaria diagnosis. Comput Vis Image Underst 114:21–32
15. Ma C, Harrison P, Wang L, Coppel RL (2010) Automated estimation of parasitaemia of plasmodium yoelii-infected mice by digital image analysis of Giemsa-stained thin blood smears. Malar J 9:348
16. Khot ST, Prasad RK (2012) Image analysis system for detection of red blood cell disorders using artificial neural network. Int J Eng Res Technol 1:1–14
17. Kumarasamy SK, Ong SH, Tan KSW (2011) Robust contour reconstruction of red blood cells and parasites in the automated identification of the stages of malarial infection. Mach Vis Appl 22:461–469

18. Suradkar PT (2013) Detection of malarial parasite in blood using image processing. Int J Eng Innov Technol 2:124–126
19. Rajesh Kanna B, Aravindan C, Kannan K (2012) Image based area estimation of any connected region using y-convex region decomposition. AEU Int J Electron Commun 66 (2):172–183. https://doi.org/10.1016/j.aeue.2011.06.010
20. Rajesh Kanna B, Aravindan C, Kannan K (2012) Development of yConvex hypergraph model for contour based image analysis. In: Proceedings of the 2nd IEEE international conference computer communication and informatics (ICCCI), vol 2, pp 1–5. https://doi.org/ 10.1109/iccci.2012.6158806
21. Ross NE, Pritchard CJ, Rubin DM, Duse AG (2006) Automated image processing method for the diagnosis and classification of malaria on thin blood smears. Med Biol Eng Comput 44:427–436
22. Springl V (2009) Automatic malaria diagnosis through microscopy imaging. Faculty of Electrical Engineering. Master thesis, Czech Technical University, In Prague, Czech Republic
23. Malihi L, Ansari-Asl K, Behbahani A (2013) Malaria parasite detection in giemsa-stained blood cell images. In: Proceedings of 8th Iranian conference on machine vision and image processing (MVIP), Zanjan, Iran, pp 360–365
24. Anggraini D, Nugroho AS, Pratama C, Rozi IE, Iskandar AA, Hartono RN (2011) Automated status identification of microscopic images obtained from malaria thin blood smears. In: Proceedings of international conference on electrical engineering and informatics (ICEEI), Bandung, Indonesia, pp 1–6
25. Vink JP, Laubscher M, Vlutters R, Silamut K, Maude RJ, Hasan MU, Haan G (2013) An automatic vision-based malaria diagnosis system. J Microsc 250:166–178
26. Maity M, Maity AK, Dutta PK, Chakraborty C (2012) A web accessible framework for automated storage with compression and textural classification of malaria parasite images. Int J Comput Appl 52:31–39
27. Materka A, Strzelecki M (2015) On the effect of image brightness and contrast nonuniformity on statistical texture parameters. Found Comput Decis Sci 40. ISSN 0867-6356

Fusion of Panchromatic Image with Low-Resolution Multispectral Images Using Dynamic Mode Decomposition

V. Ankarao, V. Sowmya and K. P. Soman

Abstract Remote sensing applications, like classification, vegetation, environmental changes, land use, land cover changes, need high spatial information along with multispectral data. There are many existing methods for image fusion, but all the methods are not able to provide the resultant without any deviations in the image properties. This work concentrates on embedding the spatial information of the panchromatic image onto spectral information of the multispectral image using dynamic mode decomposition (DMD). In this work, we propose a method for image fusion using dynamic mode decomposition (DMD) and weighted fusion rule. Dynamic mode decomposition is a data-driven model and it is able to provide the leading eigenvalues and eigenvectors. By separating the leading and lagging eigenvalues, we are able to construct modes for the datasets. We have calculated the fused coefficients by applying the weighted fusion rule for the decomposed modes. Proposed fusion method based on DMD is validated on four different datasets. Obtained results are analyzed qualitatively and quantitatively and are compared with four existing methods—generalized intensity hue saturation (GIHS) transform, Brovey transform, discrete wavelet transform (DWT), and two-dimensional empirical mode decomposition (2D-EMD).

Keywords Dynamic mode decomposition · Multispectral image fusion
DWT · 2D-EMD

V. Ankarao (✉) · V. Sowmya · K. P. Soman
Center for Computational Engineering and Networking (CEN), Amrita School
of Engineering, Amrita Vishwa Vidyapeetham, Coimbatore 641112, India
e-mail: ankarao93@outlook.com

V. Sowmya
e-mail: v_sowmya@cb.amrita.edu

© Springer Nature Singapore Pte Ltd. 2018
A. K. Nandi et al. (eds.), *Computational Signal Processing
and Analysis*, Lecture Notes in Electrical Engineering 490,
https://doi.org/10.1007/978-981-10-8354-9_31

1 Introduction

Remote sensing has emerged as powerful technology to understand the changes in our planet with the help of massive informational data. Data is collected from different sources and it is available in different formats like maps, images. Current era works on space bone datasets. Satellites are able to provide earth images by scanning the earth in various regions of electromagnetic spectrum [1]. These images can be panchromatic image of a single band or multispectral image consisting of three to seven different bands and hyper-spectral image of 100–200 spectral bands. Information present in space bone imagery plays an important role in many applications like classification, segmentation, change detection, vegetation. [2]. All these applications require high-resolution data. Many satellites are there to provide high-resolution data but, the datasets have the limitation such as, lack of spatial information in spectral bands. To overcome this limitation, a lot of methods are available like pan-sharpening, image fusion. But, all methods have tradeoff with statistical properties of images. Image fusion is emerging field which intent to give a superior perception of data by merging multiple images of same area. Due to the various constraints, several fusion methods were developed to merge two images together resulting an image of better spectral and spatial resolutions. Among the existing methods, intensity, hue, saturation (IHS) transform is the most used method from last decades [3]. But this method has the limitation with number of bands. Apart from this method, there are lot of methods were proposed for image fusion with wavelet decomposition, Brovey transform, variational mode decomposition, etc. [4, 5], in IHS transforms the multispectral bands are transformed into the IHS domain by the application of linear operators on each band. To overcome the effect of spectral distortions caused by the linear operations, Tu et al. proposed the extended IHS method [6]. The mathematical equation used in the extended IHS is: $F_L = m_L + \beta$, where F_L is the fused band of band L, m_L is the multispectral band of band L and $\beta = \text{pan} - K$ where $K = \frac{1}{L}\sum_{i=1}^{L} m_i$, $L = 1, \ldots, N$, where N is the numbers of bands in multispectral image. Later, Nunez et al. proposed a method for image fusion using wavelets by component substitution followed by the framework based on multiresolution analysis is proposed for image fusion [7]. Survey about the available image fusion methods is presented in [8–10]. Recently, Vishnu et al. proposed a method for image using variational mode decomposition [5]. All these methods have limitation with the spectral distortion.

In this work, we propose a framework for embedding spatial information onto multispectral image using image fusion with dynamic mode decomposition (DMD). Obtained results are analyzed qualitatively and quantitatively and compared with four existing methods—generalized intensity hue saturation (GIHS) transform, Brovey transform, discrete wavelet transform (DWT), and two-dimensional empirical mode decomposition (2D-EMD) [4, 6, 11].

Rest of the paper is organized as follows: Sect. 2 gives the detailed explanation about dynamical mode decomposition (DMD). Section 3 discusses about the proposed method and Sect. 4 gives the details about the results obtained by the proposed method and comparison of quality metrics with existing methods [4, 6, 11].

2 Dynamical Mode Decomposition

Dynamic mode decomposition (DMD) is a mathematical procedure that was formed and refined during the last decade. It is mainly used in simulating, controlling, and studying the variations observed in nonlinear complex systems without having much knowledge about the underlying equations that makeup the system. Informational data of these nonlinear systems collected experimentally (or) through various simulations (or data generation methods) is taken in various steps of time are given to DMD for processing. DMD tries to analyze the data so that we may get an idea of the states of system at current time or in a future context, and to identify the consistent parameters associated within the system. It can be used to ascertain spatial-temporal data analytics of a system [12]. The major capability of DMD can be understood by taking into consideration, the low dimensionality of a complicated system and then describing the system in a computationally and theoretically tractable form. The data collected for analysis using DMD is time spaced regularly. The proper orthogonal decomposition (POD) is used for the determination of the variable dynamics of the system is approximated by DMD. In this work, we separated leading and lagging eigenvalues from the eigenvalues of POD [13]. From the separated eigenvalues, we constructed DMD modes.

Let us assume an $m - 1$ sequences of frames (or vectors) separated in time by Δt, given by

$$X_1^{m-1} = \left[x_1 \quad x_2 \quad x_3 \quad \cdots \quad x_{m-1} \right]$$

Next is to find operator (or) transformation matrix A such that $AX_1^{m-1} \approx X_2^m$ where $X_2^m = \left[x_2 \quad x_3 \quad x_4 \quad \cdots \quad x_m \right]$

This implies A is such that

$$Ax_1 = x_2, A^2 x_1 = Ax_2 = x_3, \ldots, A^{m-1} x_1 = x_m$$

Let us assume that the system is evolving slowly and the matrix is of low rank, so that, by the time, A reaches to the index m, frame x_m can be represented as a linear combination of previous frames with small error. That is,

$$x_m = \sum_{j=1}^{m-1} a_j x_j + r$$

$$X_2^m = X_1^{m-1} S + r e_{m-1}^T$$

So, we write

$$X_2^m \approx X_1^{m-1} S$$

On taking SVD of X_1^{m-1}, we obtain

$$X_2^m \approx U\Sigma V^H S$$

$$S \approx V\Sigma^{-1} U^H X_2^m$$

Through similarity transformation, we obtain

$$\tilde{S} = \left(V\Sigma^{-1}\right)^{-1} S\left(V\Sigma^{-1}\right) = U^H X_2^m V\Sigma^{-1}$$

A is related with S. But the frame size of S is less than A.

If total number of frames m, then size of S is $(m - 1 \times m - 1)$. So, A has m eigenvalues, whereas S has $m - 1$ eigenvalues.

$$AX_1^{m-1} \approx X_1^{m-1} S = X_2^m$$

$$AU\Sigma V^H \approx X_2^m$$

$$AU \approx X_2^m V\Sigma^{-1}$$

$$U^H AU \approx U^H X_2^m V\Sigma^{-1} = \tilde{S}$$

$$AU \approx U\tilde{S} = U\left(W\Omega W^{-1}\right)$$

$$A(UW) \approx (UW)\Omega$$

$$A\Phi \approx \Phi\Omega$$

From the above relations, we can say eigenvalues of S are also eigenvalues of A. Here, W is of size $(m - 1) \times (m - 1)$. So, number of columns in Φ is $m - 1$. The columns of Φ represent eigenvectors of A, but there are $m - 1$ vectors. From W and Ω, we can write the relation $\tilde{S}w_j = \mu_j w_j$

$$\Phi = \begin{bmatrix} | & | & | & & | \\ \phi_1 & \phi_2 & \phi_3 & \cdots & \phi_{m-1} \\ | & | & | & & | \end{bmatrix} = UW \Rightarrow \phi_j = Uw_j$$

These ϕs are eigenvectors of A. Theoretically

$$A^k = \Phi_{full}\Omega^k_{full}\Phi^{-1}_{full}$$

But full eigenvectors of A are not available to access. Since A is of low rank, the obtained eigenvectors are enough to capture the dynamics of A. So, $m - 1$ eigenvalues and eigenvectors of A are present in the truncated version $A\Phi \approx \Phi\Omega$. The time instance between two data frame is Δt. To obtain the system vector x_t at any t, define

$$\omega_j = \log(\mu_j)/\Delta t, \quad t/\Delta t = k$$

and

$$e^{\omega_j t} = e^{t \times \log(\mu_j)/\Delta t} = e^{(t/\Delta t) \times \log(\mu_j)} = e^{k \times \log(\mu_j)} = e^{\log\left((\mu_j)^k\right)} = (\mu_j)^k$$

Modes for datasets are

$$X_{DMD}(t) = X_{DMD}(t) = A^t x_1 = \Phi_{full}\Omega^t_{full}\left(\Phi^{-1}_{full}x_1\right) \approx \Phi(\Omega_{\Delta t})^t b$$

where b is obtained by taking pseudo inverse. That is,

$$b = (\text{pinv}(\Phi)) \times x_1 = \left(\Phi^T\Phi\right)^{-1}\Phi^T x_1$$

$$X_{DMD}(t) = A^t x_1 = \Phi(\Omega_{\Delta t})^t b = \sum_{j=1}^{l < m-1} b_j \phi_j e^{\omega_j t}$$

$$X_{DMD}(t) = x_t = A^t x_1 = \Phi(\Omega_{\Delta t})^t b = b_p \phi_p e^{\omega_p t} + \sum_{j \neq p}^{l < m-1} b_j \phi_j e^{\omega_j t}, \; \|\omega_p\| \approx 0$$

$b_p \phi_p e^{\omega_p t}$ correspond to low energy samples and $\sum_{j \neq p}^{l < m-1} b_j \phi_j e^{\omega_j t}$ correspond to high energy samples.

3   Proposed Methodology

In this work, we propose a method for image fusion using dynamic mode
decomposition (DMD) and weighted fusion rule. Dynamic mode decomposition is
a data-driven model and it is able to provide the leading eigenvalues and eigenvectors. Separation of eigenvalues helps to analyze the spatial and temporal deviations in the data, which helps to figure out the maximum data presence. By
separating the leading and lagging eigenvalues, we are able to construct modes for
the datasets. Construction of modes from the eigenvalues and eigenvectors reduces
the effect occurred by the spectral distortions and helps to find the changes in data
whether in spatial or temporal resolutions further used in remote sensing applications. Reducing the effect of spectral distortions helps to increase the quality of
fused image. We have calculated the fused coefficients by applying the weighted
fusion rule for the decomposed modes. Framework for proposed method is given in
Fig. 1. In this method, DMD takes the responsibility of constructing modes.

As shown in Fig. 1, DMD is applied on individual band of multispectral image,
panchromatic image and the fusion rule is applied on DMD outcome to get fused
band. This procedure is repeated for all the bands in multispectral data. Fused bands
are stacked together for getting fused multispectral image. As preprocessing task,
input data for DMD is created by stacking the multispectral band and panchromatic
image. Let 'A' be single band of multispectral image and B is a panchromatic image,
then A-B-A-B is the input data. Then, DMD applied on the input data. Fusion rule
is given by

$$F(x) = \alpha(x)I_i(x) + \beta(x)J(x) \quad i = 1, 2, 3 \ldots m$$

where 'm' denotes the total number of bands in multispectral image, $F(x)$ denotes
the fused image, $I_i(x)$ and $J(x)$ denote the modes corresponding to each multispectral band and panchromatic image. $\alpha(x)$ and $\beta(x)$ denote the weighted coefficients satisfying the condition $\alpha(x) + \beta(x) = 1$.

Fig. 1 Proposed framework
for DMD-based panchromatic
and multispectral image
fusion

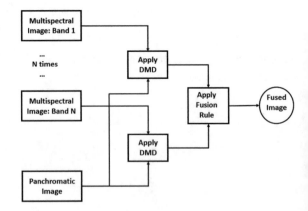

4 Results and Discussion

4.1 Dataset Description

Proposed fusion method based on DMD is validated on four different datasets. First dataset is of location—Rajasthan, India, consisting of panchromatic image of resolution 60 cm and multispectral band image of resolution 2.4 m obtained from QuickBird satellite [14]; the second dataset is the location of Capetown, South Africa, consisting of panchromatic image of resolution 50 cm and multispectral band image of 2 m resolution obtained by GeoEye-1 satellite; third dataset is the location of Adelaide, Australia, and consists of panchromatic image of resolution 30 cm and multispectral band image of resolution 1.2 m obtained from WorldView-3 satellite; and fourth dataset is the location of Sydney, Australia, and consists of panchromatic band image of resolution 50 cm and multispectral band image of resolution 2 m. Figure 2 represents the sample dataset, for experimental purpose each multispectral data is converted as low-resolution image by blurring the image using Gaussian filter.

4.2 Results

Fused image by proposed methodology using DMD is shown in Fig. 3. In Fig. 3c, we observe enhanced visualization of multispectral image with high spatial information. Figure 4c represents the fused image using 2D-EMD where there is minimal loss of edge information due to the reconstruction of modes using the residue, which is a low-frequency component and it contains spatial information. Figure 5 shows the fused images obtained using the existing techniques for image fusion. The fused image obtained using DWT (Fig. 5a) is blurred because the fused coefficients are obtained from low frequency of nth level of decomposition and reconstruction is done using high-frequency bands. The image statics are affected

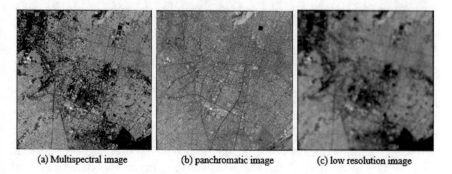

(a) Multispectral image   (b) panchromatic image   (c) low resolution image

Fig. 2 Sample dataset for panchromatic and low-resolution multispectral image fusion

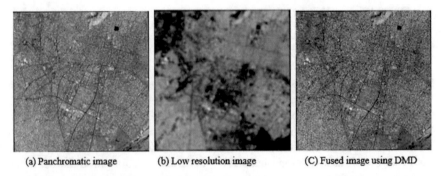

(a) Panchromatic image   (b) Low resolution image   (C) Fused image using DMD

Fig. 3 Fused image using proposed method

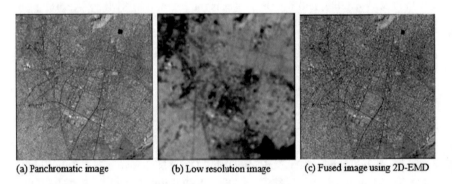

(a) Panchromatic image   (b) Low resolution image   (c) Fused image using 2D-EMD

Fig. 4 Fused image using 2D-EMD

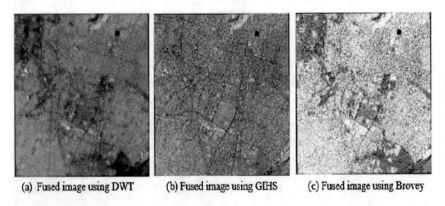

(a) Fused image using DWT   (b) Fused image using GIHS   (c) Fused image using Brovey

Fig. 5 Fused images using existing methods

for fused image obtained using GIHS transform (Fig. 5b) due to the space transformation. The level of details of the fused image obtained using Brovey transform (Fig. 5c) is less because of the linear combination of multispectral image and panchromatic image.

Table 1 Quality metric for proposed method

Dataset	SSIM index	PSNR (dB)	MSE	NAE	Correlation coefficients		
					R	G	B
1	0.46	12.67	3500	0.33	0.92	0.91	0.90
2	0.76	20.75	562.1	0.19	0.77	0.82	0.83
3	0.69	19.80	678.9	0.22	0.92	0.92	0.91
4	0.57	20.43	588.3	0.47	1.0	1.0	0.95

4.3 Quality Metrics and Comparison of Quality Metrics

Five quality metrics are calculated in qualitative analysis, and consists of mean square error (MSE), normalized average error (NAE), structural similarity index (SSIM), correlation coefficients (CC), peak signal to noise ratio (PSNR), which are used to analyze the performance of any image processing application [15–17]. Quality metrics of the proposed method is tabulated in Table 1.

PSNR for all the datasets is above 13 dB, which reflects the good image quality of fused image. The normalized average error (NAE) values are also low, which assures that fused image quality is good. The structural similarity between fused image and multispectral image is calculated by the metric structural similarity index (SSIM). The values of SSIM index are in the range of 0.45–0.8, which shows that fused image has high structural similarity. Similarly, from all the metric values, it is evident that the fused multispectral image has good quality with enhanced spatial details with spectral information. The comparison of quality metrics computed for the proposed method against the existing techniques for image fusion is tabulated in Tables 2, 3, 4 and 5.

The experimental values for standard deviation are 5, α is 0.7, and β is 0.3. Quality metrics of different methods for the first dataset are tabulated in Table 2. By comparing PSNR values of DWT, GIHS, and Brovey transforms with the PSNR of proposed method, it is evident that fused multispectral image quality of proposed method is better. Similarly, the comparison of metrics such as NAE, MSE, and SSIM shows that the fused multispectral image obtained from proposed fusion

Table 2 Quality metrics comparison of different methods for first dataset

Dataset	SSIM index	PSNR (dB)	MSE	NAE	Correlation coefficients		
					R	G	B
DMD	**0.46**	**12.67**	**3500**	**0.33**	**0.92**	**0.91**	**0.90**
2D-EMD	**0.47**	**12.8**	**3300**	**0.349**	**0.93**	**0.93**	**0.94**
DWT	0.3	4.9	21,000	223.54	0.2	0.22	0.24
GIHS	0.31	7.5	1100	0.41	0.62	0.56	0.59
Brovey	0.33	8.6	8800	0.4	0.65	0.63	0.64

Table 3 Quality metrics comparison of different methods for second dataset

Method	SSIM index	PSNR (dB)	MSE	NAE	Correlation coefficients		
					R	G	B
DMD	**0.76**	**20.75**	**562.1**	**0.19**	**0.77**	**0.82**	**0.83**
2D-EMD	**0.73**	**20.13**	**630.33**	**0.27**	**1.0**	**1.0**	**1.0**
DWT	0.3	9.45	7300	280.7	0.21	0.26	0.27
GIHS	0.69	18.35	955.86	0.2	0.68	0.76	0.78
Brovey	0.69	18.55	900	0.21	0.88	0.87	0.86

Table 4 Quality metrics comparison of different methods for third dataset

Method	SSIM index	PSNR (dB)	MSE	NAE	Correlation coefficients		
					R	G	B
DMD	**0.69**	**19.80**	**678.9**	**0.22**	**0.92**	**0.92**	**0.91**
2D-EMD	**0.70**	**20.43**	**588.5**	**0.21**	**0.98**	**0.96**	**0.97**
DWT	0.49	8.6	8000	239.9	0.24	0.24	0.23
GIHS	0.55	12.75	3400	0.36	0.65	0.65	0.63
Brovey	0.53	14.51	1800	0.299	0.75	0.75	0.75

Table 5 Quality metrics comparison of different methods for fourth dataset

Method	SSIM index	PSNR (dB)	MSE	NAE	Correlation coefficients		
					R	G	B
DMD	**0.57**	**20.43**	**588.3**	**0.47**	**1.0**	**1.0**	**0.95**
2D-EMD	**0.55**	**20.45**	**585.1**	**0.52**	**0.99**	**0.99**	**1.0**
DWT	0.39	12.40	3700	261.9	0.26	0.22	0.24
GIHS	0.45	17.67	1100	0.3	0.75	0.71	0.68
Brovey	0.42	18.4	932.4	0.59	0.82	0.82	0.83

method has better image quality. The metric correlation coefficients decide the spatial information present in fused multispectral image. Correlation coefficients obtained from the fused multispectral image using proposed method are better than DWT, GIHS, Brovey transforms. Apart from the proposed method, the 2D-EMD method also produces results which are comparable with the proposed method since the reconstruction of fused image with the residue contains the high spatial information and edge information. From the comparison of quality metrics for all datasets, it is evident that the proposed method performs better for all datasets.

5 Conclusion

The work presented in this paper proposes a new method for embedding spatial information onto multispectral image by using the image fusion with DMD and weighted fusion rule. Quality metrics obtained by proposed method are compared with the quality metrics obtained by existing methods like GIHS, Brovey transform, DWT, and 2D-EMD. From the comparison of quality metrics of proposed method with DWT, GHIS transform, Brovey transform, and 2D-EMD, each metric has better results. In case of correlation coefficients, high improvement is observed which confirms the maximum spatial information embedded on multispectral image. Detail analysis of quality metrics assures that proposed method performs well compared with DWT, GHIS transform, Brovey transform and performs comparable with 2D-EMD due to the consideration of residue in reconstruction. In case of the proposed method based on DMD, output fused image obtained using the fusion rules causes the slight spectral distortions due to the linear operation.

References

1. Richards JA, Jia X (1999) Sources and characteristics of remote sensing image data. Remote sensing digital image analysis. Springer, Berlin, Heidelberg, pp 1–38
2. Asrar G, Dozier J (1994) EOS: science strategy for the earth observing system. American Institute of Physics, Woodbury, NY
3. Carper W (1990) The use of intensity-hue-saturation transformations for merging SPOT panchromatic and multispectral image data. Photogramm Eng Remote Sens 56(4):457–467
4. Wang Z et al (2005) A comparative analysis of image fusion methods. IEEE Trans Geosci Remote Sens 43(6):1391–1402
5. Vishnu PV, Sowmya V, Soman KP (2016) Variational mode decomposition based multispectral and panchromatic image fusion. Int J Control Theor Appl 9(16): 8051–8059
6. Tu TM et al (2001) A new look at IHS-like image fusion methods. Inf Fusion 2(3):177–186
7. Nunez J et al (1999) Multiresolution-based image fusion with additive wavelet decomposition. IEEE Trans Geosci Remote Sens 37(3):1204–1211
8. Thomas C et al (2008) Synthesis of multispectral images to high spatial resolution: a critical review of fusion methods based on remote sensing physics. IEEE Trans Geosci Remote Sens 46(5):1301–1312
9. Gómez-Chova L et al (2015) Multimodal classification of remote sensing images: a review and future directions. Proceedings of the IEEE 103(9):1560–1584
10. Ghassemian Hassan (2016) A review of remote sensing image fusion methods. Inf Fusion 32:75–89
11. Wang J, Zhang J, Liu Z (2008) EMD based multi-scale model for high resolution image fusion. Geo-spatial Inf Sci 11(1):31–37
12. Brunton SL et al (2015) Compressed sensing and dynamic mode decomposition. J Comput Dynam 2(2)
13. Grosek J, Kutz JN (2014) Dynamic mode decomposition for real-time background/foreground separation in video. arXiv preprint arXiv:1404.7592
14. URL {http://glcf.umd.edu/data/quickbird/}
15. Agarwal J, Bedi SS (2015) Implementation of hybrid image fusion technique for feature enhancement in medical diagnosis. Human-centric Comput Inf Sci 5(1):1

16. Kaur S, Kaur K (2012) Study and implementation of image fusion methods. Int J Electron Comput Sci Eng 1(03):1369–1373 (IJECSE, ISSN: 2277–1956)
17. Moushmi S, Sowmya V, Soman KP (2015) Multispectral and panchromatic image fusion using empirical wavelet transform. Indian J Sci Technol 8(24)

Enhanced Scalar-Invariant Feature Transformation

S. Adithya and M. Sivagami

Abstract The proposed work enhances the feature point detection in the scalar-invariant feature transformation (SIFT). The sequence of steps in the SIFT algorithm drops most of the feature points in the low-contrast regions of the image. This paper provides a solution to this problem by adding the output of Sobel filtered image on the input image iteratively until the entropy of the input image is increased to a saturation level. The SIFT descriptors generated for this enhanced image tend to describe the redundant features in the image. To overcome this problem, affinity propagation clustering is done.

Keywords Scalar-invariant feature transformation (SIFT) · Low contrast
Entropy · Sobel · Affinity propagation clustering

1 Introduction

Scalar-invariant feature transformation (SIFT) proposed by Lowe et al. [1] is one of the most popular object recognition techniques in content-based image retrieval. The popularity of this algorithm is due to its scalar invariance, local feature description, and robustness to illumination and clutter. SIFT is being used in competition with convolution neural networks in several industrial and research applications like object detection, object recognition, 3D reconstruction, and motion recognition. Although being such a robust algorithm, SIFT discards the feature points in low-contrast areas in an image. This is due to the sequence of steps starting from difference of Gaussian which produces fine detailed response in the

S. Adithya · M. Sivagami (✉)
School of Computing Sciences and Engineering, VIT University, Chennai Campus,
Chennai, India
e-mail: msivagami@vit.ac.in

S. Adithya
e-mail: sadithya.2014bce1086@vit.ac.in

© Springer Nature Singapore Pte Ltd. 2018
A. K. Nandi et al. (eds.), *Computational Signal Processing
and Analysis*, Lecture Notes in Electrical Engineering 490,
https://doi.org/10.1007/978-981-10-8354-9_32

low-contrast area and this thin response in the low-contrast regions is removed by Taylor series expansion.

Apart from SIFT, several other feature point detectors are being used in industrial and research applications like SURF, BRISK, and FREAK. SURF stands for speeded-up robust feature which was inspired from SIFT algorithm, the SURF uses Hessian blob detection to find the feature points and whereas the feature vectors are obtained by suming up the Haar Wavelets. BRISK stands for binary robust-invariant scalable keypoints which calculates 512-bit binary feature points using average-weighted Gaussian on the feature point location. Fast Retina keypoint also known as FREAK is a binary feature detector which is a cascade of binary strings, and these strings are calculated by retinal sampling pattern and image intensities. FREAK occupies low memory and performs really fast computation, making it easier to use in real-time applications.

Section 2 represents the related work in the enhancements of the SIFT algorithm. Section 3 discusses the proposed work. Section 4 describes the experimental results obtained after application of the algorithm. The conclusion is provided in Sect. 5, and references are provided in the end.

2 Related Work

Pulung et al. [2] have used contrast limited adaptive histogram equalization as a preprocessing for SIFT feature point matching for underwater images. In this work, Rayleigh distribution is used to represent the heavy noise and the authors have compared their work with contrast stretching as a reference method.

Linagping et al. [3] have proposed histogram equalization as a preprocessing for SIFT to increase the number of feature points and detection.

Sebastian et al. [4] have improved the SIFT by focusing on the large feature vectors by performing the geometric correction, creation of orthophotos, and 3D modeling.

Guohua et al. [5] have described building SIFT-like descriptors for image registration using information of the gradient, and the results showed that these descriptors have outperformed the traditional SIFT descriptors.

3 Proposed Work

In this paper, we propose a modified SIFT technique to enhance the feature point detection in the low contrast regions of the image by iteratively adding the Sobel filtered output to the input image until the entropy of the image is saturated. The Sobel edge detector is applied on the input image to enhance the image highlighting the feature points in a thicker form, and this enhanced result is added to the input image, resulting in an increased entropy.

- Iterative Sobel based on entropy
- Scale space extrema detection
- Keypoint localization
- Orientation assignment
- Local image descriptor
- Affinity propagation clustering

3.1 Iterative Sobel Based on Entropy

Sobel is a spatial filter applied to extract the features in an image. The Sobel calculates the first-order derivative gradient at each pixel producing a thick (high) response at the location of features while a zero response in case of non-feature locations.

The Sobel operator is a good preprocessing operation for several content-based image retrieval applications. Sobel serves to reduce the noise in image by smoothening the image and produce thick response feature points. As a result Taylor series expansion does not drop these feature points especially the one in low-contrast region anymore as there they are thick in characteristic.

Based on this idea the output of Sobel filter is added to the original image so as to enhance its features in low-contrast regions.

We have used the entropy metric to decide whether image has been enhanced to maximum level after applying Sobel filter. The Sobel filter is applied iteratively till the entropy becomes constant. Entropy describes the randomness in a particular system, so it has been applied to the images to extract information or understand the characteristics of the image. Entropy is the base for several compression algorithms and is also used to find the contrast in images. Lower the entropy, lower is the contrast. Entropy of an image is calculated using (1):

$$\text{entropy} = -\sum P_i \times \log(P_i) \tag{1}$$

$$g = \sqrt{g_x^2 + g_y^2} \tag{2}$$

$$\Theta = \tan^{-1}\left(\frac{g_y}{g_x}\right) \tag{3}$$

Here, P(i) in (1) represents pixel value at the position i.

The proposed work starts with applying Sobel filter to the input image. There are two different kernels of Sobel operator: Each of them computes gradient in the x and y directions, respectively, as shown in Fig. 1, and we finally sum the gradients by taking root over the sum of squares of these gradients (2). The output is an enhanced image which is added to the original image. This enhancement process is done continuously until the entropy is saturated.

$$g_x = \begin{array}{|c|c|c|} \hline -1 & 0 & 1 \\ \hline -2 & 0 & 2 \\ \hline -1 & 0 & 1 \\ \hline \end{array} \quad g_y = \begin{array}{|c|c|c|} \hline -1 & -2 & -1 \\ \hline 0 & 0 & 0 \\ \hline 1 & 2 & 1 \\ \hline \end{array}$$

Fig. 1 Sobel filter in X and Y directions

Fig. 2 The first set of images are the input images from ImageNet dataset, and the second set of images represent enhanced output due to the iterative Sobel filter based on entropy

The result after this iterative process is an image enhanced up to saturation level, i.e., for sure the low-contrast regions are enhanced in the image where several important feature point is hidden in the set of input as shown in Figs. 2 and 9.

3.2 Scale Space Extrema Detection

The blurring is done using Laplacian of Gaussian at different scales as shown in Fig. 3. The objective of blurring at different scales is to obtain maximum information from the image by choosing the scale that has maximum response for that particular pixel. The scale used is different for different images so as to extract the best possible information from the image by subsampling. In practice, the Laplacian of Gaussian is implemented as difference of Gaussian as it is a simpler operation and resembles the way how visual cortex identifies different objects.

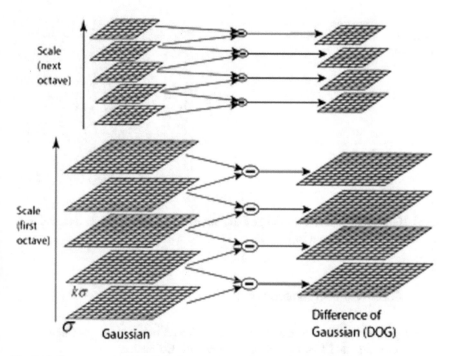

Fig. 3 Scale space generation of the image using difference of Gaussian

In the SIFT algorithm first the Gaussian filter is applied to the input image for different values of σ and represented in a pyramid format [6]. For the first octave, the input image is subsampled by half and this is done sequentially for all the subsequent octaves. In each octave, the blurring is achieved by Gaussian filter which is done with $K = k^\wedge(i) * \sigma$ where i is the number of the octaves. The same filter is applied at different scales to find the most stable scale for each pixel as the effect of the filter changes when the image is subsampled.

The empirical analysis of the SIFT feature detector was performed for different images, and it was found that the repeatability was highest for the scale 3 and σ value 1.6.

$$\frac{\partial G}{\partial \sigma} = \sigma \Delta^2 G \tag{4}$$

$$\sigma \Delta^2 G = \frac{\partial G}{\partial \sigma} = \frac{G(x, y, k\sigma) - G(x, y, \sigma)}{k\sigma - \sigma} \tag{5}$$

$$G(x, y, k\sigma) - G(x, y, \sigma) \approx (k - 1)\sigma^2 \Delta^2 G \tag{6}$$

After creating the scale space, the extremas are located by checking the 26-pixel neighborhood including one level up, one level down, and the one in the same

Fig. 4 Scale space extrema
calculation and representation
of neighbors, one level up,
down, and at the same level

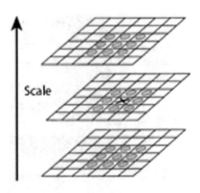

Scale

level. If the pixel is smallest or largest among its neighborhood, then it is said to be
a feature point as shown in Fig. 4.

3.3 Feature Point Localization

In this part, the outliers in the image are eliminated using techniques like Taylor
series expansion and Hessian Principal Curvature Estimation.

The Taylor series expansion is calculated for all the feature points and if the
value at the feature point location is less than the threshold, then we drop it.

$$X = (x, y, \sigma)^T \tag{7}$$

The X represents the ordered pair of x, y, and σ (7) which is used to represent the
difference of Gaussian (D) as a function as shown in (8).

$$D(X) = D + \frac{\partial D^T}{\partial X} X + \frac{1}{2} X^T \frac{\partial^2 D}{\partial X^2} X \tag{8}$$

The Taylor series expansion of the difference of Gaussian at the extrema for the
given value of x is shown in (9).

$$\hat{X} = -\frac{\partial^2 D^{-1}}{\partial X^2} \frac{\partial D}{\partial X} \tag{9}$$

For the point to be accepted, the value of difference of Gaussian should be
greater than the threshold value.

$$|D(X)| > \text{th} \tag{10}$$

The threshold value is set to 0.03.

The poorly localized points that are along the edge are removed using Hessian matrix.

In this formulation, the difference of Gaussian is treated as a surface and its principal curvature is calculated using the Hessian. If the curvature is low at one side and high at the other side, then the feature point is said to be a poorly localized feature point and these points are removed from the set of feature points.

$$H = \begin{bmatrix} D_{xx} & D_{xy} \\ D_{xy} & D_{yy} \end{bmatrix} \tag{11}$$

The Hessian matrix computes the partial double derivative along all dimensions measuring the variation of the intensity across the space as shown in (11). The point which is poorly localized satisfies the following criteria determined using trace, determinant, and ratio r of the matrix as shown in (16, 17):

$$\text{Tr}(H) = D_{xx} + D_{yy} = \lambda_1 + \lambda_2 \tag{12}$$

$$\text{Det}(H) = D_{xx}D_{yy} - D_{xy}^2 = \lambda_1\lambda_2 \tag{13}$$

$$r = \frac{\lambda_1}{\lambda_2} \tag{14}$$

$$\frac{\text{Tr}(H)^2}{\text{Det}(H)} = \frac{(\lambda_1 + \lambda_1)^2}{\lambda_1\lambda_2} = \frac{(r\lambda_1 + \lambda_1)^2}{r\lambda_2^2} = \frac{(r+1)^2}{r} \tag{15}$$

$$\frac{\text{Tr}(H)^2}{\text{Det}(H)} < \frac{(r+1)^2}{r} \tag{16}$$

$$r < 10 \tag{17}$$

3.4 Orientation Assignment

The orientation at a point is the most orientation-dominant gradient vector among all the gradient vector of that pixel (Figs. 5 and 6). This is done to achieve the rotation invariance. Based on this orientation, the descriptors are calculated.

To implement this, the gradient of the feature point is calculated and a weighted histogram is formed for each such feature point. The weighted histogram is of 36 bins with each of them having allotted weights and the maximum or peak is chosen among all the orientation.

In case there are two peaks and if the ratio of first nearest neighbor to the second nearest neighbor is less than 0.8, then the first nearest neighbor is chosen, otherwise ambiguity arises as shown in Fig. 7.

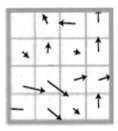

Fig. 5 Weighted direction histogram in neighborhood of the feature point

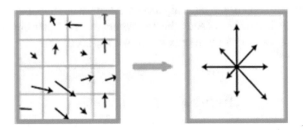

Fig. 6 A complete histogram generation

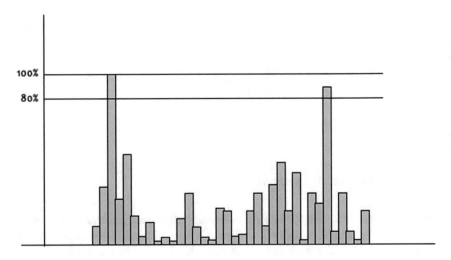

Fig. 7 Peak detection

3.5 Local Image Descriptor

To obtain the image descriptor, a gradient orientation histogram is created, making the descriptor invariant to 3D viewpoint and spatial transformation as shown in Fig. 8.

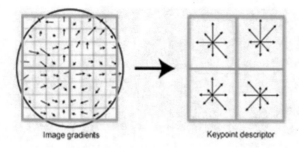

Fig. 8 Conversion of weighted orientation histogram of 8 × 8 around an interest point into a 2 × 2 descriptors

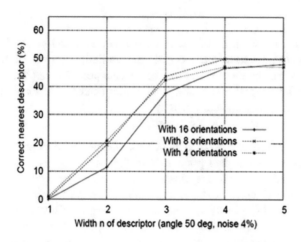

Fig. 9 Analysis of descriptor for different orientation windows

The descriptors are created by a weighted histogram around 16 × 16 neighborhood with each having 8 bins (Fig. 9). So a total of 16 × 8 = 128-dimensional vector is generated. Thus, the optimal window is reduced from 16 × 16 to 4 × 4. The final descriptors generated are shown in Fig. 10.

3.6 Affinity Propagation Clustering

Affinity propagation clustering [7] algorithm automatically determines the number of clusters and performs clustering by message passing.

Affinity propagation only requires the similarity between each data point as an input.

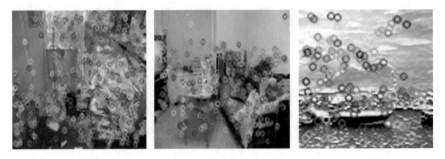

Fig. 10 The feature points detected after enhancing the input set of images

$$s(i, k) = -\|x_i - x_k\|^2 \qquad (17)$$

Here, in (17) $s(i, k)$ is the similarity between nodes i and k.

The input need not necessarily contain the similarity between all the nodes.

$s(k, k)$ describes the input preference of the data point k being the cluster center; if all of them have equal preference, then cluster center is decided using the subsequent messaging and updation activity.

There are two updations in the algorithm:

1. These are the type of messages that check whether the data point k is a perfect cluster center for the data point i; basically, the points here are fighting for adding the data points to the clusters. The message here contains the responsibility matrix R which contains responsibility between nodes i and k, and it is updated using (18). For the first iteration, the responsibility matrix is initialized using similarity measure between the each data point. Algorithm over iterations effectively assigns data point to clusters. The stability is achieved when the availability between these data points goes negative making the high similarity value ineffective and allowing the algorithm to settle to stable end rapidly. $r(k, k)$ is termed as the "self-responsibility." The self-responsibility parameter describes the preference of node k as the cluster center. So as to keep $r(k, k)$ values stable, the availability is kept negative so that the other $r(i, k)$ of positive values does not affect the decision.

2. The availability matrix A represents the availability between nodes i and k as a (i, k), and it is updated using (19). For the first iteration, the availability matrix is initialized to zero. Then, the availability is updated based on (19). Only the positive values of the responsibility are chosen as we require good cluster center to explain the data points. Similar to "self-responsibility," there is another parameter "self-availability" which indicates that how qualified the point k is to be the cluster center based on the positive message responsibility sent by other nodes. This "self-availability" is updated based on formula (20).

The cluster centers can be located using the availability and responsibility values of the node I as shown in (21). Similarly, cluster centers can be found using (21).

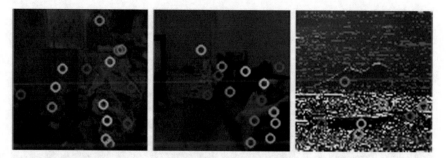

Fig. 11 Representation of the feature points obtained after affinity propagation clustering

During updations, there are many circumstances where oscillation of the data points occur. To avoid this, damping is done using a damping factor, and for each message sent, a damping factor is multiplied by the value of previous iteration plus the 1-damping factor times the prescribed update value. The default value of the damping factor is between 0.5 and can vary from 0 to 1. Figure 11 gives a visualization of how affinity clustering automatically finds the cluster centers and clusters the feature points achieved from Fig. 10.

$$r(i,k) \leftarrow s(i,k) - \max\{a(i,k') + s(i,k')\} \tag{18}$$

$$a(i,k) \leftarrow \min(0, r(k,k) + \sum \max\{0, r(i',k)\}) \tag{19}$$

$$a(k,k) \leftarrow \sum \max\{0, r(i',k)\} \tag{20}$$

$$a(i,k) + r(i,k) \tag{21}$$

4 Experimental Results

After a series of experiments on the input images from Handmetric Authentication Beijing Jiao Tong University (HA-BJTU) biometric database [8], it has been observed that the Sobel operator is an effective enhancement technique. Figure 12 shows the several palm print images to which the Sobel filter and difference of Gaussian are applied.

The Sobel filtered image tends to have thicker features as compared to the difference of Gaussian as shown in Fig. 12, reducing its chances of being dropped out by Taylor series expansion in the SIFT.

The original SIFT algorithm fails to detect feature points in the input images of the above dataset due to low-contrast problem. After entropy-based Sobel filtered enhancement, the enhanced image delivered almost 47 feature points after application of SIFT. These features are then further clustered using affinity propagation

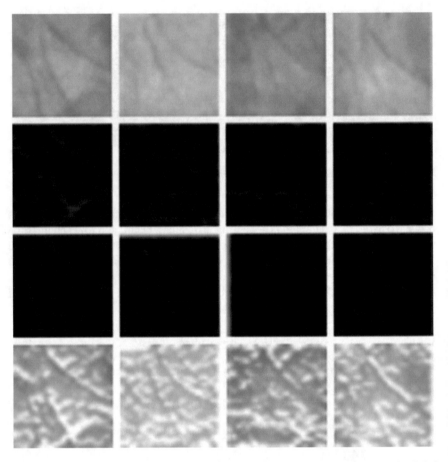

Fig. 12 First row is the input set of palm print images, the second row represents the Sobel filtered image, the third row of images is difference of Gaussian enhanced, and the fourth row of image represents proposed work enhancement

algorithm which reduces the overhead of redundant feature points by reducing it to 8 feature points from 47 for the given dataset (Fig. 13).

The comparison between several other algorithms that have been used to enhance image especially in low-contrast regions is represented in Table 1.

The advantages of our proposed method are that it does not perform excessive feature point generation and it also makes the feature point matching a lot more faster. Algorithms like histogram and contrast limited adaptive histogram equalization tend to lose features due to radical change in pixel intensity, while our method does not perform any such transformation and preserves several feature points.

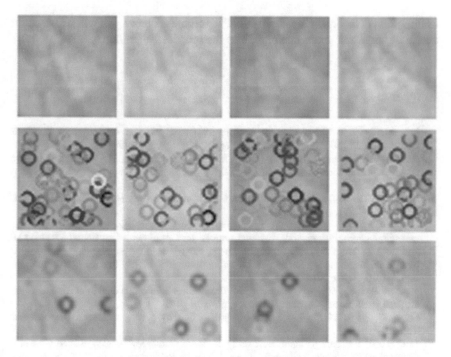

Fig. 13 Feature points for the input set of the first row represent the feature points detected using SIFT algorithm. The feature points detected in row images are after enhancement by iterative entropy-based Sobel filter. The third describes the clustered feature points

Table 1 A comparison of feature generated after applying each contrast enhancement technique

Algorithm	Feature points
SIFT	0
Unsharp masking [9]	14
OS unsharp masking [10]	24
Histogram equalization	27
Contrast stretching	30
Contrast limited adaptive histogram equalization [11]	39
Sobel-entropy enhancement	44

5 Conclusion

The experimental results show that the Sobel operator has a high response to edge compared to difference of Gaussian so it covers even the feature points in the low-contrast areas of the image. The direct application of SIFT would not be able to detect the feature points in the low-contrast region unlike the enhanced SIFT which detects more feature points in the image in the low-contrast region. It increases the feature point detection compared to SIFT algorithm. Future work will be based on improving time and space complexity of the proposed algorithm.

References

1. Lowe DG (1999) Object recognition from local scale-invariant features. Proc Int Conf Comput Vis 2:1150–1157. doi: https://doi.org/10.1109/iccv.1999.790410
2. Andono PN, Purnama I, Eddy K, Hariadi M (2013) Underwater image enhancement using adaptive filtering for enhanced SIFT-BASED image matching. J Theor Appl Inform Technol 52(3):273–280
3. Liangping Tu, Dong C (2013) In: 2013 6th international congress on image and signal processing (CISP), December 2013 pp 16–18
4. Castillo-Carrión S, Guerrero-Ginel J-E (2017) SIFT optimization and automation for matching images from multiple temporal sources. Int J Appl Earth Obs Geoinf 57:113–127
5. Guohua Lv, Teng SW, Lu G (2016) Enhancing SIFT-based image registration performance by building and selecting highly discriminating descriptors. Pattern Recogn Lett. https://doi.org/10.1016/j.patrec.2016.09.011
6. Witkin AP (1983) Scale-space filtering. In: Proceedings 8th international joint conference on artificial intelligence, Karlsruhe, Germany, pp 1019–1022
7. Dueck D (2009) Affinity propagation: clustering data by passing messages. University of Toronto
8. Li Q (2006) Research on handmetric recognition and feature level fusion method (Doctoral dissertation, PhD thesis, BeiJing JiaoTong University, Beijing)
9. Polesel A, Ramponi G, Mathews VJ (2000) Image enhancement via adaptive unsharp masking. IEEE Trans Image Process 9(3):505–510
10. Mitra SK, Li H, Lin I, Yu T (1991) A new class of nonlinear filters for image enhancement. Acoust Speech Signal Process 4:2525–2528
11. Pizer SM, Amburn EP, Austin JD, Cromartie R, Geselowitz A, Greer T, ter Haar Romeny B, Zimmerman JB, Zuiderveld K (1987) Adaptive histogram equalization and its variations. Comput vision Graph image process 39(3):355–368

Selection of a Hall Sensor for Usage in a Wire Rope Tester

Akshpreet Kaur, Aarush Gupta, Hardik Aggarwal, Manu Sharma, Sukesha Sharma, Naveen Aggarwal, Gaurav Sapra and J. K. Goswamy

Abstract This paper deals with inspection of broken strands in steel wire ropes using magnetic flux leakage (MFL) detection method-based wire rope tester. In this paper, performance of different sensors is studied experimentally in order to select suitable sensor which can appropriately detect the leakage in magnetic flux due to the defect. The Hall voltage signal corresponding to the defect gives peak at the position of the defect. Maximum and minimum signal voltages from the base signal determines the peak-to-peak amplitude verifying the capability and sensitivity of the sensor. The result of this study shows that the analog Hall sensor-Sensor G with sensitivity -90 mv/mT and range 0–10 mT is suitable for detection of defects in a wire rope.

Keywords Magnetic flux leakage · Defects · Local fault (LF)
Sensors · Hall sensor

1 Introduction

Wire ropes are extensively used in various lifting operations, cranes, elevators, cable-stayed bridges, helicopter and suspension cables etc. While deployment in field, wire ropes are subjected to dynamic loads which cause wear and tear in the rope. Some strands in the rope can fail due to excessive tensile loads occurring due to overloading. This normally results in progressive loss of strength, load sharing capacity, and service life of a wire rope. Ropes also deteriorate due to external and internal corrosion and abrasion [1]. This leads to various types of defects in a wire rope which are classified as local faults (LFs) which implies a sudden discontinuity in a rope such as a broken strand and loss in metallic area (LMA) which means that chunk of material is missing from the wire rope. The integrity of rope affects the

A. Kaur (✉) · A. Gupta · H. Aggarwal · M. Sharma · S. Sharma · N. Aggarwal · G. Sapra ·
J. K. Goswamy
UIET, Panjab University, Chandigarh 160014, India
e-mail: akshpreet9386@gmail.com

© Springer Nature Singapore Pte Ltd. 2018
A. K. Nandi et al. (eds.), *Computational Signal Processing
and Analysis*, Lecture Notes in Electrical Engineering 490,
https://doi.org/10.1007/978-981-10-8354-9_33

reliability and safety of operations for which they are in service. Visual methods of inspection are found to be unreliable. The routine retirement of ropes without inspection is costly because some ropes are discarded before they get sufficiently worn out or corroded. In order to ensure reliability and safety, proper regular inspection of broken strands in wire ropes is necessary by a wire rope tester [2]. In past few decades, several methods for inspection of wire rope have emerged such as magnetic flux leakage (MFL), X-ray detection method, acoustic emission-ultrasonic testing method. Magnetic flux leakage (MFL) testing is an efficient, economical, and most reliable inspection method for detection of defects in ferromagnetic materials [3, 4].

In this paper, experimental investigation is done to compare effectiveness of various Hall sensors in a wire rope tester. For this purpose, various sensors are interfaced to a laptop installed with MATLAB software using open-source hardware Arduino Mega 2560. In following sections, typical structure of the wire rope tester is discussed, instrumentation of wire rope tester is explained and experimental results are presented.

2 Structure of Equipment

As shown in Fig. 1, typical structure of a wire rope tester consists of magnets, sensors, ferromagnetic yoke, and liner. Magnets used in this setup can be electromagnets or temporary magnets or permanent magnets. Electromagnet is a type of magnet in which an electric current produces the magnetic field. The magnetic field vanishes when the current is turned off. The magnetic field of an electromagnet can be quickly altered by changing the amount of electric current in the windings which is an advantage over permanent magnets. However, unlike permanent magnet it requires persistent supply of current in the windings to retain the magnetic field. Temporary magnets are those which simply act like permanent magnet when they are within a strong magnetic field. They lose their magnetism in the absence of magnetic field unlike permanent magnets. They are not useful for inspection of wire ropes based on magnetic leakage principle. Permanent magnets retain their magnetism once magnetized, and lifetime is around 40 years. Therefore, permanent magnets are best suitable for inspection of wire ropes [5]. A typical wire rope tester consists of two radially magnetized ring magnets housed in a yoke as shown in Fig. 1.

The principle of detecting the defects say broken steel strands in a wire rope using a wire rope tester is shown in Fig. 2. Steel strands in a wire rope have ferromagnetic nature therefore the steel strands get magnetized adequately by permanent magnets instrumented in a wire rope tester. Magnetic field lines are

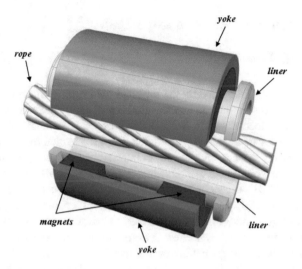

Fig. 1 Wire rope tester [16]

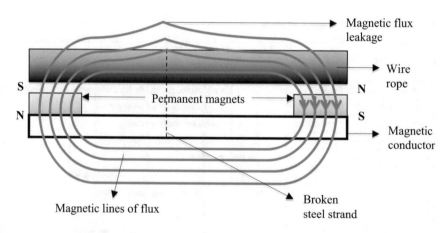

Fig. 2 Magnetic circuit developed by a wire rope tester on a wire rope

developed evenly in the wire rope and if there is any crack or broken strand, the magnetic permeability decreases at that position which results in leakage of magnetic field fluxes to the surrounding atmosphere. This magnetic flux leakage over a defect can be easily sensed by a suitable Hall sensor [6]. The signal corresponding to broken strand is shown in Fig. 2.

3   Instrumentation of Wire Rope Tester

3.1   Type of Sensors

A sensor is a device that converts the physical parameter (temperature, humidity, speed, etc.) into a signal that can be measured electrically. There are several type of sensors used to detect magnetic leakage field (as listed in Table 1) such as Hall effect sensor that can be analog or digital [5, 7], magnetoresistive sensor [8], SQUID sensor (Superconducting Quantum Interference Device) [9], fluxgate sensor [10, 11], pick-up coils [12], giant magnetoimpedance (GMI) effect sensor [13], magneto-optic (MO) sensor, anisotropic magnetoresistance (AMR) sensor, giant magnetoresistance (GMR) sensor, etc. [14]. Hall effect sensors have good accuracy, are robust, have good detection range, are easy to calibrate, and are relatively cheap [11, 14, 15]. In this paper, Hall effect sensors are used to detect the defect in the wire rope. Magnetic leakage field over a defect in a wire rope inside a wire rope tester is around 1–10 mT. Therefore, sensor used to measure magnetic leakage field should have range from about 0.1–10 mT.

Table 1 Comparison of sensors [14]

Magnetic sensor technology	Detectable field (Tesla)				
	10^{-14}	10^{-10}	10^{-6}	10^{-2}	10^{2}
Search-coil magnetometer					
Fluxgate magnetometer					
Optically pumped magnetometer					
Nuclear precession magnetometer					
SQUID magnetometer					
Hall effect sensor					
Magnetoresistive magnetometer					
Magnetodiode					
Magnetotransistor					
Fiber optic magnetometer					
Magnetic optical sensor					
Magnetoimpedance magnetometer					

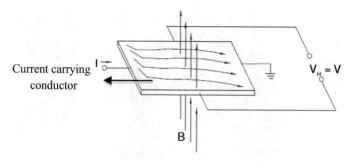

Fig. 3 Hall effect [5]

3.2 Hall Sensors

Hall sensors are based on Hall effect. They are widely used as they are cheap, efficient and have a broad temperature range (−40 to 150 °C). Consider a current carrying conductor placed in a magnetic field perpendicular to the flow of current as shown in Fig. 3. If there is no magnetic field, current follows in a straight path, whereas if magnetic field is applied perpendicular to it, the electrons accumulate on one side of the material and positive charges accumulate on the other side of the material due to which a potential difference develops which is known as the Hall voltage [5]. The voltage measured at opposite sides of the Hall plate is:

$$V_H = IB\frac{R_H}{d_H} = IBk_H \tag{1}$$

where $k_H = \frac{R_H}{d_H}$, k_H is the sensitivity of the Hall generator, d_H is thickness of the Hall plate, R_H is the Hall coefficient, I is the current flowing in the conductor, and B is the applied magnetic field [5].

Hall sensor is a three-pin device consisting of input, output, and ground pins. Hall sensor typically gives the output voltage of about 30 μV in the presence of 10^{-4} T. In order to amplify it, an amplifier is sometimes already provided in the chip. Different circuits can be integrated in the chip itself according to the requirement like amplifiers, Schmitt triggers, concentrators, comparators.

4 Hall Sensor Inspection System and Experiments

Hall sensors which may be analog or digital are used in this work to detect the magnetic flux leakage due to the defect. The selection of sensors plays a very crucial role in detection of defect. A sensor head has been fabricated to hold all the sensors in place inside the wire rope tester, perpendicular to the flux leakages. The sensor head is 3D printed using polylactic acid (PLA) material. As shown in Fig. 4,

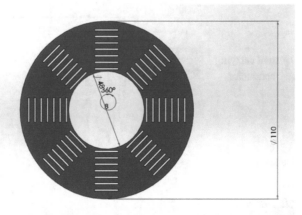

Fig. 4 CAD model of sensor head

Table 2 Specifications of sensor head

Specifications of sensor head		
S. No.	Parameters	Dimensions
1	Thickness	10 mm
2	Outer diameter	110 mm
3	Inner diameter	48 mm
4	Lift off	3–21 mm
5	No. of slots (at same lift off)	8

sensor head of thickness 10 mm has 56 slots allowing to place sensors at different lift-offs (distance between sensor and wire rope) and angular positions. The dimensions of the sensor head are tabulated in Table 2.

As shown in Fig. 5, seven sensors out of which four are analog and three are digital (as listed in Table 3) are fixed in the sensor head and connections are made with Arduino Mega 2560 and thereafter interfaced with a computer.

4.1 Performance Analysis of Hall Sensors

Performance of seven Hall sensors is observed by comparing their time responses when the sensors are subjected to a magnetic field of a small magnet. The distance between the small magnet and a sensor is measured by a sharp sensor. A small magnet is brought first close to the sensors and then moved away and the corresponding signals are displayed in Fig. 6. The sensor voltages of analog sensor-Sensor G and digital sensor-Sensor C are plotted with respect to time. The analog sensors give output voltage from 0 to 2 V and digital sensors give output voltage

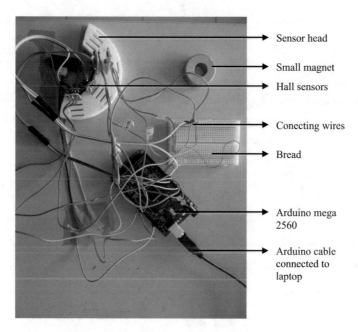

Fig. 5 Interfacing of sensors

Table 3 Specifications of sensors

S. No.	Description	Type of sensor	Sensitivity (mV/mT)	Range (mT)	Polarity
1	Sensor A	Digital	–	–	Bipolar
2	Sensor B	Digital	–	–	Unipolar (south pole high)
3	Sensor C	Digital	–	–	Omnipolar
4	Sensor D	Analog	+45	0–20	Bipolar
5	Sensor E	Analog	−23	0–40	Bipolar
6	Sensor F	Analog	−11	0–80	Bipolar
7	Sensor G	Analog	−90	0–10	Bipolar

from 0 to 5 V. The output of analog sensor is 1 V when there is no magnetic field and varies in presence of magnetic field. Among digital sensors, Sensor C is omnipolar, the output is high if either north or south pole is detected and it is low when there is no magnetic field. Other digital sensors either latch or are unipolar. Thus, analog sensors are preferred over digital sensors in this work as the digital

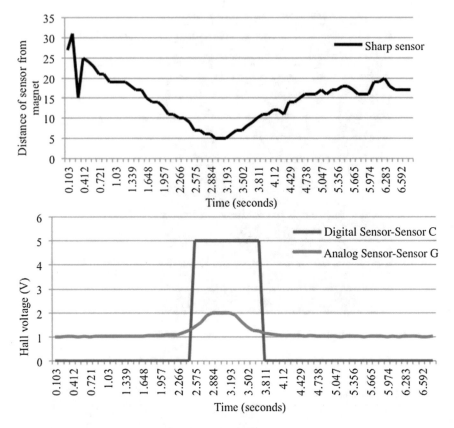

Fig. 6 Performance comparison of analog and digital sensor

sensors do not provide the required continuous information and give output as either low or high.

The analog sensors are compared on the basis of their time response and magnetic field detection range. The distance between the small magnet and sensor is first decreased and then increased and corresponding outputs are shown in Fig. 7. It is observed in Fig. 7 Sensor G with sensitivity −90 mv/mT and magnetic field range 0–10 mT (as listed in Table 3) give maximum variation in comparison to Sensor D, E and F with low sensitivity, when a magnet is brought close to the sensor head. When the magnet is very close to Sensor G, it gets saturated as the upper limit of range is 10 mT. According to magnetic field leakage measurement using Gaussmeter in this work, the field is found to lie in range 0–10 mT. So, taking into consideration this factor, Sensor G is the best choice among analog sensors.

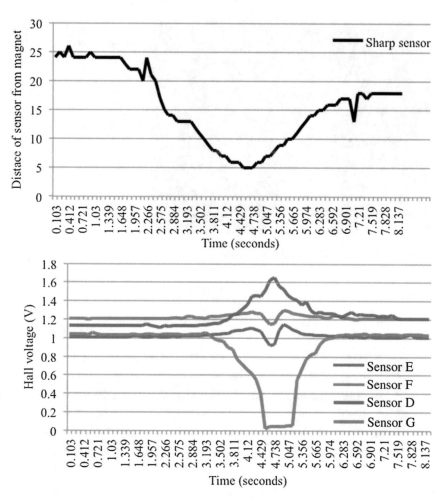

Fig. 7 Performance comparison of Hall sensors using sharp sensor

4.2 Detection of Defect Using Hall Sensor DRV5053VA in Wire Rope Tester

The Hall sensor-Sensor G is tested by in a wire rope tester on a wire of cross-section 0.5 mm^2 and length 1.5 m fixed on an aluminum core. An artificial defect is created by cutting the wire which acts like a local fault (LF). The defect is passed under the sensor as shown in Fig. 8, and the output Hall voltage signal is shown in Fig. 9.

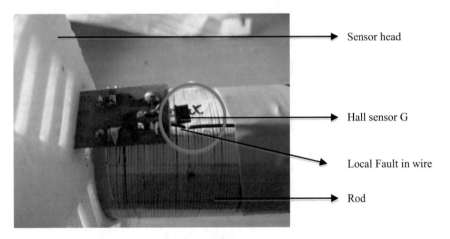

Fig. 8 Detection of broken strand using Sensor G

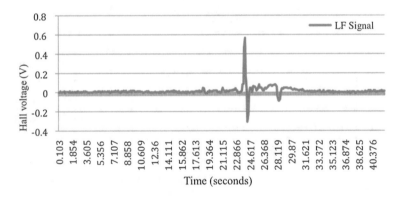

Fig. 9 Hall voltage signal using Hall Sensor G corresponding to broken strand

Hall sensor signal over a defect is an electrical impulse which gives higher amplitude corresponding to the defect than that of base signal. As observed in Fig. 9, the maximum voltage of the signal is 570 mV and minimum voltage is 310 mV. The peak-to-peak amplitude corresponding to the defect is 880 mV. Therefore, it can be concluded that the level of deterioration of a wire rope can be determined easily by analog Hall Sensor G which has sensitivity of −90 mV/mT and range 0–10 mT.

5 Conclusions

In this work, an experimental investigation is done to locate appropriate Hall sensor for usage in a wire rope tester. A wire rope is magnetically saturated by creating a magnetic circuit using permanent magnets and a ferromagnetic yoke. Eight Hall sensors were used in wire rope tester to sense magnetic leakage flux over a defect in a wire rope. Experimental investigation reveals that Hall Sensor G generates sufficient Hall voltage when subjected to magnetic flux leakage over a defect. Therefore, it is concluded that analog Hall Sensor G is appropriate for usage in a wire rope tester.

References

1. Tytko AA, Ridge IML (2003) The influence of rope tension on the LMA sensor output of magnetic NDT equipment for wire ropes. Nondest Test and Eval 19(4):153–163
2. Weischedel HR, Ramsey RP (1989) Electromagnetic testing, a reliable method for the inspection of wire ropes in service. NDT Int 22(3):155–161
3. Zawada K (1999) Magnetic NDT of steel wire ropes. J Nondestr Test Ultrason (Germany) 4(8)
4. Tian J, Zhou J, Wang H, Meng G (2015) Literature review of research on the technology of wire rope nondestructive inspection in China and Abroad. In: MATEC web of conferences, vol 22. EDP Sciences
5. A manual by Honeywell, Hall effects sensing and application by Honeywell, MICRO SWITCH Sensing and Control, USA
6. Wu B, Li G (2015) Design and simulation analysis of detection sensor of localized faults for wire rope
7. Cao Y, Zhang D, Wang C, Xu D (2006) More accurate localized wire rope testing based on Hall sensor array. Mater Eval 64(9):907–910
8. Singh WS, Rao BPC, Vaidyanathan S, Jayakumar T, Raj B (2007) Detection of leakage magnetic flux from near-side and far-side defects in carbon steel plates using a giant magneto-resistive sensor. Meas Sci Technol 19(1):015702
9. Zhang D, Zhao M, Zhou Z, Pan S (2013) Characterization of wire rope defects with gray level co-occurrence matrix of magnetic flux leakage images. J Nondestr Eval 32(1):37–43
10. Wei G, Jianxin C (2002) A transducer made up of fluxgate sensors for testing wire rope defects. IEEE Trans Instrum Meas 51(1):120–124
11. Baschirotto A, Dallago E, Malcovati P, Marchesi M, Venchi G, Rossini A (2006) Multilayer PCB planar fluxgate magnetic sensor. In: 2006 Ph.D. research in microelectronics and electronics. IEEE, pp 413–416
12. Jomdecha C, Prateepasen A (2009) Design of modified electromagnetic main-flux for steel wire rope inspection. NDT E Int 42(1):77–83
13. Tehranchi MM, Ranjbaran M, Eftekhari H (2011) Double core giant magneto-impedance sensors for the inspection of magnetic flux leakage from metal surface cracks. Sens Actuators A 170(1):55–61
14. Caruso MJ, Bratland T, Smith CH, Schneider R (1998) A new perspective on magnetic field sensing. Sensors 15:34–47
15. Tang SC, Duffy MC, Ripka P, Hurley WG (2004) Excitation circuit for fluxgate sensor using saturable inductor. Sens Actuators A 113(2):156–165
16. Sukhorukov VV (2013) MFL technology for diagnostics and prediction of object condition. In: The 12th international conference of the Slovenian society for NDT, pp 4–6

Transliteration of Braille Code into Text in English Language

K. P. S. G. Sugirtha and M. Dhanalakshmi

Abstract Braille is specially designed for visually impaired persons which is a particular system of representing information in tactile form. It acts as written communication medium between sighted people and visually impaired. This paper describes a new system to recognize the braille characters in a six-dot pattern from scanned document and thus transliterate in English language. Scanned documents are preprocessed followed by segmentation of braille cells. From the segmented braille cell, centroid of each braille dot is calculated for the computation of Euclidean distance. Later, a lookup table is constructed utilizing Euclidean distance between the braille dots. Thus, the braille characters are recognized and translated to corresponding text.

Keywords Braille characters · Image processing · Euclidean distance

1 Introduction

In 1821, Frenchman Louis Braille contrived the braille system that is used to peruse and compose for visually impaired people via raised dots. Nowadays, braille characters are widely used as the instructions in door labels in the elevators, the railway stations, currency notes, ATM, mobile phones, and also the direction signs at the public places. It is reported that less than 10% of the legally dazzle individuals can truly read braille as in [1]. This situation makes a waste of the public resources. To overcome this situation, transliteration is done. In general, braille scripts are made up of cells, and each cell contains six dots that are arranged in three rows and two columns as in [2]. These six dots can be raised or flat corresponding to braille characters. Therefore, 64 conceivable diverse specks combination are

K. P. S. G. Sugirtha (✉) · M. Dhanalakshmi
Department of Biomedical Engineering, SSN College of Engineering, Chennai, India
e-mail: kpsgsugirtha@gmail.com

M. Dhanalakshmi
e-mail: dhanalakshmim@ssn.edu.in

© Springer Nature Singapore Pte Ltd. 2018
A. K. Nandi et al. (eds.), *Computational Signal Processing and Analysis*, Lecture Notes in Electrical Engineering 490,
https://doi.org/10.1007/978-981-10-8354-9_34

Fig. 1 Braille cell dimension

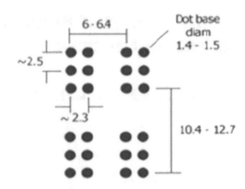

All dimensions are in Millimeters

available. The braille cell has standardized dimensions as shown in Fig. 1. A standard dimension of braille page is 11 in. by 11.5 in., and every line has a maximum of 40–43 braille cells.

The braille literacy code consists of three grades:

In grade 1, each braille cell signifies a single character, and combination of braille cells makes a word. Figure 2 signifies the grade 1 braille predominantly used by fledglings. The composition of braille cell differs for lowercase alphabets, uppercase alphabets, and numerals. Lowercase alphabets are represented by one braille cell, whereas uppercase alphabets and numerals are represented by two braille cell. Grade 2 includes some abbreviations and contractions. Figure 3 presents the instance of grade 2 braille. Grade 3 is the complex grade of braille that includes intricate phrases and sentences. Figure 4 gives some sample words in grade 3 system.

Hassan and Mohammed in [1] use feed-forward artificial neural network for conversion of braille characters. Venugopal in [3] proposed braille recognition utilizing a camera-empowered smartphone using digital image processing techniques. This framework procures 84% precision of braille character recognition. Subur et al. in [4] recommended the braille recognition system using find contour

Fig. 2 Depiction of English letters and numerals in braille

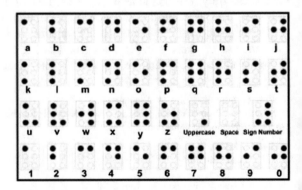

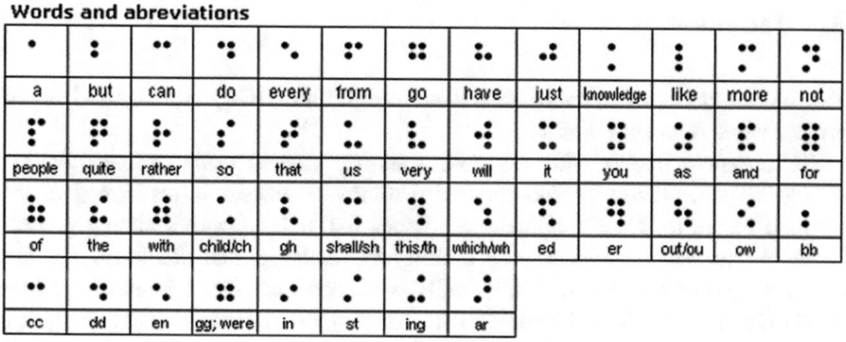

Words and abreviations

•	:	••	•:	•.	:•	::	:.	.:	: •	:	•• ••	:• •
a	but	can	do	every	from	go	have	just	knowledge	like	more	not

| :• | :: | :• | : • | :• | • | : | •: | •• | :: | : • | :: | :: |
| people | quite | rather | so | that | us | very | will | it | you | as | and | for |

| :. | :. | :: | : | :. | •• | :: | : | :• | :: | :. | .: | : |
| of | the | with | child/ch | gh | shall/sh | this/th | which/wh | ed | er | out/ou | ow | bb |

| •• | •: | •. | :: | .• | : | •. | : | | | | | |
| cc | dd | en | gg; were | in | st | ing | ar | | | | | |

Fig. 3 Illustration of Grade 2 braille

Fig. 4 Depiction of Grade 3 braille

Romanized Form	Braille	Description
account	:: :: ::	acc
acknowledge	:: :: ::	ack
acknowledgement	:: :: :: ::	ack dots 5-6 t

method. From the experimental analysis using this method, an accuracy level of 100% is be achieved on the tilted image of 0° to 0.5°. When the image is tilted greater than 1, the level of accuracy decreases and the images cannot be perceived. Li et al. in [5] have foreseen a framework which makes use of the standard distance between the braille dots in each cell. Jiang in [6] describes mandarin braille word segmentation and transforms into Chinese characters. Mennens in [7] stated a problem of shading effect in the scanned image. Padmavathi et al. in [8] emphasize on the translation of scanned braille documents to equivalent text in English, Hindi, and Tamil languages. These texts are recited out by speech synthesizer and mapping errors occurred when the braille has similar representation for the script and the accentuation.

Though all the above-mentioned techniques transliterate in different languages with different algorithms [9], most of the researchers use mobile phone [10], and scanned braille picture has more distortion effect. Therefore, this project work aims to scan the braille image and use image processing techniques to achieve enhanced braille cells. The centroids of each braille code were found, and Euclidean distances between the braille dots were computed to create a lookup table [8] that helps in recognizing the braille to corresponding text. This recognition reduces the stretch of remembering braille character and diminishes usual time taken by visually impaired people. It requires very less number of trained tutors and helps to serve a large number of users with a single document. It helps to safeguard old braille books. It benefits clique who do not know braille.

2   Methodology

The block diagram for braille character recognition (BCR) system used in this project work is as shown in Fig. 5.

BCR system begins with a simple scanning process using ordinary flat-bed scanner. Then, clamor evacuation (noise removal) process is performed on the scanned image to expel enlightenment issues that appear from scanning process. Other preprocessing includes gray scaling, thresholding, and dilation. After preprocessing, segmentation of braille cells is carried out. These braille cells have standardized dimension; therefore, a new technique is evolved using Euclidean distance. The centroids of each braille cell are calculated. The Euclidean distance is calculated between the centroids of a reference dot and centroids of corresponding dots in the braille cell. Finally, using consistent Euclidean distance, braille characters are recognized. Deciphered content in English language is stored in text file format. The system is discussed in detail as follows (Fig. 6).

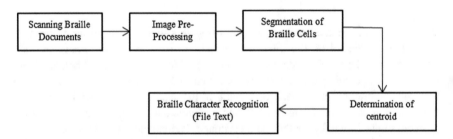

Fig. 5 System block diagram of BCR

Fig. 6 Sample of scanned image

2.1 Image Preprocessing

Firstly, the acquired braille documents are examined with Hp DESKJET 2050 scanner with resolution of 300 dpi [11]. Ensued image is in.jpeg format and has RGB scale. The image is preprocessed in order to prepare the document for recognition process. The scanned documents in RGB scale are converted to grayscale image so as to reduce reminiscence space. Then, thresholding is performed to enhance the dots of the braille cells and suppress the background of the scanned document. Enrichment of braille dots is performed by dilation, a morphological operator that enlarges braille dots as in [5].

2.2 Segmentation of Braille Cells

In order to recognize the braille character, the braille cells are extracted using segmentation. Every braille cell comprises of two sections (columns) and three lines (rows), i.e., 3×2 dimension dots. Each cell ought to be extricated for recognition process. This cell has standardized dimension in millimeter (mm) that is converted to pixel-wise dimension 127×72. A mask is created for this aspect using this pixel-wise dimension that slides over the entire braille document. However, the mask size varies for numerals and uppercase letters. Since they contain two braille cells for representing single letter or number, the mask size should be doubled.

2.3 Estimation of Centroids

Each braille cell is now segmented with braille dots enhanced. The centroid [12] of each braille dot is to be estimated so as to find the Euclidean distance between them. The center of mass of a region is called centroid. The horizontal coordinate (or x-coordinate) of the center of mass is specified as the first element of the centroid and the vertical coordinate (or y-coordinate) as second element.

$$\text{Centre } X = \frac{\min X + \max X}{2} \tag{1}$$

$$\text{Centre } Y = \frac{\min Y + \max Y}{2} \tag{2}$$

where center X in Eq. (1) as in [4] is center coordinate x of each braille dot; minimum value and maximum value of coordinate x are represented as min X and max X, respectively; center Y in Eq. (2) as in [4] is center coordinate y of each braille dot; minimum value and maximum value of coordinate y are given by min Y and max Y, respectively. The computation of Euclidean distance is between the centroid of dots in each braille cell. Based on the number of dots in each cell, a lookup table is created using the Euclidean distance.

2.4 Braille Character Recognition

Braille character recognition is the conversion of scanned document into machine encoded text. Each letter is transliterated using the consistent distance from the lookup table. Braille character recognition helps to reduce the storage space as the natural braille code consumes more space than the normal English language.

3 Results and Discussion

3.1 Image Preprocessing

Image preprocessing incorporates the accompanying strides. Grayscaling process is used to convert the original RGB scale image to grayscale images. Gray scale has the variance of shades from 0 to 255. Adaptive thresholding method is used to convert the grayscale image into binary image. The resultant of the threshold image is further enriched by dilation, a morphological operator as shown in Fig. 7.

3.2 Determination of Centroids

Centroids are computed for each braille dot in a braille cell. Using these centroids, Euclidean distance is determined. It shows the reckoning of centroids of dots in each braille cell. In Fig. 8, asterisk depicts the centroid of each dot in braille cell (Fig. 6).

Fig. 7 Preprocessed image

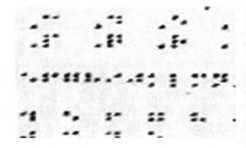

Fig. 8 Centroid estimation

3.3 Computation of Euclidean Distance

A lookup table is constructed based on the consistent Euclidean distance among each braille cell. The consistent distances are listed based on the number of dots available in each braille cell. For instance, Fig. 9a represents braille cell of letter 'd' with three dots and Fig. 9b represents two braille cell of number '8' with seven dots. Considering Fig. 9a, from the left side, with respect to first cell, first dot of column 1 is considered as reference dot. Euclidean distance is calculated from the reference dot to other corresponding dots. D1 represents distance between the reference dot and second dot in column 2. D2 represents distance between the reference dot and first dot in column 2. The Euclidean distance D1 and D2 are computed as 40 and 30, respectively. These Euclidean distance values were entered in the lookup table. Therefore, if the BCR systems scan a braille cell with this Euclidean distance, it is compared with the lookup table and the scanned braille cell is recognized as letter 'd.' Figure 9b shows Euclidean distance D1, D2, D3, D4, D5, and D6 from the reference braille dot for the number 8.

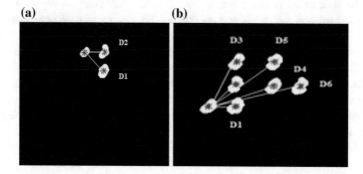

Fig. 9 **a** Euclidean distance recognizes the lowercase letter d. **b** Euclidean distance recognizes the number 8

(a)   **(b)**

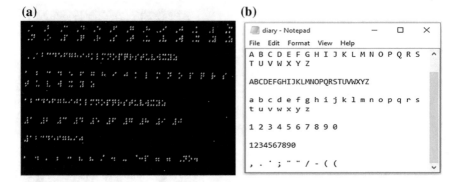

Fig. 10 **a** Braille grade 1 characters. **b** Transliterated characters

3.4 Braille Character Recognition

The Euclidean distances of each braille cell are entered in a lookup table with its corresponding alphabet or numerals. When the braille document is scanned, the braille cell is automatically segmented and is compared with the lookup table to recognize the corresponding text. Figure 10a consists the basics of grade 1 characters which include uppercase letters, lowercase letters, numerals with space and without space, and punctuations. Figure 10b represents transliterated text as stored in text file.

4 Conclusion

This project focuses mainly on the conversion of scanned braille documents of grade 1 system to corresponding text in English language. This system uses flat-bed scanner and avoids the usage of complicated hardware. After identifying the start of the braille code in the scanned braille documents, braille cells are segmented based on the standard measurements and computing the number of dots available in each braille cell. Using the consistent distance, braille cells are identified and matched with its corresponding text in English language. Subsequently, these recognized alphabets are saved in text file format. During preprocessing of the scanned braille document, the extracted dots vary in shape due to resolution of scanner and therefore, centroid values differ for few characters resulting in mapping errors. Also, when two braille characters have similar representation, the mapping error persists. Future work could be extended for grade 2 of braille system by creating a lookup table for contracted words and by eliminating these mapping errors. braille character recognition (BCR) system described here shows the feasibility of cost-effective system.

Acknowledgements Braille documents are acquired from National Institute for Empowerment of Persons with Multiple Disorders (NIEPMD) with consent of Dr. Himangsu Das, Director, NIEPMD. We would also like to thank Mr. Stalin Arul Regan, Mr. Jayakumar, Ms. Meenachi, and Mr. Mahadevan, special teachers for giving detailed insight into braille and its documents acquisition.

References

1. Hassan MY, Mohammed AG (2011) Conversion of English characters into braille using neural network. IJCCCE 11(2):30–37
2. Grades of Braille [online]. Available http://www.acb.org/tennessee/braille
3. Venugopal-Wairagade G (2016) Braille recognition using a camera-enabled smartphone. Int J Eng Manuf 4:32–39
4. Subur J, Sardjono TA, Mardiyanto R (2015) Braille character recognition using find contour method. In: 5th international conference on electrical engineering and information, vol 4(4), pp 48–53
5. Yin J, Wang L, Li J (2010) The research on paper-mediated braille automatic recognition method. In: Fifth international conference on frontier of computer science and technology
6. Jiang M (2002) Braille to print translations of Chinese. Int J Adv Eng Res 91–100
7. Mennens J (1993) Optical recognition of Braille writing. IEEE Int J Softw Innov 428–431
8. Padmavathi S, Manojna KSS, Sphoorthy Reddy S, Meenakshy D (2013) Conversion of braille to text in English, Hindi and Tamil languages. Int J Comput Sci Eng Appl 3(3)
9. Shreekanth T, Udayashankara V (2013) A review on software algorithms for optical recognition of embossed braille Characters. Int J Comput Appl (0975-8887) 81(3):421–429
10. Zhang S, Yoshino K (2007) A braille recognition system by the mobile phone with embedded camera. In: Dengel A, Spitz AL (eds) World Scientific Publishing Co, IEEE 413–421
11. Antonacopoulos A, Bridson D (2004) A robust braille recognition system. In: Dengel A, Marinai S (eds) Document analysis system VI. Lecture notes in computer science, LNCS, vol 3163. Springer, Berlin, pp 533–545
12. Nian-feng L, Li-rong W (2011) A kind of braille paper automatic marking system. Mechatronic science electric engineering and computer (MEC) international conference, pp 664–667

Semi-blind Hyperspectral Unmixing Using Nonnegative Matrix Factorization

R. Subhashini, N. Venkateswaran and S. Bharathi

Abstract In hyperspectral imaging applications, spectral unmixing aims at identifying the constituent materials of a remotely sensed data and estimates its corresponding spectral signature for data exploitation. In this paper, the unmixing is primarily based on a linear mixture version in which every pixel is considered as a sum of definite number of absolutely clear spectra or endmembers, in accordance with means of abundance. Firstly, the number of endmembers in a given scene is determined using hyperspectral signal subspace identification by minimum error (Hysime) algorithm. Then, a vertex component analysis (VCA) method is used for unsupervised endmember extraction. Based on the observation that a negative reflectance is not possible, it is supportive and significant to constrain with non-negativity. Thus, a nonnegative matrix factorization is applied for decomposing a given scene into its endmembers and abundance matrix. The successfulness of the researched technique is served using the simulated knowledge supported by USGS laboratory collected by the AVIRIS on mineral mining district, Nevada.

Keywords Hyperspectral unmixing · Nonnegative matrix factorization (NMF) Spectral signatures · Blind source separation

1 Introduction

Recently, hyperspectral imaging techniques have been applied in various fields due to the significant information that it carries across the spectral bands. A key requirement in a remotely sensed image is digital signal processing. It has potential

R. Subhashini (✉) · N. Venkateswaran · S. Bharathi
Department of ECE, SSN College of Engineering, Chennai, India
e-mail: subhashini.rajendiran@gmail.com

N. Venkateswaran
e-mail: venkateswarann@ssn.edu.in

S. Bharathi
e-mail: bharathikarthi92@gmail.com

© Springer Nature Singapore Pte Ltd. 2018
A. K. Nandi et al. (eds.), *Computational Signal Processing
and Analysis*, Lecture Notes in Electrical Engineering 490,
https://doi.org/10.1007/978-981-10-8354-9_35

383

applications in defense and security and civilian purposes such as military surveillance, mineral exploration, surveying of resources, environmental monitoring, satellite weather prediction images. Due to mixed pixels in hyperspectral image, the different materials get combined into a single homogenous form due to their lower spatial resolution. In such a case, the spectral range of a solitary pixel is a blend of a number of endmember spectra, valued by their fragmentary abundances. Spectral unmixing (SU) [1] is a common problem of decomposing the mixed pixels. SU plays the role of segregating the endmembers and its corresponding abundance matrices, given only the spectral mixture data. This mathematical limitation is considered as blind source separation (BSS) problem. BSS [2] is one such procedure of evaluating singular input segments from their blends. It is referred as blind since we do not use any other information besides the mixtures. In hyperspectral imaging, the observation matrix is the collected mixtures, mixing matrix is the endmember signatures, and spectral unmixing contains abundances as sources. For rectifying the SU/BSS issue specified above, there are two techniques: (1) the two-stage method (TSM) and (2) the single-stage method. The TSM extracts the endmembers and afterward assesses the abundance in view of the extracted endmembers. An extensive aggregate error will happen in the estimation of the abundances, when the assessment exactness of minerals is less. Single-stage method is an alternate method that obtains the endmembers and their abundances simultaneously. Independent component analysis (ICA) [3] is a factual and computational approach for uncovering concealed components that underlie units of irregular factors, estimations, or signs. Abundance sum to one, a limitation that occurs in SU, is a weakness and the utilization of ICA is hence confined. But nonnegative matrix factorization (NMF) achieves nonnegative results using the needed values. Therefore, NMF serves as a better tool for solving spectral unmixing.

2 Background

The number of substances and their reflectance are unknown in most of the cases leading it into a class of blind source separation problems. ICA [4] is on the view that sources are not dependent on each other which is an unseen one in hyperspectral data, and the result of the total value of the abundance fraction remains unchanged, shows statistical dependence exists in them. Because of the limitation created by dependency factor, using ICA over hyperspectral images is prohibited. The ICA detects endmembers ground truth by the product of spectral indices and unmixing array that gives needed exclusive values from the channels by restricting the mutuality that exists between them. Approaches based on the geometric features of hyperspectral mixtures are mentioned in [5]. The minimum volume transforms (MVT) [6] algorithm produces a much easier minimum volume from the given data. A convex hull is produced by the algorithm from the data results, and then it will be processed to obtain a simpler minimum volume of the data. With the intention of reducing the computational difficulties, pixel purity index (PPI) and N-FINDR [7] are used. It produces the easier minimum volume from the data cloud, independence exists in minimum one pixel per endmember is taken. In this paper, we introduce the Hysime algorithm for subspace identification to iteratively project the data with right-angle directions to the subspace. The algorithm processes a loop in order to find the needed and much better endmembers than PPI and N-FINDR.

The paper is ordered like this. Section 3 explains the mixing model and Hysime algorithm. Section 4 evaluates the algorithm using the simulated data, and Sect. 5 ends the paper with concluding remarks.

3 Hyperspectral Unmixing

NMF assures that all the entries of the resulted matrix to be positive which itself is a natural property of the measured quantities.

3.1 Linear Mixing Model

This model requires a pixel in a mining imagery dataset Y as a linear combination of K recognized mineral identification values, called endmembers: $A = [a_1, a_2, a_3, ... a_K]$, in which a_i is the spectral signature of ith endmember whose proportion is the abundance source denoted by S. In the linear mixture model, we have

$$Y = AS + N$$
$$\text{s.t } A \geq 0, \ S \geq 0$$

where A is called the mixing matrix containing the constituent materials and S is a matrix that holds fractional abundance and N denotes noise.

3.2 NMF—Nonnegative Matrix Factorization

In NMF, the input image is depicted as a mixture data matrix, which is divided into two nonnegative matrices collected with endmembers and abundances, respectively. In NMF, the squared Euclidean distance is adopted as the cost function. So as to increase the rate of convergence, we use an optimized cost function. The objective is to minimize this function and is represented as:

$$\min(A, S)(1/2)\| Y - AS\|_F^2; \quad \text{s.t } A \geq 0, \ S \geq 0$$

where $\| \ \|_F$ stands as Frobenius norm of the matrix.

3.3 Dimensionality Reduction

Heavy function processors and memory capabilities are demanded by hyperspectral sensors for processing. Dimensionality reduction refers to the process of converting a set of data having multi-proportions into data with lesser proportions making sure that it conveys the necessity information concisely. The Hysime strategy is used to find out the endmembers in a given picture where signal and the error-linking lattices are estimated initially. A division of eigenvectors is chosen next, that excellently displays signal. By using this, two-term objective function can be made to a least value. First term associates to the signal projection error and the second term associates to the power of the noise projection error.

Let a pixel z in a hyperspectral imagery can be shown as a vector in R^P, P represents the bands across the spectrum, i.e.,

$$z = x + e$$

where x and e are vectors with P-dimensions of source and additive noise, respectively.

As a first step, set of right-angle directions in the signal region is found out as noise estimation. This division is then obtained by the MMSE between x, the source signal, and e error that appears over it. We supposed that the noise is distributed in Gaussian fashion with covariance matrix R_n and that of signal correlation matrix R_x be

$$R_x = E\Sigma E^T$$

where E^T represents the transpose of E.

3.4 Work Flow

The operating steps required in unmixing chain are atmospheric correction, dimensionality reduction, and unmixing as appeared in Fig. 1. Beneath, we give a concise portrayal of each of these means:

(1) *Atmospheric correction*: Scattering and assimilation impacts from the environment are expelled by changing over radiance into reflectance, a fundamental character of the materials.

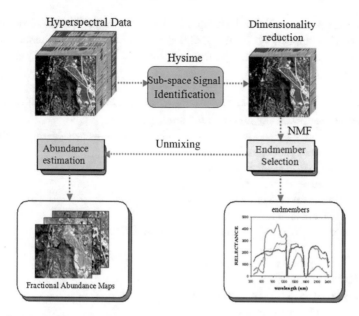

Fig. 1 Flowchart of hyperspectral unmixing steps

(2) *Dimensionality reduction*: Selecting and obtaining meaningful subspaces which best represents the information of the image. The dimensionality is reduced and improves data storage with the application of Hysime method over the optimal subspaces.

(3) *NMF Unmixing*: It is choosing the endmembers in the scene and the partial abundances at every pixel. Nonnegative matrix factorization is used to unmix the data. Availability of spectral libraries makes this framework to put into practical usage and solutions.

3.5 Unmixing Algorithm

The hyperspectral signal subspace identification by minimum error (Hysime) comes about with the signal elements and decides its dimensionality. Noise estimation part and the signal subspace identification part are involved in it. The complex factor is much reduced in this algorithm when compared to other algorithms. It seems to be a great advantage for noise estimation in hyperspectral data. Noise correlation matrix R_n is figured first and afterward processes the signal correlation lattice R_x. At last, a minimization objective function is connected to get an estimate p, the quantity of endmembers in information data.

The main purpose of this technique is the bases mean-squared error strategy. Then, NMF is applied on hyperspectral data to obtain the endmember [A] and abundance matrices [S]. Finally, the initialized A and S matrices are updated iteratively.

Algorithm: Semi Blind Hyperspectral Unmixing

---

1. **Input:**   Observed mixture data **Y**
                Regularization parameter γ
2. **Output:** Number of endmembers k
                Endmember signature matrix **A**
                Abundance matrix **S**.
3. **Estimate k**   – Hysime /*subspace signal identification*/
                – Noise estimation
                – Signal subspace identification
4. **Initialize A** and **S**
5. **Factorization:** **Y = AS**   / *NMF */
6. **Repeat** until convergence
7. Update **A**
$$A \leftarrow A. * YS^T./ASS \quad\text{———— (1)}$$
8. Update **S**
$$S \leftarrow S. * A^TY./ \left(A^TAS+\gamma\right) \quad\text{———— (2)}$$

---

4 Experiments and Results

4.1 Description of the Datasets

In this area, the projected methodology is connected to a sub-picture of the Cuprite mineral information collection obtained by the AVIRIS [8] sensing element. The AVIRIS sensor, flying at a height of 20 km having IFOV of 20 m and it sees a width more than 10 km. It has 224 groups in the picture, covering the wavelength of range 0.37–2.48 μm, with a ghastly determination of 10 nm. For our analysis, a square with the span of 250 × 190 was expelled from the information data. The noisy groups (1–3 and 221–224 bands) and water ingestion groups (104–115 and 148–170 bands) were dispensed with, leaving 182 groups (Fig. 2).

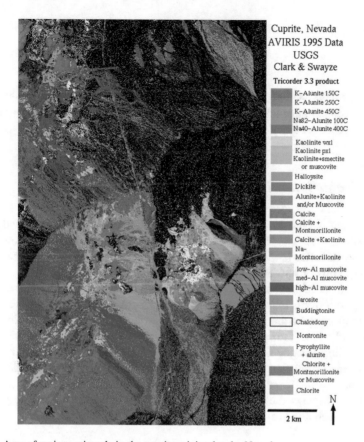

Cuprite, Nevada
AVIRIS 1995 Data
USGS
Clark & Swayze

Tricorder 3.3 product

K-Alunite 150C
K-Alunite 250C
K-Alunite 450C
Na82-Alunite 100C
Na40-Alunite 400C

Kaolinite wxl
Kaolinite pxl
Kaolinite+smectite
or muscovite

Halloysite

Dickite

Alunite+Kaolinite
and/or Muscovite

Calcite

Calcite +
Montmorillonite

Calcite +Kaolinite

Na-
Montmorillonite

low-Al muscovite
med-Al muscovite
high-Al muscovite

Jarosite

Buddingtonite

Chalcedony

Nontronite

Pyrophyllite
+ alunite

Chlorite +
Montmorillonite
or Muscovite

Chlorite

N

2 km

Fig. 2 Area of various minerals in the cuprite mining locale, Nevada

4.2 Performance Measures

Assessment of the shape closeness between the evaluated endmember signature A and the genuine endmember signature $\bar{A}$ is decided by SAD, characterized as:

$$\text{SAD}(A, \bar{A}) = \cos^{-1} \frac{A\bar{A}}{\|A\|\|\bar{A}\|}$$

Spectral angle distance between two endmember signatures is given by the SAD; smaller the values better the estimation result. Evaluation of the abundance estimations is done by root mean square error (RMSE). For the ith endmember, RMSE is defined as

$$\sqrt{\frac{1}{M} \sum_{j=1}^{M} (X_{ij} - X'_{ij})^2}$$

where X_{ij} represents the real abundances, and X'_{ij} denotes the calculated values. The average estimation of all endmembers' RMSEs is figured out. Smaller the RMSE shows a better accuracy of the estimation.

4.3 Evaluation with Experimental Data

Cuprite data is used to investigate the properties of the algorithm correctly. The scene consists of $n = 47,750$ pixels. There can be high quantity of endmembers (p) in real scenarios in the imagery. It is not good to have substantial minerals count in a single pixel itself. Resultant data loaded in .mat file is shown in Fig. 3.

The number of endmembers obtained is shown in Fig. 4, that is, $p = 10$ since the mean square error is minimum at this point. Impact of the mixing matrix is analyzed in order to perform the estimation of endmembers and abundance fractions. The minerals obtained through the NMF algorithm are displayed in Fig. 5 and its corresponding fractional abundance in Fig. 6. Visual interpretations are used to establish the association between an endmember and the respective USGS library signature.

Table 1 displays the SAD values of individual minerals illustrating that the introduced technique can yield better outcomes for minerals extraction.

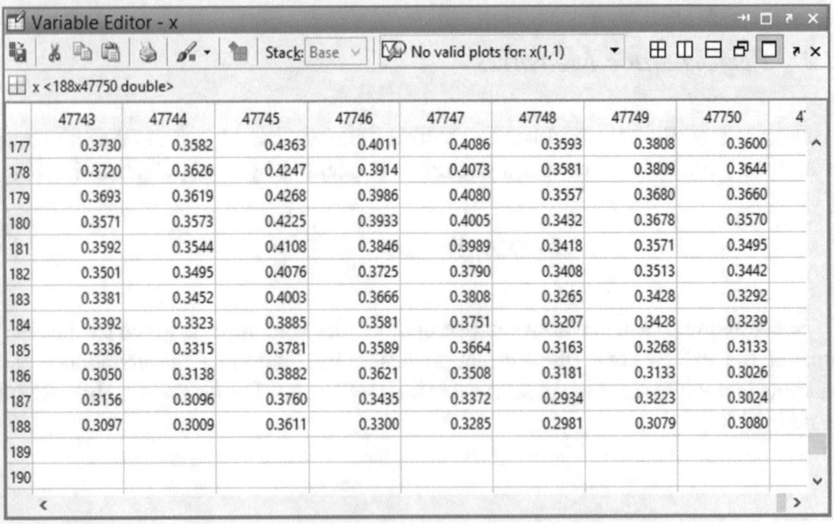

Variable Editor - x

x <188x47750 double>

	47743	47744	47745	47746	47747	47748	47749	47750	4
177	0.3730	0.3582	0.4363	0.4011	0.4086	0.3593	0.3808	0.3600	
178	0.3720	0.3626	0.4247	0.3914	0.4073	0.3581	0.3809	0.3644	
179	0.3693	0.3619	0.4268	0.3986	0.4080	0.3557	0.3680	0.3660	
180	0.3571	0.3573	0.4225	0.3933	0.4005	0.3432	0.3678	0.3570	
181	0.3592	0.3544	0.4108	0.3846	0.3989	0.3418	0.3571	0.3495	
182	0.3501	0.3495	0.4076	0.3725	0.3790	0.3408	0.3513	0.3442	
183	0.3381	0.3452	0.4003	0.3666	0.3808	0.3265	0.3428	0.3292	
184	0.3392	0.3323	0.3885	0.3581	0.3751	0.3207	0.3428	0.3239	
185	0.3336	0.3315	0.3781	0.3589	0.3664	0.3163	0.3268	0.3133	
186	0.3050	0.3138	0.3882	0.3621	0.3508	0.3181	0.3133	0.3026	
187	0.3156	0.3096	0.3760	0.3435	0.3372	0.2934	0.3223	0.3024	
188	0.3097	0.3009	0.3611	0.3300	0.3285	0.2981	0.3079	0.3080	
189									
190									

Fig. 3 Cuprite data loaded in .mat file

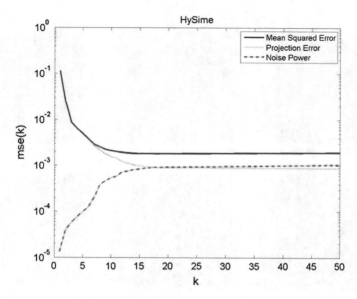

Fig. 4 Mean square error obtained from Hysime

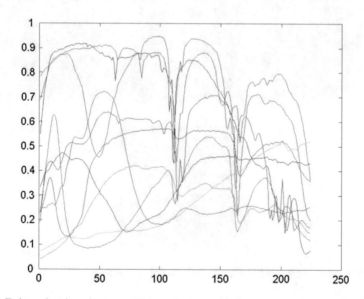

Fig. 5 Estimated endmembers providing good match with their corresponding spectral library

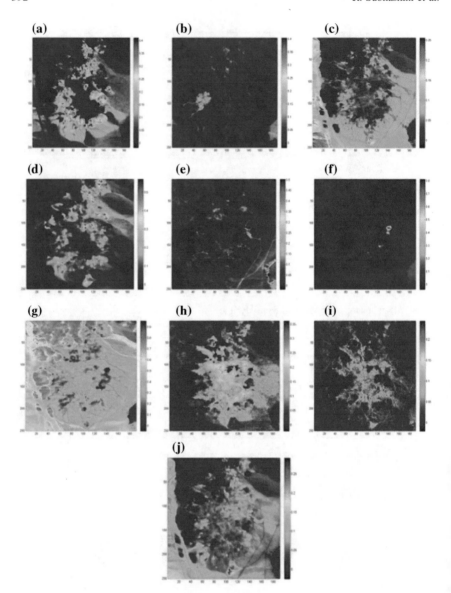

Fig. 6 Abundance maps corresponding to the extracted endmembers

Table 1 SAD values of the extracted minerals

Minerals	SAD
(a) Sphene	4.4922
(b) Nontronite	5.3521
(c) Chalcedony	7.3028
(d) Kaolinite	6.0039
(e) Dumortierite	4.3013
(f) Buddingtonite	3.7302
(g) Alunite	4.6978
(h) Andradite	4.1750
(i) Montmorillonite	3.0785
(j) Pyrope	6.6389

5 Conclusion

Hyperspectral imagery contains mixed pixels which are factorized based on NMF is presented. The projected technique is run on the dataset and is evaluated with the help of USGS spectral library. Here, the Hysime algorithm is used to estimate the signal and noise correlation to represent a subset of eigenvectors efficiently in signal subspace by minimum mean-squared error sense. Simulation results are then provided using SAD metrics depicting effective unmixing. Future work includes algorithm implementation of cost function using other weighting function.

References

1. Keshava N, Mustard JF (2002) Spectral unmixing. IEEE Signal Process Mag 19(1):44–57
2. Ma W et al (2014) Signal processing perspective on hyperspectral unmixing. IEEE Signal Process Mag 31(1):67–81
3. Bayliss J, Gualtieri JA, Cromp R (1997) Analysing hyperspectral data with independent component analysis. Proc. SPIE 3240:133–143
4. Comon P, Jutten C, Herault J (1991) Blind separation of sources, part II: problem statement. Signal Process 24:11–20
5. Boardman J (1993) Automating spectral unmixing of AVIRIS data using convex geometry concepts. In: Summaries 4th annual JPL Airborne geoscience workshop, vol 1, pp. 11–14. JPL Publication 93–26
6. Craig MD (1994) Minimum-volume transforms for remotely sensed data. IEEE Trans Geosci Remote Sens 32(1):99–109
7. Winter ME (1999) N-findr: an algorithm for fast autonomous spectral end-member determination in hyperspectral data. In Proceedings of the SPIE conference on imaging spectrometry V, pp 266–275
8. Aviris Cuprite Nevada Data set. [Online]. Available: http://aviris.jpl.nasa.gov/data/free_data.html

Comparison of Butterworth and Chebyshev Prototype of Bandpass Filter for MRI Receiver Front End

Shraddha Ajay Joshi, Thyagarajan Jayavignesh and Rajesh Harsh

Abstract **Background/Objectives**: Filter in MRI front-end receiver chain plays an important role in rejecting the undesired image frequencies. These filters are designed for RF and IF stages in the front-end chain. **Methods/Statistical analysis**: Bandpass filters are designed for obtaining particular frequency band. The MRI frequency lies in this frequency band selected by the bandpass filter. Different types of prototypes for design of bandpass filter are discussed here. The major two prototypes of filter are designed and simulated. **Findings**: There are various practical filter prototypes in which the major two filter prototypes: Butterworth and Chebyshev are analyzed. These filters give better response and results for the MRI frequency. They are designed at novel frequency of 63.87 MHz which is the frequency or the MRI receiver chain. The filter is designed and simulated at different orders for obtaining better performance with respect to the insertion loss and return loss. The novelty of the results obtained is that the insertion loss is minimum at the frequency of the MRI receiver front-end chain. This will reduce the loss in the MRI system, increases gain of system, and thus image obtained will be more clear and accurate. **Application/Improvements**: Better filter can be designed by reducing the passband bandwidth which will provide more accuracy to the system.

Keywords Butterworth prototype · Chebyshev prototype · Insertion loss
Return loss · MRI receiver chain

S. A. Joshi · T. Jayavignesh (✉)
School of Electronics Engineering, VIT University, Chennai, India
e-mail: jayavignesh.t@vit.ac.in

S. A. Joshi
e-mail: shradhajoshi9403@gmail.com

R. Harsh
Scientist at SAMEER, Mumbai, India
e-mail: rajesh1@sameer.gov.in

© Springer Nature Singapore Pte Ltd. 2018
A. K. Nandi et al. (eds.), *Computational Signal Processing and Analysis*, Lecture Notes in Electrical Engineering 490,
https://doi.org/10.1007/978-981-10-8354-9_36

1 Introduction

The MRI is used for the noninvasive diagnosis of a body. The microwave frequencies are used in MRI system for detection uneven tissues in our body. MRI technique is specially used for medical diagnosis where the body is not directly exposed to any harmful radiation. As magnetic resonance imaging (MRI) is a technique used to produce the images of the soft tissues present in our body, it is essential to design a receiver which can yield clear visible images where the defects hidden in our body can be detected. Therefore, the bandpass filter plays a key role in filtering the frequency of required signals which will in turn help in discarding the image frequencies.

The following block diagram (Fig. 1) illustrates the MRI receiver front-end chain. The bandpass filter stage next to the LNA is used to let on the required band of frequencies to the succeeding stage.

For various frequencies, the microwave interdigital bandpass filter is designed. It is designed to allow band of required frequencies by which the frequency response of the filter gets monitored. Using ADS software, the design and simulation of equi-ripple, i.e., Chebyshev filter and interdigital bandpass filter, is described in paper [1].

To obtain the desired frequencies, the parameters such as length, spacing as well as width and L, C values are varied for interdigital and lumped bandpass filter, respectively.

A microstrip filter with Chebyshev response having 0.5 dB ripple is designed for the frequency of 2.4 GHz. First, the low-pass filter prototype is designed and transformed into bandpass filter by applying scaling transformations.

Further MCLIN and MCFIL microstrip lines are used to design the respective filter and theoretical and simulated results are compared using ADS software. The observations for this paper are made in favor of microstrip bandpass filter as compared to that of the lumped component filter design [2].

A parallel-coupled microstrip BPF is designed at frequency of 2.44 GHz having fractional bandwidth of 3.42%. Filter parameter, insertion loss was considered to be a major factor for optimizing results [3].

Fig. 1 Block diagram of MRI receiver chain

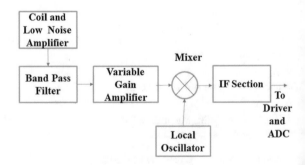

The comprehensive treatment of RF and microwave circuit elements used for the designing the circuits using the lumped elements is described. The topics discussed in the book are related to the materials, fabrication, analyses, design, modeling, and physical, electrical, and thermal practical considerations of the design. The entire design of the bandpass filter can be calculated theoretically using the formulas [4].

The passband of ideal filter has zero insertion loss and linear phase response in the passband and infinite attenuation in stopband. There is no any clear way to improve the design in image parameter method as it yields unstable filter response. Therefore, insertion loss method is used and it allows the high degree of control over the passband and stopband amplitude and phase characteristics [5–8].

2 Filter Design

The filter schemed in this paper is designed for the frequency of 63.87 MHz which is resonant frequency in the MRI signaling. The filter is designed using insertion loss method with the help of lumped components such as L and C. The low-pass filter design is executed first and then converted to bandpass filter using scaling transformation technique. To design a filter, its bandwidth and design frequency should be known.

Usually, Butterworth and Chebyshev prototypes are availed for microwave frequencies. The aspects to be taken into consideration while designing the filter is insertion loss, return loss, passband and stopband bandwidth, quality factor, and attenuation. The desired specifications for conception of bandpass filter are tabulated in Table 1.

2.1 Low-Pass Filter Design

The low-pass filter is designed for the aforementioned specifications. The input and output impedance of the filter is matched to standard value of 50 Ω. The component values of the circuit given in Fig. 2 are designed using the formulae given below:

$$L'_k = \frac{R_0 L_k}{\omega_c}, \quad C'_k = \frac{C_k}{R_0 \omega_c}. \tag{1}$$

Table 1 Specifications of bandpass filter

Center frequency (f_c)	63.87 MHz
3-dB bandwidth (BW)	2 MHz
Fractional bandwidth (Δ)	0.03

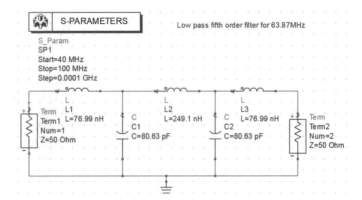

Fig. 2 Schematic of fifth-order low-pass filter

Table 2 Element values for Butterworth low-pass filter prototype

Order	3	5	7	9
$g1$	1.0000	0.6180	0.4450	0.3473
$g2$	2.0000	1.6180	1.2470	1.0000
$g3$	1.0000	2.0000	1.8019	1.5321
$g4$	1.0000	1.6180	2.0000	1.8794
$g5$		0.6180	1.8019	2.0000
$g6$		1.0000	1.2470	1.8794
$g7$			0.4450	1.5321
$g8$			1.0000	1.0000
$g9$				0.3473
$g10$				1.0000

The values given in Tables 2 and 3 are element values for Butterworth and Chebyshev prototype, respectively, substituted in place of L_k and C_k for afore-mentioned formulae.

The schematic of the low-pass filter is shown in Fig. 2, and the results are shown in Fig. 3.

2.2 Frequency Scaling of the Filter

The filter is designed for the required resonance frequency and with the standard impedance for matching. To design a bandpass filter, frequency scaling is to be done for the low-pass filter circuit. The low-pass filter is transformed to the bandpass filter with the scaling technique. In the transformation process:

The series inductor of the low-pass filter is converted to the parallel circuit of L and C with component values as given below:

Table 3 Element values for 0.5 dB equi-ripple low-pass filter prototype

Order	3	5	7	9
$g1$	1.5963	1.7058	1.7372	1.7504
$g2$	1.0967	1.2296	1.2583	1.2690
$g3$	1.5963	2.5408	2.6381	2.6678
$g4$	1.0000	1.2296	1.3444	1.3673
$g5$		1.7058	2.6381	2.7239
$g6$		1.0000	1.2583	1.3673
$g7$			1.7372	2.6678
$g8$			1.0000	1.2690
$g9$				1.7504
$g10$				1.0000

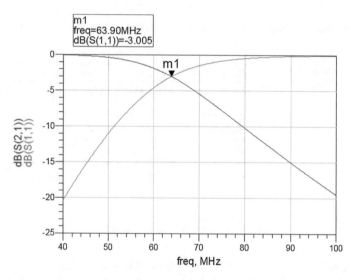

Fig. 3 Return loss and insertion loss results for the fifth-order low-pass filter

$$L' = \frac{R_0 L}{\omega_c \text{BW}} \quad C' = \frac{\text{BW}}{R_0 L \omega_c} \tag{2}$$

The shunt capacitor of low-pass circuit is converted to the series circuit of L and C with the component values given below.

$$L' = \frac{R_0 \text{BW}}{\omega_c C}, \quad C' = \frac{C}{R_0 \text{BW} \omega_c} \tag{3}$$

The fraction bandwidth and center frequency of the BPF can be calculated as:

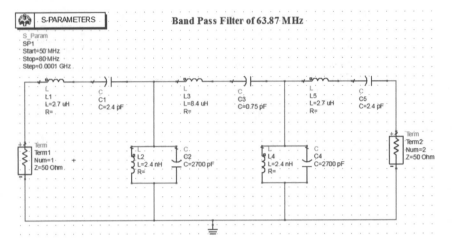

Fig. 4 Schematic of fifth-order Butterworth bandpass filter

Table 4 Component values for the Butterworth prototype

Components	Value	Components	Value
L_1	2.7 μH	L_4	2.4 nH
C_1	2.4 pF	C_4	2.7 nF
L_2	2.4 nH	L_5	2.7 μH
C_2	2.7 nF	C_5	2.4 pF
L_3	8.4 μH	Z_{in}	50 Ω
C_3	0.75 pF	Z_{out}	50 Ω

$$\Delta = \frac{\omega_2 - \omega_1}{\omega_0} \text{ and } \omega_0 = \sqrt{\omega_1 \omega_2}. \tag{4}$$

where ω_2 and ω_1 are the two cutoff frequencies of bandpass filter. The values of the components in bandpass filter for both the prototypes are calculated using the formulae mentioned above. According to the values, the schematic is designed for the two types of filter prototypes.

1. Design of Butterworth Bandpass filter

The schematic of the Butterworth bandpass filter is shown in Fig. 4. The values of the components are tabulated in Table 4. The results of return loss and insertion loss are shown in Figs. 5 and 6, respectively.

2. Design of Chebyshev bandpass filter

The schematic of the Chebyshev bandpass filter is shown in Fig. 7. The values of the components are calculated and tabulated in Table 5. The results of return loss and insertion loss of the filter design are shown in Figs. 8 and 9, respectively.

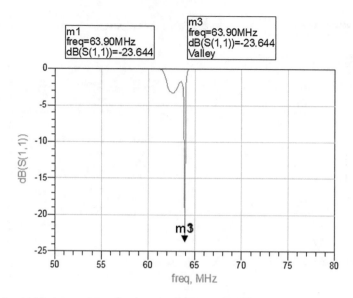

Fig. 5 Return loss results for the fifth-order Butterworth bandpass filter

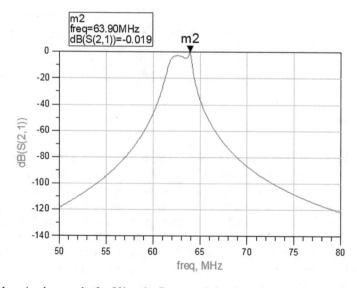

Fig. 6 Insertion loss results for fifth-order Butterworth bandpass filter

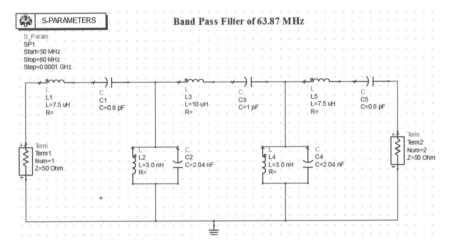

Fig. 7 Schematic of fifth-order Chebyshev bandpass filter

Table 5 Component values for the Chebyshev prototype

Components	Values	Components	Values
L_1	7.5 µH	L_4	3 nH
C_1	0.8 pF	C_4	2.04 pF
L_2	3 nH	L_5	7.5 µH
C_2	2.04 pF	C_5	0.8 pF
L_3	10 µH	Z_{in}	50 Ω
C_3	1 pF	Z_{out}	50 Ω

3 Results and Analysis

The results of two prototypes are tabulated below which shows that the Butterworth filter gives superior output in terms of return loss and insertion loss. Also, the order of the filter has an effective impact on these two measuring parameters. So, it is important to be selective in terms of the order and the prototype for the filter.

The Chebyshev filter produces ripples in the passband as compared to Butterworth filter, thus affecting the insertion loss parameter resulting to a higher value in turn declining the return loss. Therefore, the fifth-order Butterworth filter is selected for designing the bandpass filter in the receiver front-end chain. The values are shown in Table 6.

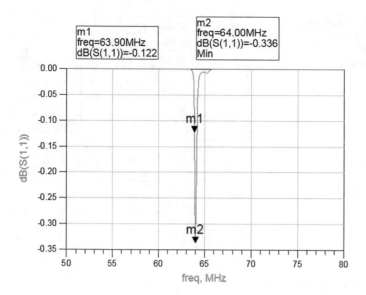

Fig. 8 Return loss results for the fifth-order Chebyshev bandpass filter

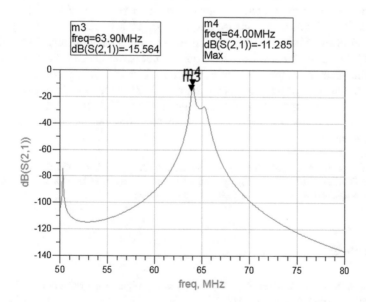

Fig. 9 Insertion loss results for fifth-order Chebyshev bandpass filter

Table 6 Simulation results of BPF for **1** Butterworth filter prototype and **2** Chebyshev filter prototype

(1) *Butterworth filter*

Order	Maximally flat prototype	
	Return loss (dB)	Insertion loss (dB)
3	0.1	14.7
5	23.64	0.01
7	1.6e–7	74.2
9	3.85e–15	89.2

(2) *Chebyshev filter*

Order	Chebyshev prototype	
	Return loss (dB)	Insertion loss (dB)
3	0.15	14.75
5	0.122	15.56
7	2.04	4.25
9	0.77	7.82

4 Conclusion

The simulation result divulges that Butterworth filter gives better results as compared to the Chebyshev filter. The novelty of the paper is the filter is designed for 63.87 MHz which is the resonance frequency of the MRI receivers RF stage in the front-end chain. The return loss of the Butterworth filter is nearly 23.64 dB and insertion loss of 0.01 dB which yields quite better performance than the Chebyshev filter, thus making us to select the Butterworth filter for receiver chain. The values of the circuit component were calculated and the filter was simulated in ADS software. Further, the circuit can be fabricated on PCB using same component values. There might be slight variations in the simulated values and practically measured results.

Acknowledgements I would like to thank Mr. Rajesh Harsh, Head of Technological Innovation Division (TID) and Mrs. Twisha Pandey, Senior Research Scientist (TID) for their continuous help and encouragement to work on this ongoing research topic.

References

1. Indira ND, Nalini K, Khan H (2013) Design of interdigital bandpass filter. Int J Eng Adv Technol (IJEAT) 2(4). ISSN: 2249–8958
2. Khandelwal KS, Kureshi AK (2014) Realization of microstrip band-pass filter design. Int J Adv Res Comput Eng Technol (IJARCET) 3(12):4242–4247
3. Srivastava S, Manjunath RK, Shanthi P (2014) Design, simulation and fabrication of a microstrip bandpass filter. Int J Sci Eng Appl (IJSEA) 3(5):154–158
4. Bahl Inder (2003) Lumped elements for RF and microwave circuits. Artech House Inc., Bosten, London

5. Rohde UL, Newkirk DP (2000) RF/microwave circuit design for wireless applications. Wiley, New York
6. Leenaerts D, Tang J, Vaucher CS (2003) Circuit design for RF transceivers. Wiley, New York
7. Laskar J, Matinpour B, Chakraborty S (2004) Modern receiver front-ends system, circuit and integration. Wiley, New York
8. Pozar DM (2012) Microwave engineering, 3rd edn. Wiley, New York, pp 396, 401–404

Object Tracking Based on Position Vectors and Pattern Matching

V. Purandhar Reddy and A. Annis Fathima

Abstract Object tracking systems using camera have become an essential requirement in today's society. In-expensive and high-quality video cameras, availability and demand for analysis of automated video have produced a lot of interest for numerous fields. Almost all conventional algorithms are developed based on background subtraction, frame difference, and static background. They fail to track in environments such as variation in illumination, cluttered background, and occlusions. The image segmentation based object tracking algorithms fail to track in real-time. Feature extraction of an image is an indispensable first step in object tracking applications. In this paper, a novel real-time object tracking based on position and feature vectors is developed. The proposed algorithm involves two phases. The first phase is extraction of features for region of interest object in first frame and nine position features of second frame in video. The second phase is similarity estimation of extracted features of two frames using Euclidean distance. The nearest match is considered by minimum distance between first frame feature vectors and nine different feature vectors of second frame. The proposed algorithm is compared with other existing algorithms using different feature extraction techniques for object tracking in video. The proposed method is simulated and evaluated by statistical, discrete wavelet transform, Radon transform, scale-invariant feature transform and features from accelerated segment test. The performance evaluation shows that the proposed algorithm can be applied for any feature extraction technique and object tracking in video depends on tracking accuracy.

Keywords Object detection · Object tracking · Pattern matching
SIFT · FAST

V. Purandhar Reddy (✉) · A. Annis Fathima
School of Electronics Engineering, Vellore Institute of Technology, Chennai, India
e-mail: vpurandhar.reddy2015@vit.ac.in

A. Annis Fathima
e-mail: annis.fathima@vit.ac.i

© Springer Nature Singapore Pte Ltd. 2018
A. K. Nandi et al. (eds.), *Computational Signal Processing and Analysis*, Lecture Notes in Electrical Engineering 490,
https://doi.org/10.1007/978-981-10-8354-9_37

1   Introduction

Object tracking plays an important role in activity recognition, motion analysis, traffic monitoring and video surveillance attracts great interest deal in computer vision community. Even though many tracking methods are developed for object tracking from past decade, it still be a challenging task because of many variable factors like illumination variation, occlusions, and cluttered background in video data sets.

Without proper knowledge about the target, many model-free trackers are developed to obtain accurate generic object tracking [1, 2]. Effective tracker design is difficult task due to various parameters, such as dynamic background, occlusions, and illumination variations. Object tracking techniques based on pattern matching use euclidean distance for feature vector matching and tracking [3, 4]. In recent years, an extensive number of object tracking algorithms is used solve tracking problems in crowded backgrounds.

The conventional object tracking methods are generally classified as either a one-filter-per-target approach [5–8] or a one-state-per-target [9–12]. Existing approaches converts the single object feature and position vectors to global vectors and performs minimum distance search on overall state space [13]. To withstand object tracking with appearance change, the algorithms are there with online learning appearance models [5, 6, 9–11]. However, all these approaches majorly depend on the online appearance model to estimate the object from the background, which leads to mistracking with same appearance.

In this paper, we proposed an algorithm for real-time object tracking. Pattern matching is key role in this method to get better accuracy in object tracking. Pattern matching is done based on feature vectors matching using minimum distance. The pattern matching is used in order to measure the similarity is complex task in many computer vision applications. Many applications like robot localization, object recognition, navigation, image registration, and activity recognition real-time pattern matching is the key role. In general, the conventional pattern matching based on feature vectors can be classified based on global and local features. Global feature-based algorithms are used to recognize an object in whole image. To get this image segmentation and object extraction is required. This type of algorithm is particularly suitable for object tracking, where the object is easily extracted from the image background using image segmentation. For the segmented object, eigenvector covariance matrix [14] or Hu moments [15] can be used as global features. The algorithms based on global features are simple and fast, but there is drawback in object recognition under illumination changes and object pose. In this paper, we used DWT, DTCWT, and Radon transform and SIFT as global features.

In this paper, we proposed object tracking algorithm based on position vectors and pattern matching. Our algorithm is based on position vector estimation and region of interest pattern matching. The tracking method between successive frames proposed in this paper uses region of interest pattern matching. As a result, the algorithms can be applied to track simultaneously various still objects and moving objects in video pictures. This paper is organized as follows: Sect. 2 describes the feature extraction methods; Sect. 3 describes the position vectors estimation; Sect. 4 describes in detail the modified object tracking.

2 Feature Vectors

Object feature extraction in video is the initial step of proposed object tracking technique. First object is cropped based on region of interest and then object feature vector is formed by combining subbands of applied transform. Four transforms DWT, RT, SIFT, and FAST are applied for feature vectors estimation in eight different directions $[-45°, +45°, +90°, -90°, +180°, -180°, +270°, -270°]$.

2.1 DWT—Discrete Wavelet Transform

Two-level DWT decomposition using Haar wavelet is used. In the proposed work, only the second level LL image is used for calculation of feature vectors. For initial frame, the selected area feature vector is calculated in nine different position directions by Eq. (1).

$$F = [LL1, LL2, LL3, LL4, LL5, LL6, LL7, LL7, LL8] \quad (1)$$

2.2 RT—Radon Transform

The Radon transform is utilized to separate the image intensity along the radial line oriented at a specific angle. Out of this object signature, the Gray-Level Covariance Matrix is estimated from which the accompanying statistical elements are extricated; Correlation, Contrast, Energy, and Homogeneity which are connected in column to form feature vector by Eq. (2) for a particular ROI object.

$$F = [Correlation, Contrast, Energy, Homogeneity] \quad (2)$$

2.3 Scale-Invariant Feature Transform

Lowe [10] proposed SIFT algorithm having major stages of computation for image features extraction:

1. *Scale-space extrema detection*: Image location scales are identified in this stage. This can be implemented effectively by Gaussian function variation to identify key interest-invariant points to scale and orientation.
2. *Key point localization*: To decide location and scale, every candidate location is utilized to fit point-by-point model. Selected of key points is based on similarity measures with their stability.
3. *Orientation assignment*: For orientations assign at each key point location, local image gradient directions are used. Position, rotation and scaling features are extracted in image to avoid invariance transformations.
4. *Key point descriptor*: To avoid illumination effects, local image gradients are used and measured at the selected scale at each key point region. All gradients are converted into a representation to change in illumination and allow minimum levels of shape distortion.

In the last step, descriptor vectors effectively extracted are group of histograms estimated from orientations and gradient magnitudes of neighbor points in window at each and around key point. To find similarity areas in two images, SIFT feature vectors can be used by distance matching in many applications. Not matched key points are discarded in their locations not to be affected for object tracking.

2.4 FAST—Features from Accelerated Segment Test

FAST algorithm is feature-based only among other feature extractions explained in this section. Since, variation in feature extraction by FAST identifies feature points differ significantly from SIFT and SURF. FAST using corner response function for robust detection in a given scene. FAST uses multigrid algorithm to identify corners that fast up the process normally [16]. The three-step multigrid algorithm used for corner detection by step-by-step procedure is as follows.

Step 1: Estimate the simple CRF in a low-resolution image, at every pixel location. Based on threshold greater than defined threshold pixels are classified as "potential corners."

Step 2: In similar way in step 1, compute the simple CRF in a high-resolution image, at every potential pixel location. Non-corner is detected if the response is lower than threshold detected already. If non-corner is not detected then use interpixel approximation and compute new response. Again non-corner pixel can be identified if the response is lower than the second threshold T2.

Step 3: Corner pixels are identified by a locally maximal CRF. This step is needed since the vicinity of a corner will have high CRF if more number of points exists, and that's why largest CRF is confirmed to be corner point. The process is also called as non-maximum suppression (NMS).

3   Position Vectors Estimation

In general, position vectors of object can be estimated by using image segmentation and object extraction. In the proposed algorithm, first select the area of an object which is to be tracked and then by using axis parameters of a block, we can extract the location points of the object by Eqs. (3) and (4).

P_{xmx} (P_{xmn}) has the maximum (minimum) x-component as shown in Fig. 1.

$$P_{xmax} = (X_{mx}, x, X_{mx}, y), \tag{3}$$

$$P_{xmin} = (X_{mn}, x, X_{mn}, y), \tag{4}$$

where X_{mx}, x, X_{mx}, y, X_{mn}, y and X_{mn}, x, are x and y coordinates of the leftmost and rightmost boundary of the block, respectively.

Similarly in y-direction position vectors defined as by Eqs. (5) and (6)

$P_{ymn} = (Y_{mn}, x, Y_{mn}, y)$, $P_{ymx} = (Y_{mx}, x, Y_{mx}, y)$. Then we estimate the height h and the width of the block by

$$w_i(t) = X_{mx}, x - X_{mn}, x, \tag{5}$$

$$h_i(t) = Y_{mx}, y - Y_{mn}, y. \tag{6}$$

The positions block in each frame is defined as follows by Eqs. (7) and (8).

$$P = (X_1, Y_1)$$

$$X_1(t) = (X_{mx}, x + X_{mn}, x)/2 \tag{7}$$

$$Y_1(t) = (Y_{mx}, y + Y_{mn}, y)/2 \tag{8}$$

Position vectors of other object to be tracked in successive frames are estimated by using object movement in nine different directions by shift on m point in X-Y plane as shown in Table 1.

Fig. 1 Position vector estimation of ROI block in image

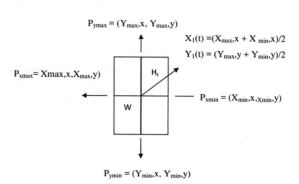

Table 1 Position vectors estimation in different directions

Direction	Position vector
+90°	$P1 = (X_1 - m, Y_1)$
−90°	$P2 = (X_1 + m, Y_1)$
+180°	$P3 = (X_1, Y_1 + m)$
−180°	$P4 = (X_1, Y_1 - m)$
+125°	$P5 = (X_1 - m, Y_1 - m)$
−125°	$P6 = (X_1 + m, Y_1 + m)$
+45°	$P7 = (X_1 - m, Y_1 + m)$
−45°	$P8 = (X_1 + m, Y_1 - m)$
+0°	$P9 = (X_1, Y_1)$

4 Proposed Object Tracking Technique

Object tracking in frame by feature vector matching is done by minimum distance search among feature vectors as shown in Fig. 2. Using ROI position of the present frame object, first we extract the feature vector of the ROI image of previous frame. After we perform match by the minimum distance between present and previous frames for all ROI images in the next frame using estimation position vectors. Finally the ROI image present frame is matched with the image in the next frame by minimum distance search. By repeating the matching procedure for all the frames with stored ROI cropped images, all regions can be identified one by one and can maintain track of the regions in between the frames. Further refinements of developed algorithm are as follows: The distance measure is not specified for matching yet. By experimental results, we confirm that the Manhattan distance is simpler and enough for object tracking applications than Euclidean distance.

Fig. 2 Proposed object tracking technique

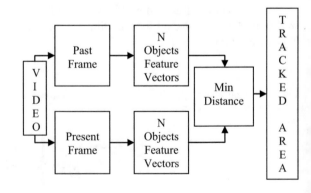

5 Simulation Results

We have performed simulation by using two different videos of dynamic background on a laptop with 2.8 GHz Intel processor CPU and a 2 GB memory, and implemented using MATLAB software. Pattern matching is done by using DWT, RT, SIFT, and FAST and calculated accuracy for performance evaluation (Tables 2 and 3).

$$\text{Accuracy} = (N_1 - N_2)/N1 \times 100$$

N_1 Number of frames in video
N_2 Number of mistracked frames in video

Table 2 Object tracking accuracy result of sample video-1

Number of frames	Accuracy			
	DWT	RT	SIFT	FAST
10	60	70	80	90
20	75	85	90	95
30	80	86.666667	90	93.333333
40	77.5	80	85	90
50	72	78	82	92
60	70	68.333333	71.666667	73.333333
70	70	65.714286	70	71.428571
80	70	68.75	70	73.75
90	67.777778	65.555556	66.666667	75.555556
100	63	61	63	78

Table 3 Object tracking accuracy result of sample video-2

Number of frames	Accuracy			
	DWT	RT	SIFT	FAST
10	50	60	80	90
20	55	60	80	90
30	56.666667	60	80	90
40	57.5	60	82.5	92.5
50	62	62	84	92
60	65	63.333333	85	86.666667
70	67.142857	64.285714	85.714286	88.571429
80	70	65	87.5	90
90	72.222222	66.666667	88.888889	91.111111
100	74	68	84	87

Fig. 3 Overall tracking
accuracy results of sample
video-1

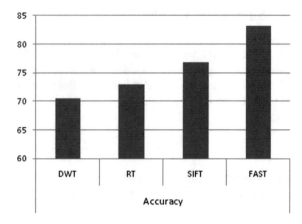

Fig. 4 Overall tracking
accuracy results of sample
video-2

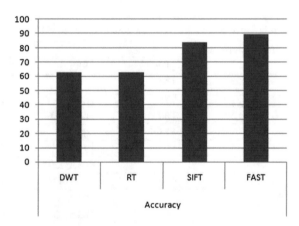

The tracked accuracy results are shown in Figs. 3 and 4. Based on calculated accuracy results in Tables 1 and 2. FAST feature extraction technique is better. Given the detection responses, for the two video data sets, our current implementation runs at 1.8 and 1.7 frames/s, respectively. The cost of good training time is not included. Based on Fig. 5 simulation results we affirm that the algorithm execution is better in time at least comparable with conventional techniques.

6   Conclusion

Fast object tracking framework in video is presented in this paper. By performance evaluation, we conclude FAST based feature vector estimation gets good accuracy. Simulation results of successive frames with moving objects verify that the algorithm is essential for accurate tracking of object in motion. We additionally affirmed that the algorithm performs exceptionally well for more crowded video pictures

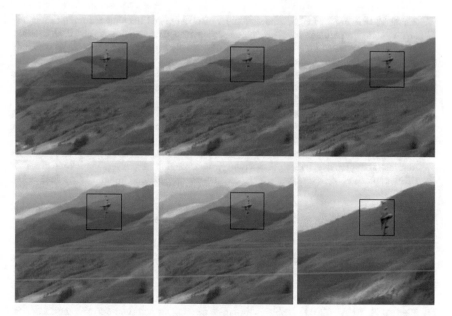

Fig. 5 Sample video-1 object tracking simulation results using SIFT

including occlusion and rotation of objects. To overcome the limitation, real-time feature matching FAST features are taken to explicit fast matching and introduced target interactions to fill the gaps of locations by retrieving the mistracked areas. Two sample data sets are used for experimental verification, and a sufficient improvement in object tracking performance has been obtained. For future enhancements, we plan to develop an efficient very fast moving object tracking algorithm, which is explicit to track multiple moving objects, and missed detections by recovering using same framework. Finally we conclude that block matching technique should be designed, and more reliable feature vectors and minimum distance strategy should be considered to attain accurate and fast object tracking in video for desirable practical applications.

References

1. Li X, Hu W, Shen C, Zhang Z, Dick A, Van Den Hengel A (2013) A survey of appearance models in visual object tracking. ACM Trans Intell Syst Technol 4(4). Article ID 58
2. Wu Y, Lim J, Yang M-H (2013) Online object tracking: a benchmark. In: Proceedings of IEEE conference on computer vision and pattern recognition, Portland, OR, USA, June 2013, pp 2411–2418
3. Mei X, Ling H (2011) Robust visual tracking and vehicle classification via sparse representation. IEEE Trans Pattern Anal Mach Intell 33(11):2259–2272

4. Zhang T, Ghanem B, Liu S, Ahuja N (2012) Robust visual tracking via multi-task sparse learning. In: Proceedings of IEEE conference on computer vision and pattern recognition, Providence, RI, USA, June 2012, pp 2042–2049. Nicole R (in press) Title of paper with only first word capitalized. J Name Stand Abbrev

5. Breitenstein MD, Reichlin F, Leibe B, Koller-Meier E, Van Gool L (2011) Online multiperson tracking-by-detection from a single, uncalibrated camera. IEEE Trans Pattern Anal Mach Intell 33(9):1820–1833

6. Yang M, Lv F, Xu W, Gong Y (2009) Detection driven adaptive multi-cue integration for multiple human tracking. In: Proceedings of IEEE international conference on computer vision, Kyoto, Japan, Sept 2009, pp 1554–1561

7. Xing J, Ai H, Lao S (2009) Multi-object tracking through occlusions by local tracklets filtering and global tracklets association with detection responses. In: Proceedings of IEEE conference on computer vision and pattern recognition, Miami, USA, June 2009, pp 1200–1207

8. Khan Z, Balch T, Dellaert F (2006) MCMC data association and sparse factorization updating for real time multitarget tracking with merged and multiple measurements. IEEE Trans Pattern Anal Mach Intell 28(12):1960–1972

9. Kuo CH, Huang C, Nevatia R (2010) Multi-target tracking by on-line learned discriminative appearance models. In: Proceedings of IEEE conference on computer vision and pattern recognition, San Francisco, USA, June 2010, pp 685–692

10. Lowe DG (2004) Distinctive Image Features from Scale-Invariant Keypoints. Int J Comput Vision. 60(2):91–110. doi:https://doi.org/10.1023/B:VISI.0000029664.99615.94

11. Benfold B, Reid I (2011) Stable multi-target tracking in real-time surveillance video. In: Proceedings of IEEE conference on computer vision and pattern recognition, Colorado Springs, USA, June 2011, pp 3457–3464

12. Yang B, Nevatia R (2012) Multi-target tracking by online learning of non-linear motion patterns and robust appearance model. In: Proceedings of IEEE conference on computer vision and pattern recognition, Providence, USA, June 2012, pp 1918–1925

13. Poiesi F, Mazzon R, Cavallaro A (2013) Multi-target tracking on confidence maps: an application to people tracking. Comput Vis Image Underst 117(10):1257–1272

14. Lee Y, Lee K, Pan S (2005) Local and global feature extraction for face recognition. Springer, Berlin

15. Vuppala SK, Grigorescu SM, Ristic D, Gräser A (2007) Robust color object recognition for a service robotic task in the system FRIEND II. In: 10th international conference on rehabilitation robotics—ICORR'07, 2007

16. Trajkovic M, Hedley M (1998) FAST corner detector. Image Vis Comput 16:75–87

17. Juan L, Gwun O (2009) A comparison of SIFT, PCA-SIFT, and SURF. Int J Image Proc (IJIP) 3(4):143–152

Printed in the United States
By Bookmasters